KITCHEN & BATHROOM PLUMBING

TIME
LIFE
BOOKS®

Other Publications:

FITNESS, HEALTH & NUTRITION
SUCCESSFUL PARENTING
HEALTHY HOME COOKING
UNDERSTANDING COMPUTERS
THE ENCHANTED WORLD
THE KODAK LIBRARY OF CREATIVE PHOTOGRAPHY
GREAT MEALS IN MINUTES
THE CIVIL WAR
PLANET EARTH
COLLECTOR'S LIBRARY OF THE CIVIL WAR
THE EPIC OF FLIGHT
THE GOOD COOK
WORLD WAR II
HOME REPAIR AND IMPROVEMENT
THE OLD WEST

KITCHEN & BATHROOM PLUMBING

TIME-LIFE BOOKS
ALEXANDRIA, VIRGINIA

Fix It Yourself was produced by
ST. REMY PRESS

MANAGING EDITOR	Kenneth Winchester
MANAGING ART DIRECTOR	Pierre Léveillé

Staff for *Kitchen & Bathroom Plumbing*

Editor	Valerie J. Marchant
Art Director	Odette Sévigny
Research Editor	Katherine Zmetana
Contributing Writers	Susan Bryan Reid, Rosemary Collins, Cathleen Farrell, Kent Farrell, Judee Ganten, Buzz Gibbs, Maryellen Kennedy, Michael Kleiza
Contributing Illustrators	Gérard Mariscalchi, Jacques Proulx
Technical Illustrator	Robert Paquet
Cover	Robert Monté
Index	Christine M. Jacobs

Administrator	Denise Rainville
Coordinator	Michelle Turbide
Systems Manager	Shirley Grynspan
Systems Analyst	Simon Lapierre
Studio Director	Daniel Bazinet

Time-Life Books Inc. is a wholly owned subsidiary of
TIME INCORPORATED

FOUNDER	Henry R. Luce 1898-1967
Editor-in-Chief	Henry Anatole Grunwald
Chairman and Chief Executive Officer	J. Richard Munro
President and Chief Operating Officer	N. J. Nicholas Jr.
Chairman of the Executive Commitee	Ralph P. Davidson
Corporate Editor	Ray Cave
Group Vice President, Books	Kelso F. Sutton
Vice President, Books	George Artandi

TIME-LIFE BOOKS INC.

EDITOR	George Constable
Director of Design	Louis Klein
Director of Editorial Resources	Phyllis K. Wise
Acting Text Director	Ellen Phillips
Editorial Board	Russell B. Adams Jr., Dale M. Brown, Roberta Conlan, Thomas H. Flaherty, Lee Hassig, Donia Ann Steele, Rosalind Stubenberg, Kit van Tulleken, Henry Woodhead
Director of Photography and Research	John Conrad Weiser

PRESIDENT	Christopher T. Linen
Chief Operating Officer	John M. Fahey Jr.
Senior Vice President	James L. Mercer, Leopoldo Toralballa
Vice Presidents	Stephen L. Bair, Ralph J. Cuomo, Terence J. Furlong, Neal Goff, Stephen L. Goldstein, Juanita T. James, Hallett Johnson III, Robert H. Smith, Paul R. Stewart
Director of Production Services	Robert J. Passantino

Editorial Operations

Copy Chief	Diane Ullius
Editorial Operations	Caroline A. Boubin
Production	Celia Beattie
Quality Control	James J. Cox (director)
Library	Louise D. Forstall

Correspondents	Elisabeth Kraemer-Singh (Bonn); Maria Vincenza Aloisi (Paris); Ann Natanson (Rome).

THE CONSULTANTS

Consulting Editor **David L. Harrison** is Managing Editor of Bibliographics Inc. in Alexandria, Virginia. He served as an editor of several Time-Life Books do-it-yourself series, including *Home Repair and Improvement, The Encyclopedia of Gardening* and *The Art of Sewing.*

Richard Day, a do-it-yourself writer for nearly a quarter century, is a founder of the National Association of Home and Workshop Writers and the author of four books on plumbing. He has built two houses from the ground up, and now lives in southern California.

John Dinolfo has taught plumbing for 17 years to vocational students in the C.E.C.M., Montreal's Catholic school system.

The nonprofit **International Association of Plumbing and Mechanical Officials** in Los Angeles provided valuable information on plumbing codes and standards.

Elliot Levine, special consultant for Canada, is a mechanical engineer and third-generation plumber. He operates Levine Brothers Plumbing in Montreal.

Mark M. Steele, a professional home inspector in the Washington, D.C. area, is an editor of home improvement articles and books.

Library of Congress Cataloguing-in-Publication Data
Kitchen & bathroom plumbing.
(Fix it yourself)
Includes index.
1. Kitchens. 2. Bathrooms. 3. Plumbing–Amateurs' manua
I. Time-Life Books. II. Title: Kitchen & bathroom plumbing.
III. Series.
TH6507. K58 1987 696'.1 87-1931
ISBN 0-8094-6208-7
ISBN 0-8094-6209-5 (lib. bdg.)

For information about any Time-Life book, please write:
Reader Information
541 North Fairbanks Court
Chicago, Illinois 60611

First printing. Printed in U.S.A.
Published simultaneously in Canada.
School and library distribution by Silver Burdett Company, Morristown, New Jersey.

CONTENTS

HOW TO USE THIS BOOK

Kitchen & Bathroom Plumbing is divided into three sections. The Emergency Guide on pages 8-13 provides information that can be indispensable, even lifesaving, in the event of a household emergency. Take the time to study this section *before* you need the important advice it contains.

The Repairs section—the heart of the book—is a system for troubleshooting and repairing faucets, sinks, bathtubs, showers, water heaters, drains and septic systems. Pictured below are four sample pages from the chapter on bathtubs and showers, with captions describing the various features of the book and how they work. If you have a leaking double-handle faucet, for example, the Troubleshooting Guide will offer a number of possible causes. If the problem is a worn O-ring, you will be directed to page 74 for detailed, step-by-step directions for removing and replacing the faulty part.

Each job has been rated by degree of difficulty and the average time it will take for a do-it-yourselfer to complete. Keep in mind that this rating is only a suggestion. Before deciding whether you should attempt a repair, first read all the instructions carefully. Then be guided by your own confidence, and the tools and time available to you. For complex or time-consuming repairs, such as replacing a faucet set or exca-

Introductory text
Describes proper use and care of home plumbing, most common breakdowns and basic safety precautions.

"Exploded" and cutaway diagrams
Locate and describe the various parts of the fixture or system.

Troubleshooting Guide
To use this chart, locate the symptom that most closely resembles your plumbing problem, review the possible causes in column 2, then follow the recommended procedures in column 3. Simple fixes may be explained on the chart; in most cases you will be directed to an illustrated, step-by-step repair sequence.

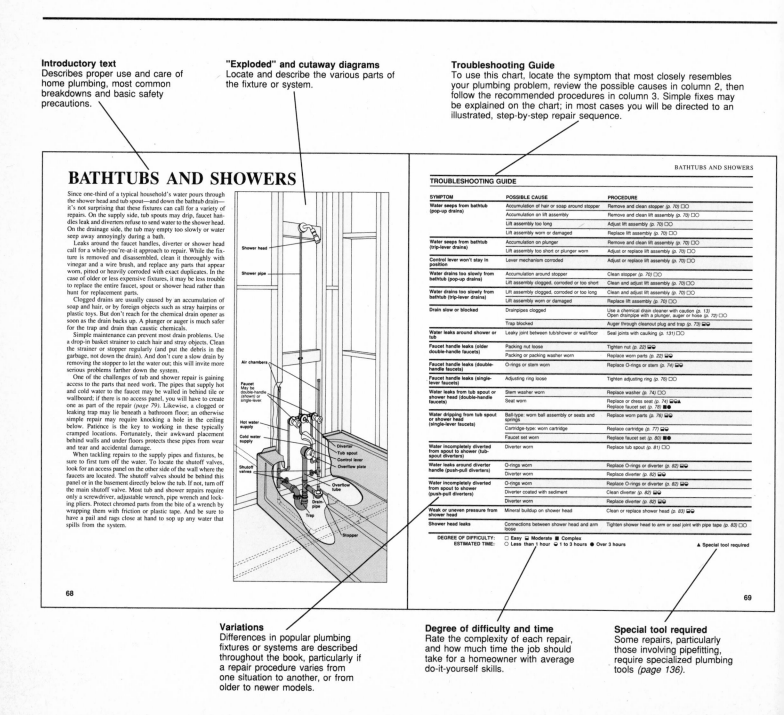

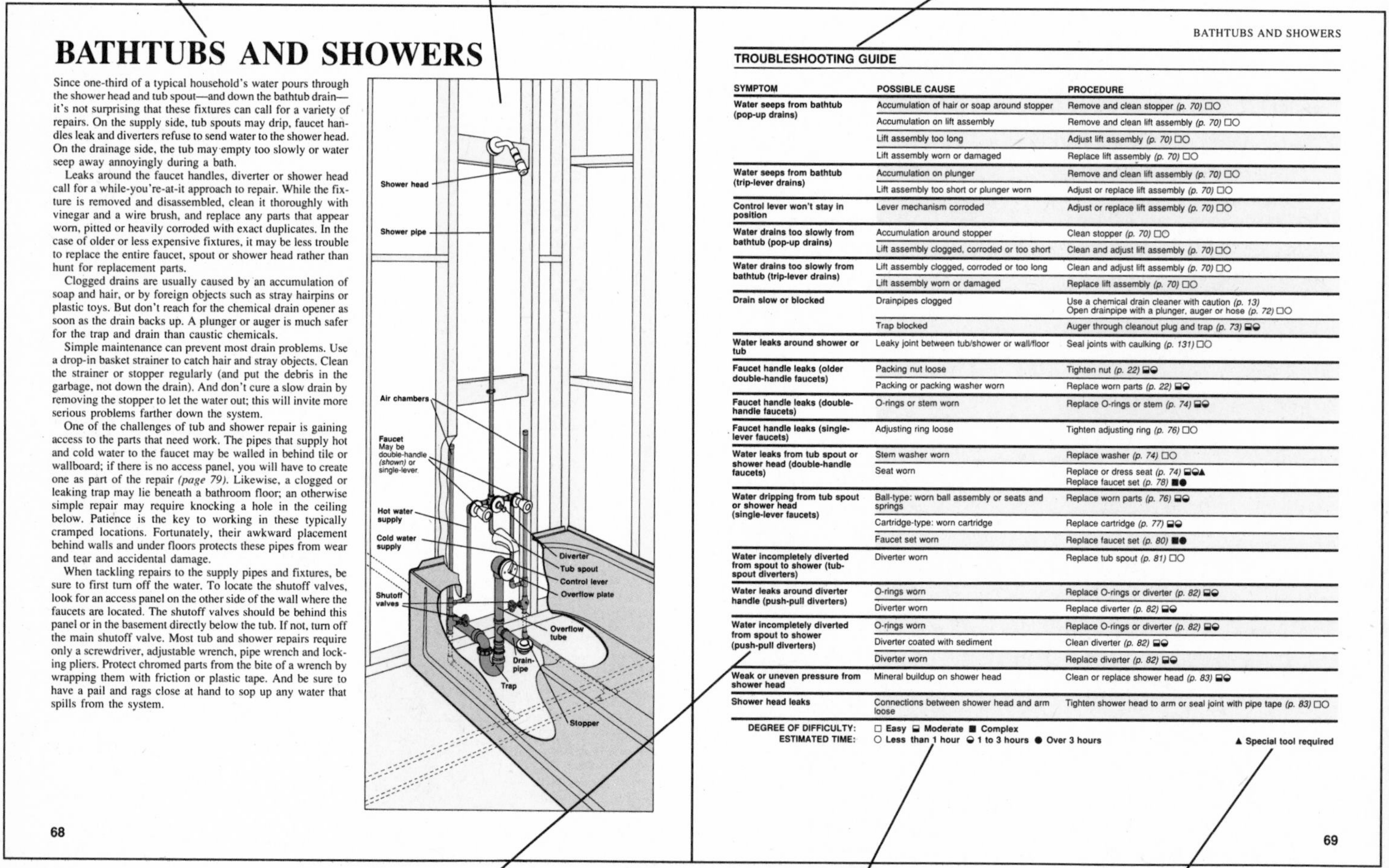

BATHTUBS AND SHOWERS

Since one-third of a typical household's water pours through the shower head and tub spout—and down the bathtub drain—it's not surprising that these fixtures can call for a variety of repairs. On the supply side, tub spouts may drip, faucet handles leak and diverters refuse to send water to the shower head. On the drainage side, the tub may empty too slowly or water seep away annoyingly during a bath.

Leaks around the faucet handles, diverter or shower head call for a while-you're-at-it approach to repair. While the fixture is removed and disassembled, clean it thoroughly with vinegar and a wire brush, and replace any parts that appear worn, pitted or heavily corroded with exact duplicates. In the case of older or less expensive fixtures, it may be less trouble to replace the entire faucet, spout or shower head rather than hunt for replacement parts.

Clogged drains are usually caused by an accumulation of soap and hair, or by foreign objects such as stray hairpins or plastic toys. But don't reach for the chemical drain opener as soon as the drain backs up. A plunger or auger is much safer for the trap and drain than caustic chemicals.

Simple maintenance can prevent most drain problems. Use a drop-in basket strainer to catch hair and stray objects. Clean the strainer or stopper regularly (and put the debris in the garbage, not down the drain). And don't cure a slow drain by removing the stopper to let the water out; this will invite more serious problems farther down the system.

One of the challenges of tub and shower repair is gaining access to the parts that need work. The pipes that supply hot and cold water to the faucet may be walled in behind tile or wallboard; if there is no access panel, you will have to create one as part of the repair *(page 79)*. Likewise, a clogged or leaking trap may lie beneath a bathroom floor; an otherwise simple repair may require knocking a hole in the ceiling below. Patience is the key to working in these typically cramped locations. Fortunately, their awkward placement behind walls and under floors protects these pipes from wear and tear and accidental damage.

When tackling repairs to the supply pipes and fixtures, be sure to first turn off the water. To locate the shutoff valves, look for an access panel on the other side of the wall where the faucets are located. The shutoff valves should be behind this panel or in the basement directly below the tub. If not, turn off the main shutoff valve. Most tub and shower repairs require only a screwdriver, adjustable wrench, pipe wrench and locking pliers. Protect chromed parts from the bite of a wrench by wrapping them with friction or plastic tape. And be sure to have a pail and rags close at hand to sop up any water that spills from the system.

68

BATHTUBS AND SHOWERS

TROUBLESHOOTING GUIDE

SYMPTOM	POSSIBLE CAUSE	PROCEDURE
Water seeps from bathtub (pop-up drains)	Accumulation of hair or soap around stopper	Remove and clean stopper *(p. 70)* □○
	Accumulation on lift assembly	Remove and clean lift assembly *(p. 70)* □○
	Lift assembly too long	Adjust lift assembly *(p. 70)* □○
	Lift assembly worn or damaged	Replace lift assembly *(p. 70)* □○
Water seeps from bathtub (trip-lever drains)	Accumulation on plunger	Remove and clean lift assembly *(p. 70)* □○
	Lift assembly too short or plunger worn	Adjust or replace lift assembly *(p. 70)* □○
Control lever won't stay in position	Lever mechanism corroded	Adjust or replace lift assembly *(p. 70)* □○
Water drains too slowly from bathtub (pop-up drains)	Accumulation around stopper	Clean stopper *(p. 70)* □○
	Lift assembly clogged, corroded or too short	Clean and adjust lift assembly *(p. 70)* □○
Water drains too slowly from bathtub (trip-lever drains)	Lift assembly clogged, corroded or too long	Clean and adjust lift assembly *(p. 70)* □○
	Lift assembly worn or damaged	Replace lift assembly *(p. 70)* □○
Drain slow or blocked	Drainpipes clogged	Use a chemical drain cleaner with caution *(p. 13)* Open drainpipe with a plunger, auger or hose *(p. 72)* □○
	Trap blocked	Auger through cleanout plug and trap *(p. 73)* ⬓◒
Water leaks around shower or tub	Leaky joint between tub/shower or wall/floor	Seal joints with caulking *(p. 131)* □○
Faucet handle leaks (older double-handle faucets)	Packing nut loose	Tighten nut *(p. 22)* ⬓◒
	Packing or packing washer worn	Replace worn parts *(p. 22)* ⬓◒
Faucet handle leaks (double-handle faucets)	O-rings or stem worn	Replace O-rings or stem *(p. 74)* ⬓◒
Faucet handle leaks (single-lever faucets)	Adjusting ring loose	Tighten adjusting ring *(p. 76)* □○
Water leaks from tub spout or shower head (double-handle faucets)	Stem washer worn	Replace washer *(p. 74)* □○
	Seat worn	Replace or dress seat *(p. 74)* ⬓◒▲ Replace faucet set *(p. 78)* ■●
Water dripping from tub spout or shower head (single-lever faucets)	Ball-type: worn ball assembly or seats and springs	Replace worn parts *(p. 76)* ⬓◒
	Cartridge-type: worn cartridge	Replace cartridge *(p. 77)* ⬓◒
	Faucet set worn	Replace faucet set *(p. 80)* ■●
Water incompletely diverted from spout to shower (tub-spout diverters)	Diverter worn	Replace tub spout *(p. 81)* □○
Water leaks around diverter handle (push-pull diverters)	O-rings worn	Replace O-rings or diverter *(p. 82)* ⬓◒
	Diverter worn	Replace diverter *(p. 82)* ⬓◒
Water incompletely diverted from spout to shower (push-pull diverters)	O-rings worn	Replace O-rings or diverter *(p. 82)* ⬓◒
	Diverter coated with sediment	Clean diverter *(p. 82)* ⬓◒
	Diverter worn	Replace diverter *(p. 82)* ⬓◒
Weak or uneven pressure from shower head	Mineral buildup on shower head	Clean or replace shower head *(p. 83)* ⬓◒
Shower head leaks	Connections between shower head and arm loose	Tighten shower head to arm or seal joint with pipe tape *(p. 83)* □○

DEGREE OF DIFFICULTY: □ Easy ⬓ Moderate ■ Complex
ESTIMATED TIME: ○ Less than 1 hour ◒ 1 to 3 hours ● Over 3 hours
▲ Special tool required

69

Variations
Differences in popular plumbing fixtures or systems are described throughout the book, particularly if a repair procedure varies from one situation to another, or from older to newer models.

Degree of difficulty and time
Rate the complexity of each repair, and how much time the job should take for a homeowner with average do-it-yourself skills.

Special tool required
Some repairs, particularly those involving pipefitting, require specialized plumbing tools *(page 136)*.

vating a broken sewer pipe, you may wish to call for profesional service. You will still have saved time and money by diagnosing the problem yourself.

Most of the repairs in *Kitchen & Bathroom Plumbing* can be made with wrenches, screwdrivers and a plunger or auger. Replacing a broken pipe or faucet set may require a propane torch, hacksaw or tube cutter. Basic plumbing tools—and the proper way to use them—are presented in the Tools & Techniques section starting on page 136. If you are a novice when it comes to home repair, read this section (and the pipefitting chapter on page 93) in preparation for a major job.

Plumbing requires patience; take your time and observe basic precautions. Before working the supply pipes or fixtures, always close the main shutoff valve *(page 10)* and drain the supply lines. Keep a bucket and rags handy to catch runoff from disassembled faucets, traps and cleanouts. Cover the sink or bathtub drain with a bath mat to protect its finish and prevent loss of parts. Check with local authorities about plumbing, building or sanitation codes that might apply to your repair, especially those involving drainage and septic systems. Most important, follow all safety tips and **Caution** warnings throughout the book.

Name of repair
You will be referred by the Troubleshooting Guide to the first page of a specific repair job.

Step-by-step procedures
Follow the numbered repair sequence carefully. Depending on the result of each step, you may be directed to a later step, or to another part of the book, to complete the repair.

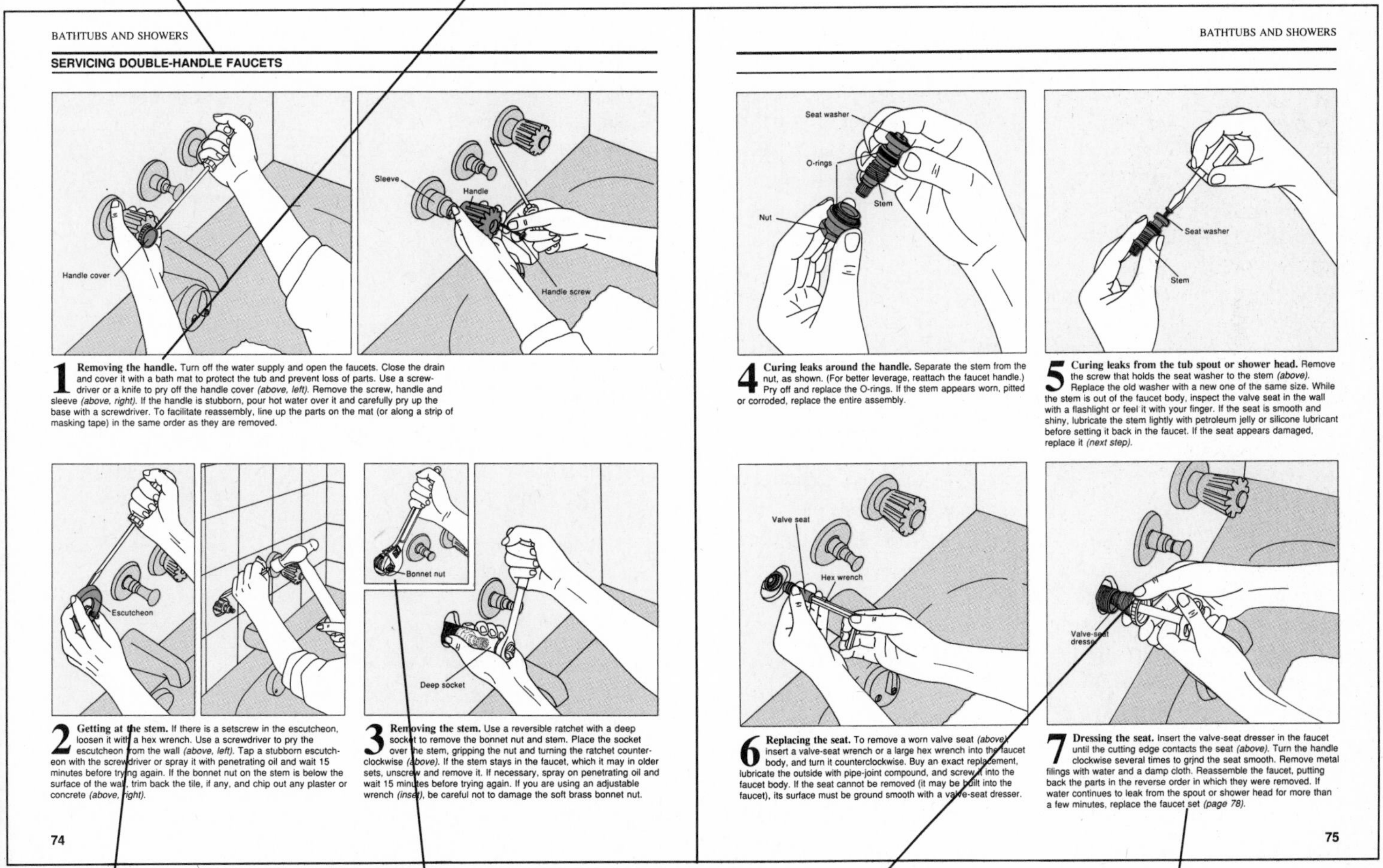
BATHTUBS AND SHOWERS

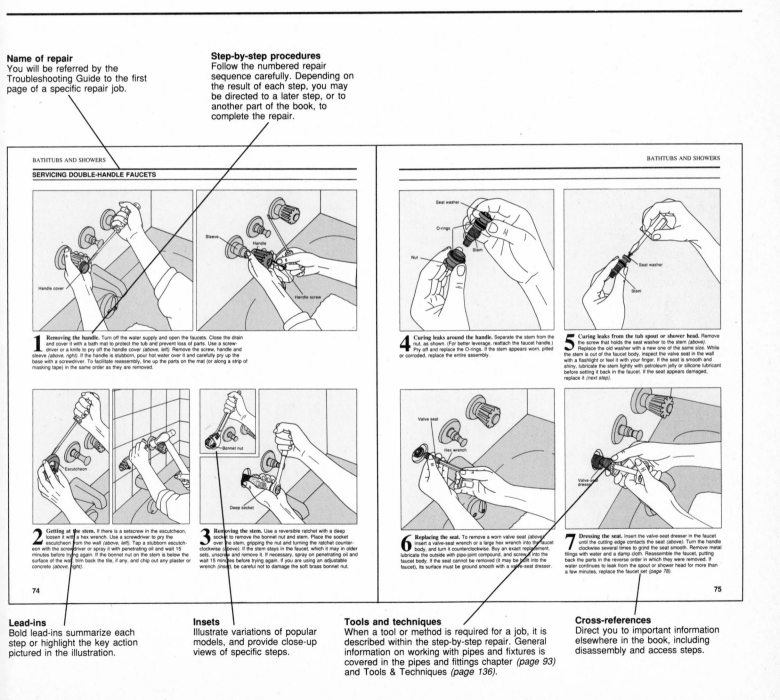

SERVICING DOUBLE-HANDLE FAUCETS

1 Removing the handle. Turn off the water supply and open the faucets. Close the drain and cover it with a bath mat to protect the tub and prevent loss of parts. Use a screwdriver or a knife to pry off the handle cover *(above, left)*. Remove the screw, handle and sleeve *(above, right)*. If the handle is stubborn, pour hot water over it and carefully pry up the base with a screwdriver. To facilitate reassembly, line up the parts on the mat (or along a strip of masking tape) in the same order as they are removed.

2 Getting at the stem. If there is a setscrew in the escutcheon, loosen it with a hex wrench. Use a screwdriver to pry the escutcheon from the wall *(above, left)*. Tap a stubborn escutcheon with the screwdriver or spray it with penetrating oil and wait 15 minutes before trying again. If the bonnet nut on the stem is below the surface of the wall, trim back the tile, if any, and chip out any plaster or concrete *(above, right)*.

3 Removing the stem. Use a reversible ratchet with a deep socket to remove the bonnet nut and stem. Place the socket over the stem, gripping the nut and turning the ratchet counterclockwise *(above)*. If the stem stays in the faucet, which it may in older sets, unscrew and remove it. If necessary, spray on penetrating oil and wait 15 minutes before trying again. If you are using an adjustable wrench *(inset)*, be careful not to damage the soft brass bonnet nut.

74

BATHTUBS AND SHOWERS

4 Curing leaks around the handle. Separate the stem from the nut, as shown. (For better leverage, reattach the faucet handle.) Pry off and replace the O-rings. If the stem appears worn, pitted or corroded, replace the entire assembly.

5 Curing leaks from the tub spout or shower head. Remove the screw that holds the seat washer to the stem *(above)*. Replace the old washer with a new one of the same size. While the stem is out of the faucet body, inspect the valve seat in the wall with a flashlight or feel it with your finger. If the seat is smooth and shiny, lubricate the stem lightly with petroleum jelly or silicone lubricant before setting it back in the faucet. If the seat appears damaged, replace it *(next step)*.

6 Replacing the seat. To remove a worn valve seat *(above)*, insert a valve-seat wrench or a large hex wrench into the faucet body, and turn it counterclockwise. Buy an exact replacement, lubricate the outside with pipe-joint compound, and screw it into the faucet body. If the seat cannot be removed (it may be built into the faucet), its surface must be ground smooth with a valve-seat dresser.

7 Dressing the seat. Insert the valve-seat dresser in the faucet until the cutting edge contacts the seat *(above)*. Turn the handle clockwise several times to grjnd the seat smooth. Remove metal filings with water and a damp cloth. Reassemble the faucet, putting back the parts in the reverse order in which they were removed. If water continues to leak from the spout or shower head for more than a few minutes, replace the faucet set *(page 78)*.

75

Lead-ins
Bold lead-ins summarize each step or highlight the key action pictured in the illustration.

Insets
Illustrate variations of popular models, and provide close-up views of specific steps.

Tools and techniques
When a tool or method is required for a job, it is described within the step-by-step repair. General information on working with pipes and fixtures is covered in the pipes and fittings chapter *(page 93)* and Tools & Techniques *(page 136)*.

Cross-references
Direct you to important information elsewhere in the book, including disassembly and access steps.

EMERGENCY GUIDE

Preventing plumbing problems. The relatively high pressure that delivers fresh water throughout the house—and the relatively low pressure that carries it away—is responsible for the majority of plumbing emergencies. On the supply side, the same pressure that sends water to a shower head or lawn sprinkler also forces water through a leaky faucet or defective pipe. On the drainage side, the pull of gravity may not be enough to draw waste water through drains that are clogged by debris. A greater force, usually provided by a plunger or auger, may be needed to clear the blockage.

Even a slow leak can waste about 15 gallons of water per day—enough for a comfortable bath. Left unchecked, that annoying drip will eventually become a steady stream. Replacing an inexpensive washer or O-ring, perhaps the simplest and most routine of all home plumbing tasks, could save your faucet set or sink. For less than a dollar you can buy a package of assorted washers to keep on hand for just such an emergency.

Do not rinse foods, fats or coffee grounds down a kitchen sink, and avoid using the toilet as a waste basket. If hair and soap frequently clog bathroom drains, install a strainer in the bathtub or sink. It is not advisable to use caustic chemical drain openers—particularly those containing lye—in a drain that is completely blocked. If the chemical does not clear the drain, you will be exposed to it as you plunge, auger or open the trap. However, a homemade concoction of 1 cup baking soda, 1 cup salt and 1/4 cup cream of tartar can help keep drains open. Every few weeks, spoon 1/4 cup of this mixture into the drain and add a quart of water. The sputtering chemical reaction will loosen accumulations in the pipes.

The repair of plumbing fixtures need not be any more dangerous than their daily use. In fact, proper repairs will prevent hazardous conditions caused by wear and neglect. The list of safety tips at right covers basic guidelines for safe service and use of your home's plumbing. See the chapters on individual fixtures, pipes and appliances for more specific advice.

The Troubleshooting Guide on page 9 puts emergency procedures at your fingertips. It lists the immediate steps to take, and refers you to pages 10-13 for more detailed information. Read this section before you need it, and familiarize yourself with the basic tools you will need for successful troubleshooting and repair on page 136.

When in doubt about the safety of your plumbing, or your ability to handle an emergency, don't hesitate to call for help. Post the telephone numbers for the fire department, gas company, electric utility and water department near the telephone. Even in non-emergency situations, they can answer questions concerning the proper use of plumbing and household utilities.

SAFETY TIPS

1. Before attempting any repair in this book, read the entire repair procedure. Familiarize yourself with the specific safety information presented in each section.

2. Know where the water shutoff valves are for the washer, dishwasher and icemaker, as well as the house's main shutoff valve. Label them.

3. Before servicing a water-using appliance, unplug the power cord or disconnect power at the service panel *(page 10)*. Leave a note on the panel so that no one reconnects the power while you are working.

4. Light and ventilate the work area well and do not reach into any area you cannot see clearly.

5. Beware of sharp metal edges and pointed screws; pad them with masking tape or wear heavy work gloves to prevent cuts.

6. Wear eye protection when using a propane torch, hammering, sawing, filing and working with overhead pipes.

7. Wear rubber gloves to handle caustic chemical drain openers.

8. When using a propane torch for loosening rusted fittings or sweat soldering, have a fire extinguisher at hand and protect flammable materials with a fireproof shield.

9. Be careful when using power tools in damp areas or around metal pipes. If possible, plug the tool into a grounded or GFCI-protected outlet.

10. Do not use a power auger for clearing drains unless you are confident in working with power tools.

11. Except for temporary repairs in an emergency, use only materials permitted by local plumbing codes. If in doubt, check with local authorities before working on your system.

12. Complete the repair before reconnecting the water, electricity or gas. Open the supply lines slowly at first to allow air to escape, then fully to flush debris from the pipes and faucet.

13. If you replace a section of metal pipe with plastic, check to see if the original pipe was part of the house's electrical grounding system. If so, have an electrician install a grounding jumper to maintain continuity.

14. If hot water is not used for 2 weeks or more, hydrogen can build up in the water heater and pipes. Before turning on appliances that use water, run all hot water taps in the house to clear out the gas.

15. Let a water heater cool before starting repairs, and never light a flame while working on a gas water heater.

16. If in doubt about the safety of a repair, call a plumber.

17. Post emergency, utility company and repair service numbers near the telephone.

18. Install smoke detectors and fire extinguishers in your home.

TROUBLESHOOTING GUIDE

SYMPTOM	PROCEDURE
Small object dropped down sink	Do not run water in sink
	Carefully remove trap under sink *(pp. 36, 46)*; have a bucket handy
Fixture leaking or overflowing	Turn off faucets
	Close shutoff valves at fixture or close main shutoff valve *(p. 10)*
Faucet bursts	Close shutoff valves at fixture or close main shutoff valve *(p. 10)*
Supply pipe leaks	Close main shutoff valve *(p. 10)*
	Patch hole temporarily with electrical tape or hose clamps and bicycle inner tube *(p. 11)*
Supply pipe bursts	Close main shutoff valve *(p. 10)*
	Drain pipe
	Repair or replace pipe *(p. 93)*
Supply pipe freezes	Turn up heat in house
	Close main shutoff valve *(p. 10)* and open nearest faucet
	Thaw with a hair dryer or heating tape *(p. 13)*
Toilet blocked	Do not flush; use plunger or closet auger to remove blockage *(p. 56)*
Toilet overflowing	Do not flush; bail out half the water, then dislodge clog and flush *(p. 56)*
Water heater, washing machine, dishwasher or garbage disposer leaking	Unplug power cord without touching the machine, or shut off power at service panel *(p. 10)*
	Turn off water supply at machine or close main shutoff valve *(p. 10)*
	Bail out or siphon out machine if full of water
Water on floor from appliance leak or overflowing appliance	If you must stand in water to mop it up, first unplug power cord or turn off power *(p. 10)*
	Dam area around water with washable, absorbent rags; clean up with mop or towels *(p. 12)*
Appliance or electrical outlet submerged	Do not enter room; if it is safe, turn off power at service panel *(p 10)*
Basement flooded	Close main shutoff valve *(p. 10)*
	Mop up, bail out or shovel water *(p. 12)*
	For 1/2 inch or less, use a wet-and-dry shop vacuum; for more than 1/2 inch, use a sump pump *(p. 12)*
Sewage fumes	Check fixture traps to be sure they have not run dry; run water to refill traps
	Pour a bucket of water in basement floor drain to replenish house trap, if any
Sewage backs up into house	Call municipality
Main shutoff valve broken	Call municipality to turn off water at the curb valve
	Replace main shutoff valve *(p. 110)*
Possibility of water contamination from back-siphonage or pollution	Do not attempt to flush out the supply lines yourself
	Have water tested for contaminants *(p. 16)*
	Call local health department
Ceiling sagging from weight of flooded water	Poke a small hole in ceiling with a large nail and catch water in a bucket *(p. 12)* When dry, repair the ceiling *(p. 139)*
Pilot light out in gas water heater	Relight pilot *(p. 114)*
Odor of escaping gas	Ventilate room
	Do not touch electrical outlets or switches
	Extinguish all flames
	Check pilots of all gas appliances and relight if necessary
Persistent odor of gas with all pilots lit	Ventilate room
	Turn off gas supply to all gas appliances *(p. 10)*
	Leave house and call gas company

SHUTTING OFF THE WATER SUPPLY

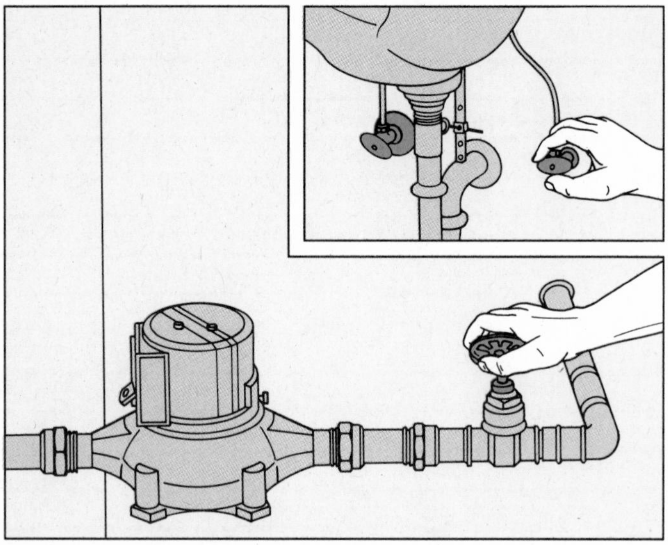

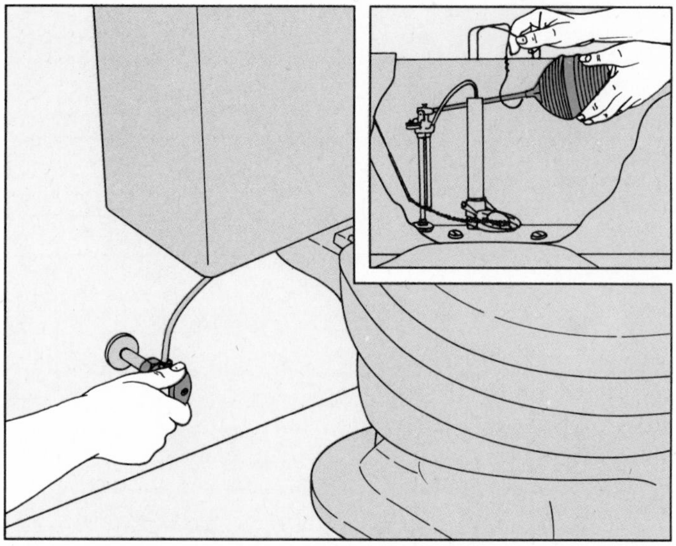

Locating shutoff valves. If water is spraying out of a broken faucet, or pouring from a burst pipe, the first step before attempting any repair is to turn off the water supply. Look for shutoff valves directly beneath the broken appliance or fixture where the water supply connects to it *(inset)*. If you don't see any handles there, or if the problem is a leaking pipe, look for the main shutoff valve supplying water to the entire house. This is usually a gate valve located near the water meter in the basement, utility room, crawlspace or even ouside in mild climates *(above, left)*. Turn the valve clockwise to shut off the water supply. To stop an overflowing toilet, close its single shutoff valve *(above, right)* or support the float ball and arm with a bent coat hanger *(inset)*.

TURNING OFF THE GAS AND ELECTRICITY

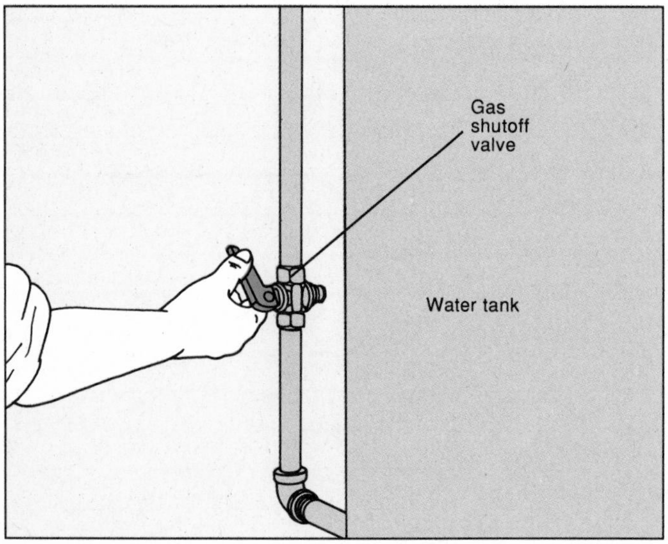

Turning off the gas supply. If the gas water heater has a valve on its supply pipe, turn the handle perpendicular to the pipe to shut off the gas *(above)*. If the gas in the room does not dissipate, leave the house and call the gas company. The gas supply to the entire house may be turned off at the meter. Using a wrench, turn the main valve so that its handle is perpendicular to the pipe.

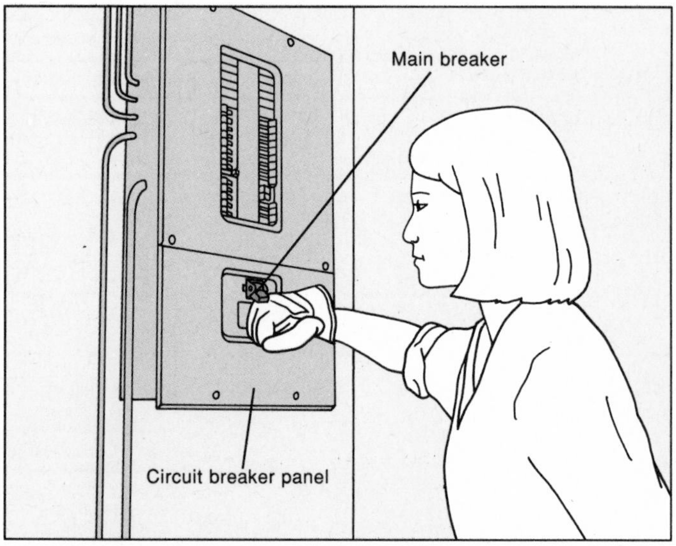

Disconnecting power at the service panel. If the floor is wet, stand on a dry board or rubber mat, wear heavy gloves and put one hand behind your back. Flip off the breaker, or unscrew the fuse, that controls the affected circuit. If you are unsure of the circuit, shut off the main power breaker. Use the back of your hand *(above)*; any shock will jerk your hand away from the panel.

STOPPING A SMALL LEAK

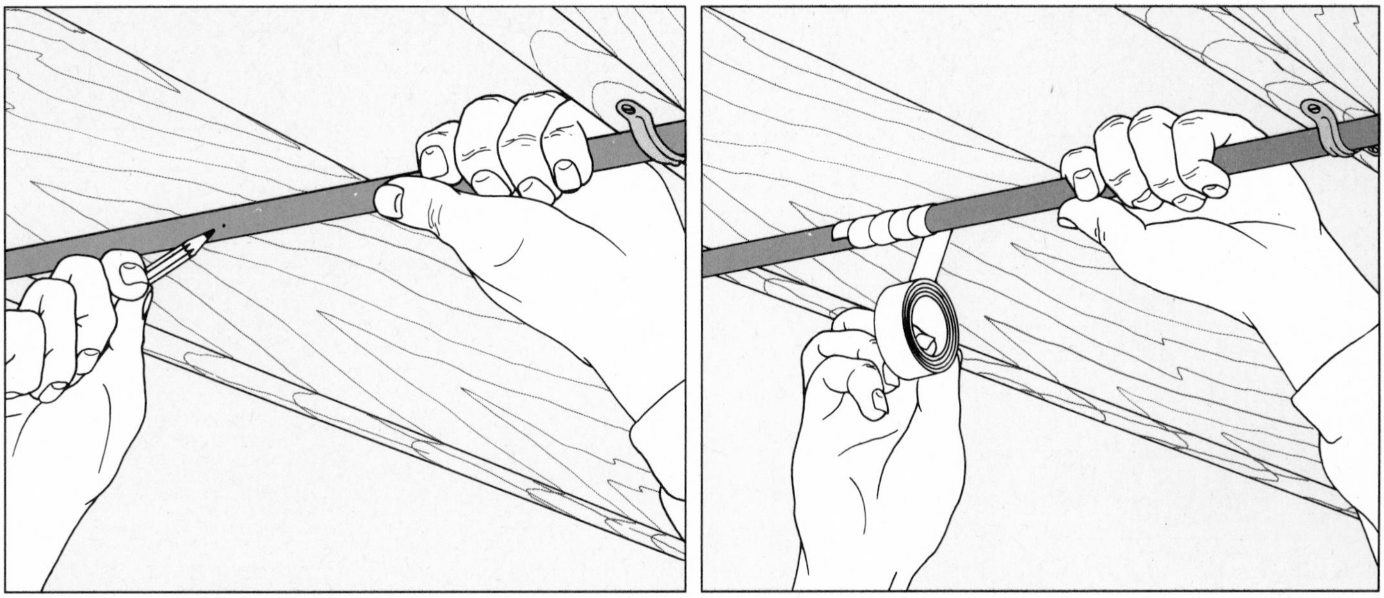

Quick fix for a pinhole leak. If you suspect a leak in the supply line, first check the water meter. If its needle or dial is moving and no one in the house is using a faucet or appliance, there is a leak somewhere. To locate it, look for stains on ceilings and walls, and listen carefully along supply pipes. As a temporary repair, turn off the water supply at the main shutoff valve, jam a pencil point into the hole and break it off *(above, left)*. A toothpick may also work, but for steel pipes, the graphite in the pencil lead will better seal the leak. To secure the plug, dry the surface of the pipe and wrap two or three layers of plastic electrical tape around the pipe for three inches on each side of the leak, overlapping each turn by half *(above, right)*. For a more permanent solution, you will have to replace the damaged section of pipe *(page 93)*.

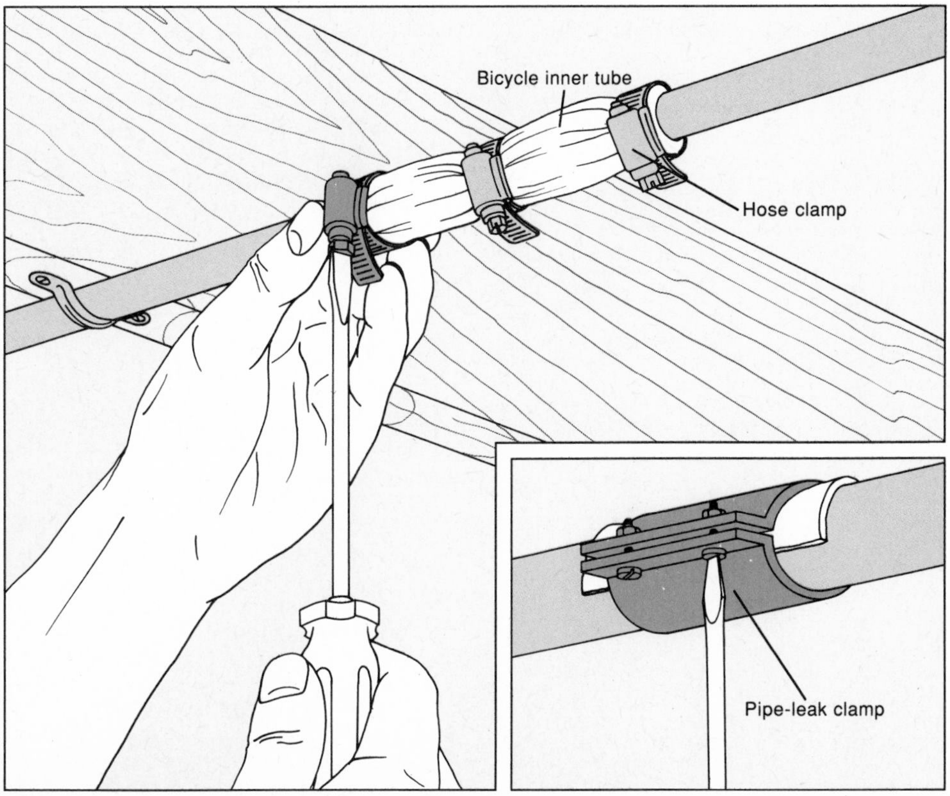

Sealing cracks or punctures. For a larger crack or puncture, close the main shutoff valve and drain the supply line. Wrap the pipe with an old bicycle inner tube and secure it with hose clamps *(left)*. Turn the water back on, slowly at first, to test for leaks. Replace the damaged section as soon as possible *(page 93)*.

To install a commercial pipe-leak clamp, remove the screws that hold the two halves of the clamp together, then fit them over the damaged pipe so that the rubber cushion seals the leak. Insert and tighten the screws *(inset)*.

Small leaks can also be patched using pipe cement or epoxy. The pipe must be drained of water and dried thoroughly for the cement to set properly. Roughen the damaged area with emery cloth, apply a thick coat of cement and allow it to dry overnight. Point a heat lamp or 100-watt bulb at the pipe to speed curing.

COPING WITH A FLOOD

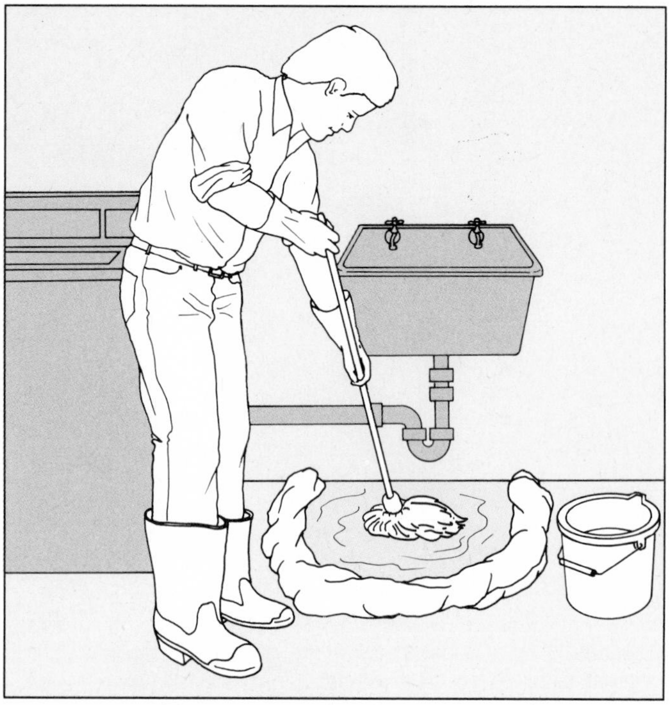

Damming the flood. Caution: If an appliance or outlet is submerged, do not enter the room; disconnect power at the service panel if it is safe to do so *(page 10)*. Turn off water to the leaking or overflowing fixture at its shutoff valves, or close the main shutoff valve. To keep water on the floor from spreading, surround it with a dam of washable, rolled-up rugs or towels. Then mop up the water within the dammed area *(above)*.

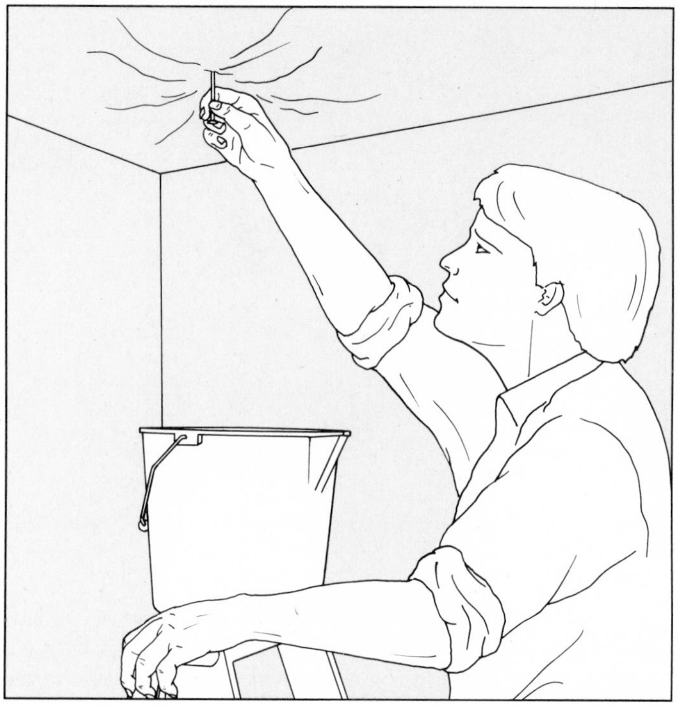

Puncturing a waterlogged ceiling. If flood water has collected above the ceiling and caused it to sag, stand on a stepladder and puncture it with a heavy nail. Place a bucket directly under the hole to catch the water as it pours out. When the water has drained, place a dehumidifier or space heater close to the damaged area to speed drying and prevent mildew. Patch the hole *(page 139)* when the ceiling is completely dry.

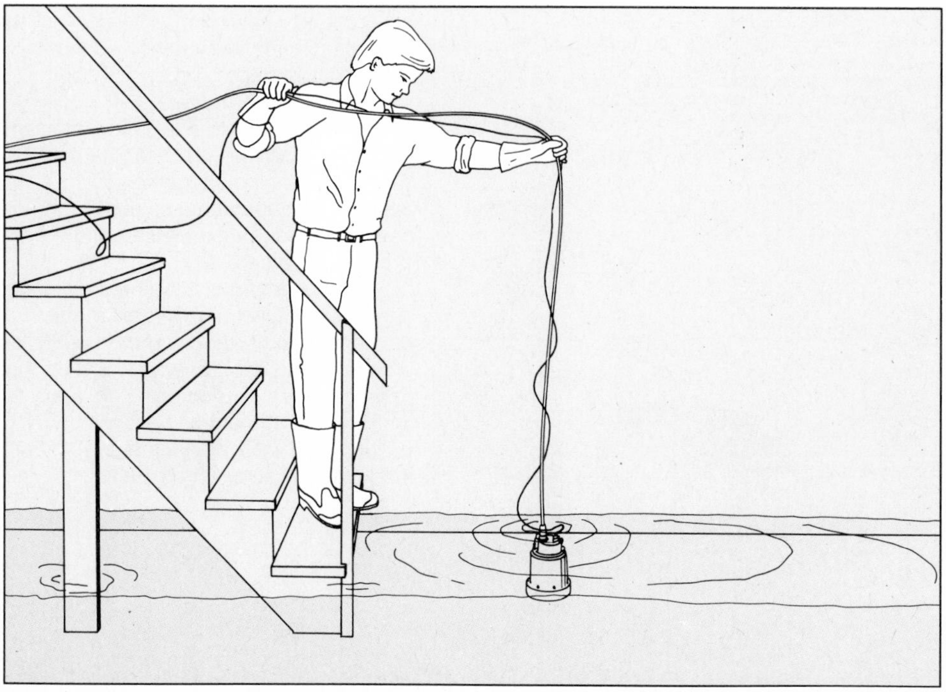

Using a sump pump. To get rid of water following a major flood, rent an electric or gas-powered sump pump from a tool-rental dealer. Flood water should be more than 1/2-inch deep to reach the pump impeller. Connect a long garden hose or 1 1/4-inch flexible pipe to the discharge fitting, then run the hose outside or to a working drain in the house. Next, lower the sump pump into the water to be drained *(left)*. To avoid electrical shock, never stand in the water while the sump pump is operating. While you are high and dry, plug the pump into a grounded outlet. Drain as much water as you can, but do not run the pump without water or you may damage its water-lubricated bearing. Stop the pump when it no longer sucks up water and remove any remaining water on the floor by bailing and mopping. Run clean water through the pump to flush its mesh filter and hose.

THAWING FROZEN PIPES

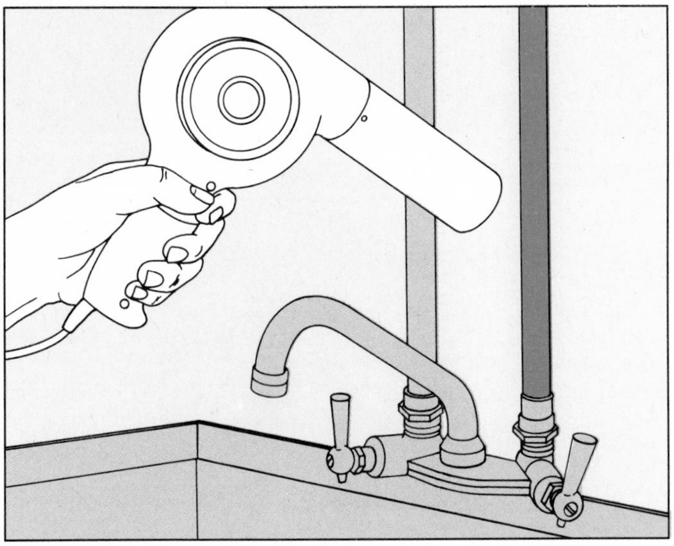

Using a hair dryer. When thawing frozen pipes, always open the nearest faucet to allow melting ice to drain, and close the main shutoff valve about 3/4 off. If you still have electricity, one of the safest remedies is to aim a hair dryer 3 to 4 inches from the affected area. Apply heat to the open faucet first, then work back along the pipe, as shown. When water begins to trickle from the tap, open the main shutoff; the flow of water will speed thawing.

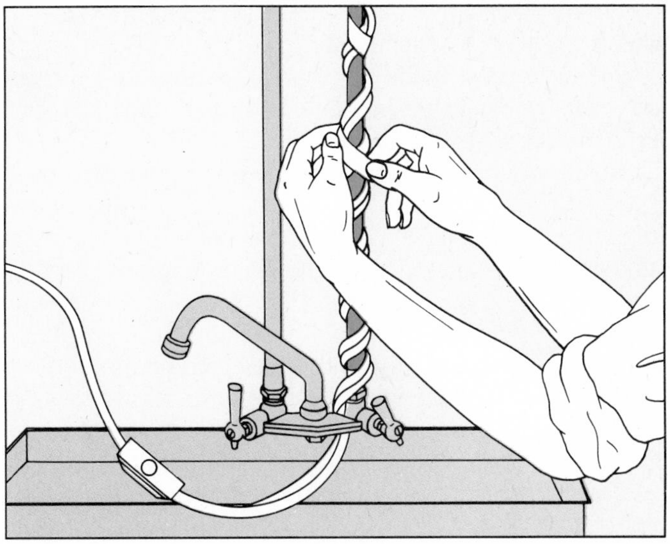

Using heating tape. Electric heating tape draws only a small amount of current to keep tap water safely above the freezing point. Some models are equipped with a thermostat that turns the tape on and off as needed; these may be permanently plugged in. Starting at a faucet or fixture, wrap the tape tightly around the exposed pipe, taking 6 to 8 turns per foot. Secure the spirals with plastic tape every 6 inches.

CHEMICAL DRAIN OPENERS

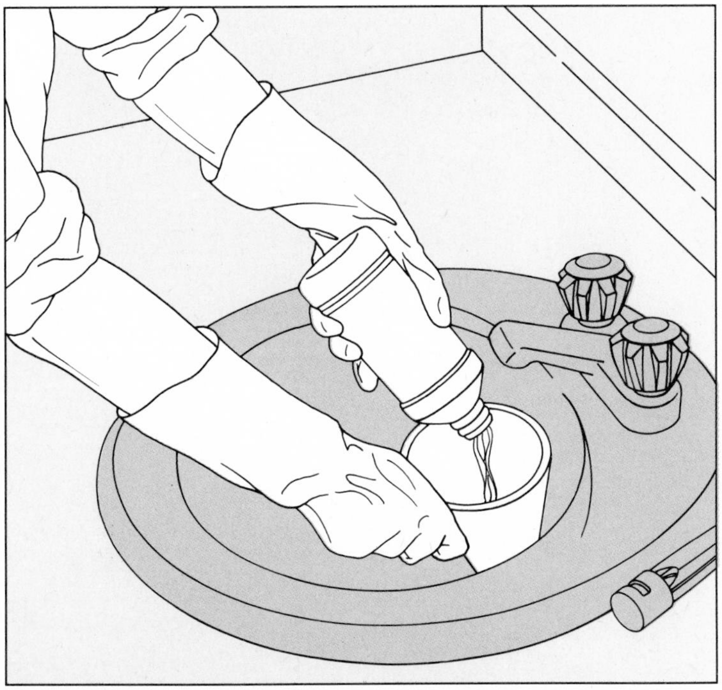

Clearing clogs with chemicals. To unblock a clogged or sluggish drain with a chemical drain opener, use a liquid alkali-based brand (with the ingredient sodium hydroxide—lye—listed on the label). Bail out any standing water in the sink or bathtub and place a funnel in the drain. Wearing rubber gloves (and safety goggles for extra protection), pour a small amount of drain opener into the funnel, being careful not to splash any of the chemical on your skin or in your eyes *(left)*. Use it only once on the clog. After 15 minutes, test the drain by flushing with cold water. If the clog remains, work directly on the trap. Because it reacts with water to produce toxic fumes, caustic soda is not recommended for household drains. It may also crystallize, inhibiting the use of a plunger or auger. Acid-based drain openers can corrode pipes and should never be used after an alkali-based variety has failed. Remember that all chemical drain openers are dangerous. They can burn eyes and skin, and must be kept out of reach of children.

YOUR HOME'S PLUMBING

Supply line
Carries water to house from private well or municipal supply.
Curb valve
Utility-owned shutoff valve.
Main shutoff valve
Allows water supply to
home to be opened or

Domestic plumbing consists of three basic systems: supply lines, fixtures and drainpipes. Although it may appear a random puzzle of pipes and fittings, the system and its various components work in a logical way.

The supply system is made up of pipes, fittings and valves that carry potable water throughout the house. Water enters your property under pressure from a reservoir, municipal water system or, in rural areas, from the pump-and-tank system of a private well. Typically, the water passes through a curb valve near the street (owned by the water utility), a water meter and the main shutoff valve. The meter is usually on the street side of the house, and the pipe enters a basement, crawlspace or, if your house is built on a slab, a utility center. Water pressure in the supply lines ranges from 35 to 100 pounds per square inch; the ideal pressure is 40-50 psi. Lower pressure may cause insufficient flow at fixtures; higher pressure may encourage water hammer or burst pipes.

From the cold water main, one supply pipe branches off to the water heater to begin a second, parallel run called the hot water main. From there, secondary branches of hot and cold water spaced about 6 inches apart snake through walls and ceilings to the various fixtures. In a well-designed system, each branch contains a shutoff valve near the point where it leaves the main line. Thus you can turn off an individual run without cutting off water to the entire house.

Plumbing fixtures include sinks, bathtubs, showers, toilets, sprinkler systems and appliances that use water and connect either permanently or temporarily to the supply and drainage systems. Not all fixtures need both of the supply lines; a toilet tank has only a cold water line, a dishwasher only a hot one. Once again, in a good system, shutoff valves control each fixture. Behind the wall at most fixtures are air chambers—capped vertical pipes that trap a column of air to cushion onrushing water when the faucet is turned off. Without an air chamber, an abrupt turn-off might create several hundred pounds of pressure within the supply system and result in water hammer.

The DWV (drain-waste-vent) system is the least visible part of the plumbing system, but the most strictly regulated by plumbing codes. The system is not under pressure, but depends on gravity to carry waste water out of the house. Each fixture is connected to a drainpipe by a P- or S-shaped trap filled with water that prevents harmful sewer gas from entering the home. When a toilet is flushed or a sink emptied, the water in the trap is replaced.

Branch drains lead to a larger vertical pipe called a stack, which drops to the level of the outgoing sewer line, and projects up through the roof to vent sewer gas and maintain atmospheric pressure in the system. (Larger dwellings may have two or more stacks.) At ground level (or below if there is a basement), the stack makes a near-45-degree turn to become the main drain, then slopes away from the house to enter a public sewer line or private septic system.

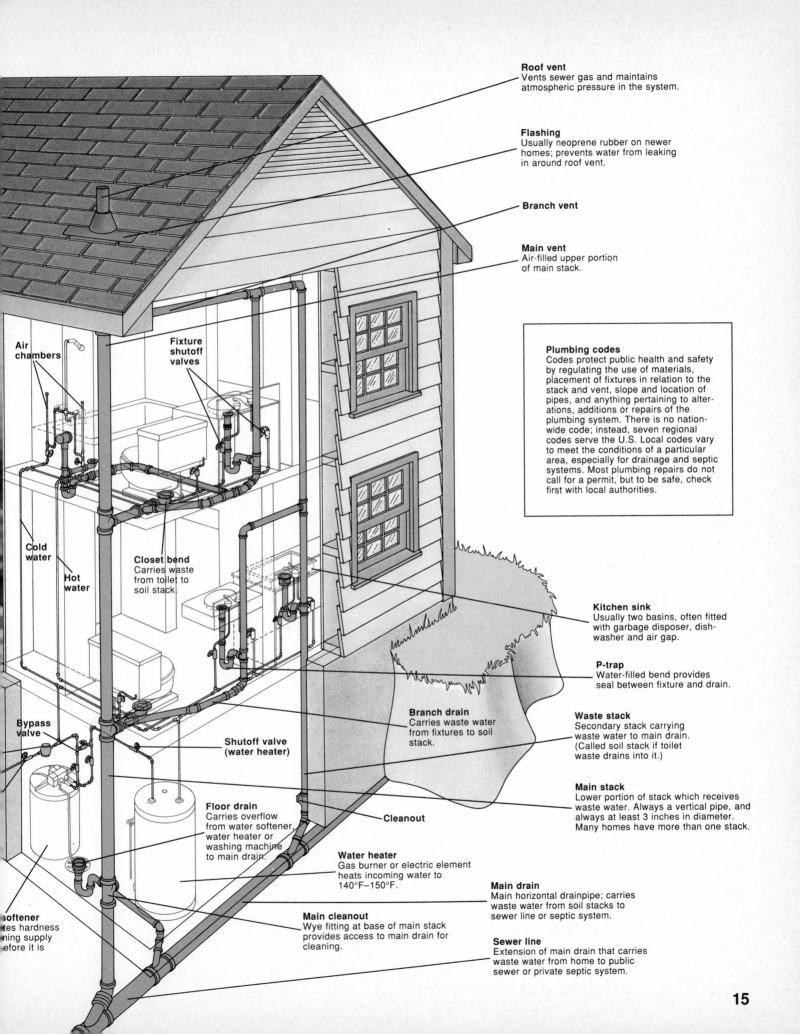

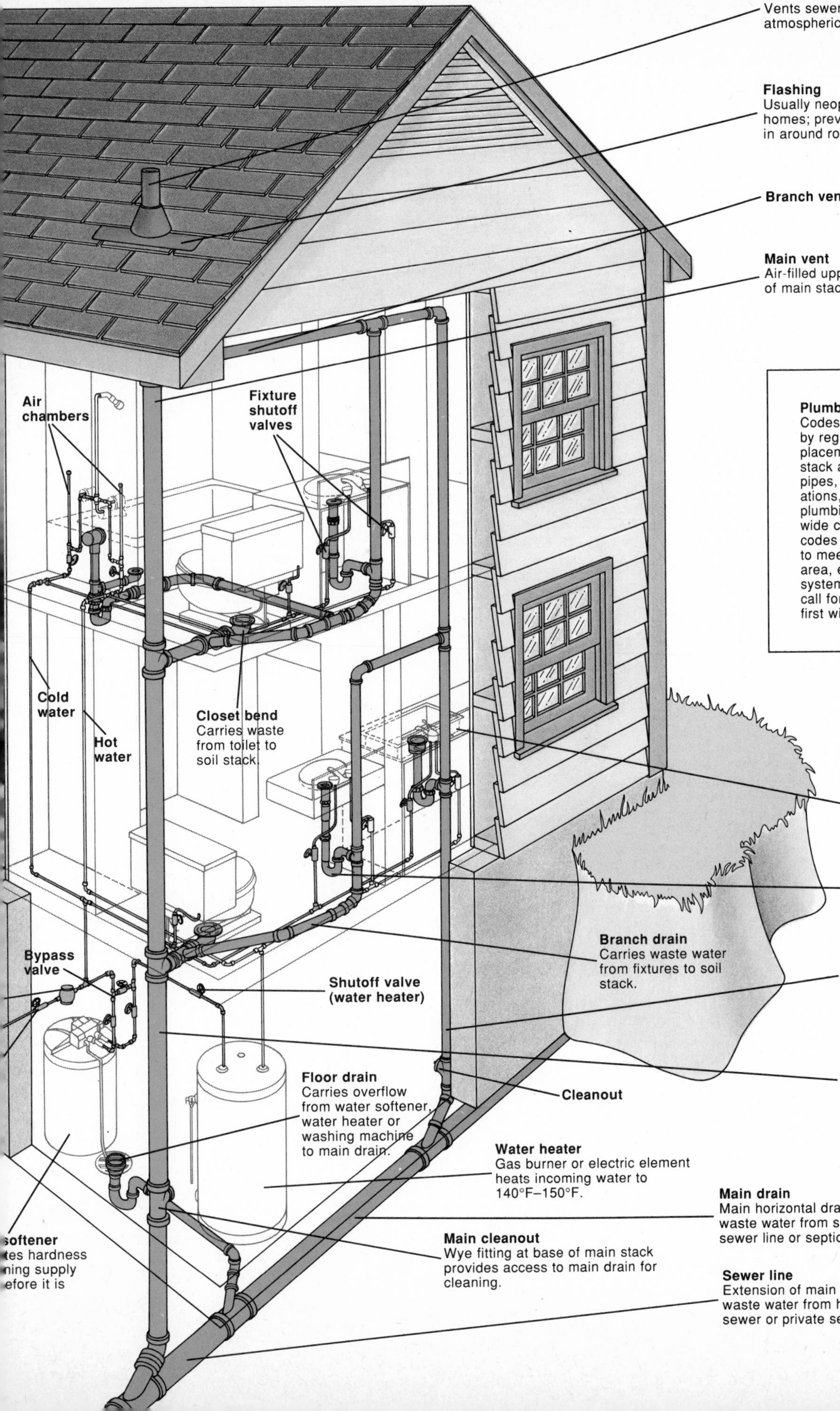
Roof vent
Vents sewer gas and maintains atmospheric pressure in the system.
Flashing
Usually neoprene rubber on newer homes; prevents water from leaking in around roof vent.
Branch vent
Main vent
Air-filled upper portion of main stack.
Air chambers
Fixture shutoff valves
Cold water
Hot water
Closet bend
Carries waste from toilet to soil stack.
Kitchen sink
Usually two basins, often fitted with garbage disposer, dishwasher and air gap.
P-trap
Water-filled bend provides seal between fixture and drain.
Branch drain
Carries waste water from fixtures to soil stack.
Waste stack
Secondary stack carrying waste water to main drain. (Called soil stack if toilet waste drains into it.)
Bypass valve
Shutoff valve (water heater)
Main stack
Lower portion of stack which receives waste water. Always a vertical pipe, and always at least 3 inches in diameter. Many homes have more than one stack.
Floor drain
Carries overflow from water softener, water heater or washing machine to main drain.
Cleanout
Water heater
Gas burner or electric element heats incoming water to 140°F–150°F.
Main drain
Main horizontal drainpipe; carries waste water from soil stacks to sewer line or septic system.
softener
tes hardness
ning supply
efore it is
Main cleanout
Wye fitting at base of main stack provides access to main drain for cleaning.
Sewer line
Extension of main drain that carries waste water from home to public sewer or private septic system.

Plumbing codes
Codes protect public health and safety by regulating the use of materials, placement of fixtures in relation to the stack and vent, slope and location of pipes, and anything pertaining to alterations, additions or repairs of the plumbing system. There is no nationwide code; instead, seven regional codes serve the U.S. Local codes vary to meet the conditions of a particular area, especially for drainage and septic systems. Most plumbing repairs do not call for a permit, but to be safe, check first with local authorities.

HOW SAFE IS YOUR WATER?

There are 60,000 public water supply systems in the United States, and the U.S. Environmental Protection Agency estimates that as many as 5,000 of these provide water that is potentially hazardous to health. The most common complaints regarding water quality include hardness, the presence of dissolved calcium and magnesium; reddish stains on plumbing fixtures caused by high levels of iron; the foul taste and smell of hydrogen sulfide; cloudy water caused by suspended particles of mud, clay or silt; and contamination by disease-carrying bacteria.

Not all chemicals found in tap water are bad. Used in moderation, chlorine is effective in killing harmful bacteria, and fluoride reduces tooth decay. But the fact is that 400 million pounds of toxic chemicals pour into America's waters each year in the form of agricultural pesticides, industrial waste, radioactive substances, gasoline from leaky storage tanks, even salt from road de-icing.

To ensure an adequate supply of potable drinking water, almost every town and city in the country now chlorinates its water. (Much chlorination is done to ensure that pure water arrives at the tap without becoming contaminated en route.) But if your water has a peculiar taste, smell or color, or if there is frequent illness in your household that may be water-borne, have your water tested. In some areas, the local health department may test well water; otherwise, have tap water tested by a commercial laboratory. (Your state department of health or environmental protection should be able to advise you on certified testing labs in your area.) There are two types of standard lab tests: a bacteria or coliform analysis, which indicates the presence of bacteria; and a chemical test that determines the levels of toxic, corrosive and "nuisance" substances. Maximum safe contaminant levels are determined by the Environmental Protection Agency *(page 17)*.

Once an impurity has been identified, one or more water treatment devices can be installed to correct it. The most common is a water softener *(page 128)*, which reduces water hardness by replacing calcium and magnesium with sodium in an ion-exchange process. Sand or turbidity filters trap suspended particles of silt, algae and organic material. Neutralizing filters use limestone chips (calcium carbonate) to raise the pH level of water and reduce acidity. Oxidizing filters remove iron, which can interfere with water softeners, and reduce hydrogen sulfide, which gives water a rotten-egg smell and tarnishes silverware. All these filters must be backflushed and regenerated on a regular basis to restore the active medium. A chlorinator kills harmful bacteria by metering a small amount of household bleach into the supply water. Such chemical feeders are often used to treat well water, which is particularly susceptible to contamination.

If a water test reveals a high level of lead in your drinking water, check the supply pipes. Older houses may contain lead plumbing, which scratches easily and is identifiable by its dull gray color. Lead solder is also used to join copper pipes, and lead ions may leach into supply water from this source. For this reason, the use of low-lead or silver solder is recommended for soldering copper pipes and fittings.

TROUBLESHOOTING GUIDE

SYMPTOM	POSSIBLE CAUSE	PROCEDURE
Scale in pipes and water heaters; soap residue on dishes and fabrics; bathtub rings	Hard water	Install ion-exchange water softener
Reddish stains on plumbing fixtures and fabrics	Iron	Install water softener, iron filter or both
Foul taste and rotten-egg smell; tarnished silverware	Hydrogen sulfide	Install oxidizing filter
Corrosion on pipes and fixtures; red stains on fixtures with galvanized pipe, blue-green stains on fixtures with copper pipe	Acidic water	Install neutralizing filter
Dirty or cloudy water	Suspended particles of mud, silt or clay	Install sand (turbidity) filter
Water salty	Alkaline water	Install reverse-osmosis purifier
Illness or disease from water	Bacteria in water supply	Have water tested and follow advice of local health department; one solution may be to install a chlorinator
Swampy odor, taste; brown or green color	Algae	Install chlorinator to kill algae and activated-carbon filter to improve taste
Lead poisoning	Lead supply pipes or lead solder in copper supply pipes	Replace lead pipes; remake all lead-soldered joints with silver solder or replace with compression fittings; install neutralizing filter

MAXIMUM SAFE CONTAMINANT LEVELS

SUBSTANCE	MAXIMUM LEVEL (mg/l)	SUBSTANCE	MAXIMUM LEVEL (mg/l)
INORGANIC CHEMICALS		**ORGANIC CHEMICALS**	
Arsenic	0.05	Endrin	0.0002
Barium	1.0	Lindane	0.004
Cadmium	0.01	Methoxychlor	0.1
Chloride	250	Toxaphene	0.005
Chromium	0.05	Total trihalomethanes - THM	0.1
Copper	1.3		
Detergents	0.5	**PESTICIDES**	
Fluoride	4.0	2,4,D	0.10
Iron	0.3	2,4,5-TP silvex	0.01
Lead	0.05		
Manganese	0.05	**RADIOACTIVE SUBSTANCES**	
Mercury	0.002	Radium 226 and 228	5 pCi/l
Nitrites and nitrates	10.0	Gross alpha particle activity	15 pCi/l
Selenium	0.01		
Silver	0.05	**OTHER**	
Sulfate	250.0	Recommended pH	6.5 - 8.5
Total dissolved solids	500.0	Total coliform bacteria	<1/ 100 ml

HOME FILTERING SYSTEMS

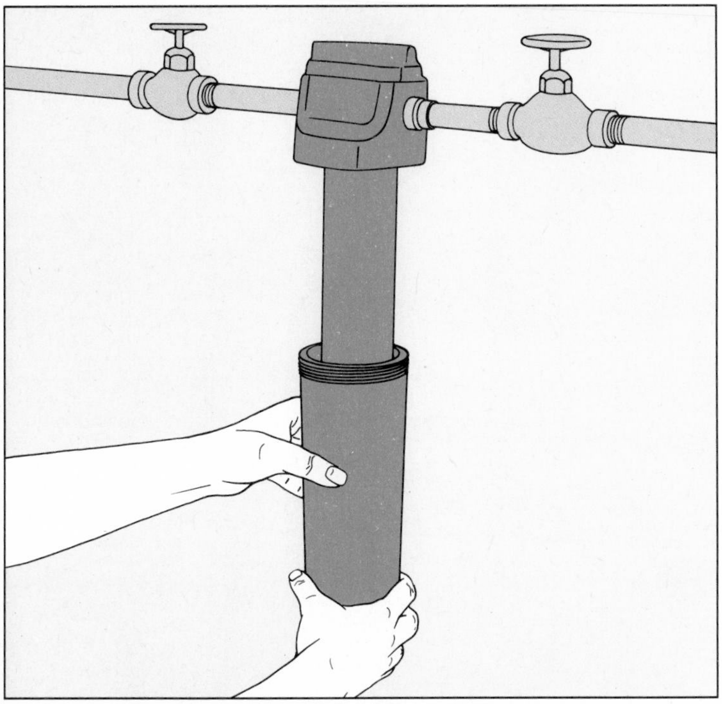

In-line cartridge filters. In-line water filters are usually installed in the supply line, where they trap tiny water-borne particles in a porous ceramic or activated-carbon cartridge. Because it becomes clogged with use, the cartridge should be replaced periodically—usually twice a year. To change it, turn off the inlet and outlet valves on each side of the filter. (Some filters have a built-in shutoff valve.) Unscrew the filter housing from the body *(above)*, remove the old cartridge and slip a new one into place. Replace the housing and open the valves.

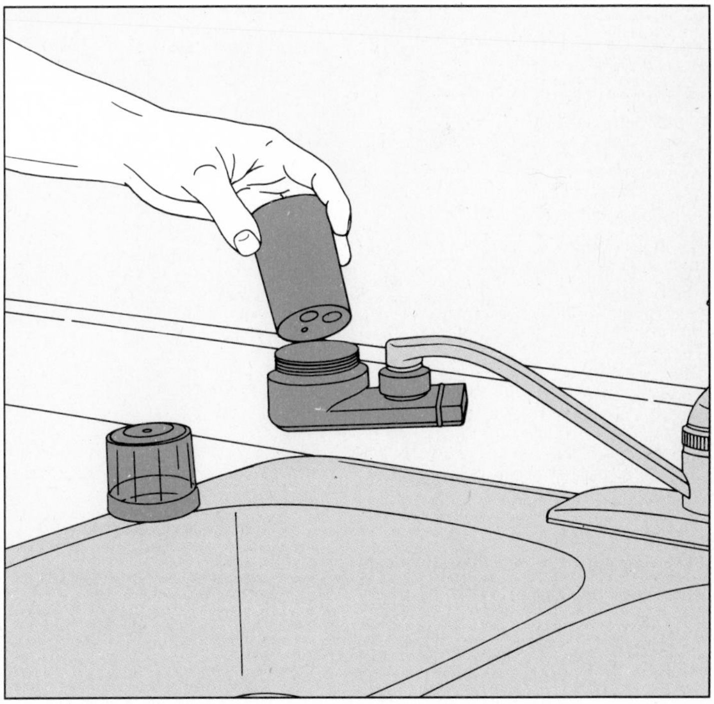

Faucet-mounted filters. These inexpensive treatment devices improve the taste and clarity of drinking water by trapping suspended particles, and remove the objectionable taste left by chlorine purification treatment. A control on the unit allows you to draw either filtered or nonfiltered water. To replace a cartridge, which typically lasts two to three months, simply unscrew the cover and lift out the old cartridge filter, as shown. Install the new cartridge, then replace the cover. Flush a gallon of water through the filter before using it.

FAUCETS

Sink faucets are simple valves that control the flow of thousands of gallons of water each year in the kitchen and bathroom. Faucet trouble usually announces itself as a steady drip, drip, drip from the spout, or as a slow leak from around the handle or collar. To solve the problem, you must first identify what kind of faucet you have so that you can buy the exact replacement parts. In most cases, this means disassembling the faucet, then comparing it to those illustrated in this chapter.

Compression and *reverse-compression* faucets, always double-handle, have a washer that rests on a seat at the bottom of the stem. When a compression faucet is turned on, the washer rises to allow water to flow to the spout. When a reverse-compression faucet is turned on, the stem lowers to create a space between the washer and seat, allowing water up. Simply changing the stem washer will often stop the spout from dripping, but on older faucets the seat may also need replacement. Replacing the O-ring or packing in the stem will usually stop leaks from the handle.

A *diaphragm* faucet, another type of double-handle faucet, is easily repaired. A change of O-ring stops most leaks from the handle. Replacing the diaphragm, which controls water flow, stops leaks from both the spout and the handle.

A *disc* faucet, double-handle or single-lever, has a pair of plastic or ceramic discs that move up and down to regulate the volume of water, and rotate to control temperature. The disc assembly rarely needs changing, but the inlet ports can become clogged, and the seals can wear out.

A *cartridge* faucet regulates water flow by means of a cartridge controlled by a single lever. Repairs involve changing the O-rings or replacing the entire cartridge.

A *rotating-ball* faucet, another single-lever faucet, employs a slotted plastic or brass ball set atop a pair of spring-loaded rubber seats. The handle rotates the ball to adjust water temperature and flow. When this faucet leaks from the spout, its springs and seats probably need replacing. Seepage around the handle points to worn O-rings or a loose adjusting ring.

As the body of a faucet set wears out, it may start to leak under the sink and must be replaced. You then have the choice to keep a double-handle faucet or to install a single-lever faucet, which usually needs less frequent repair.

Kitchen faucets often come with a sink spray, which requires a diverter inside the faucet body. Repairs for this accessory and the aerator attached to most spouts appear on page 34.

DOUBLE-HANDLE FAUCET

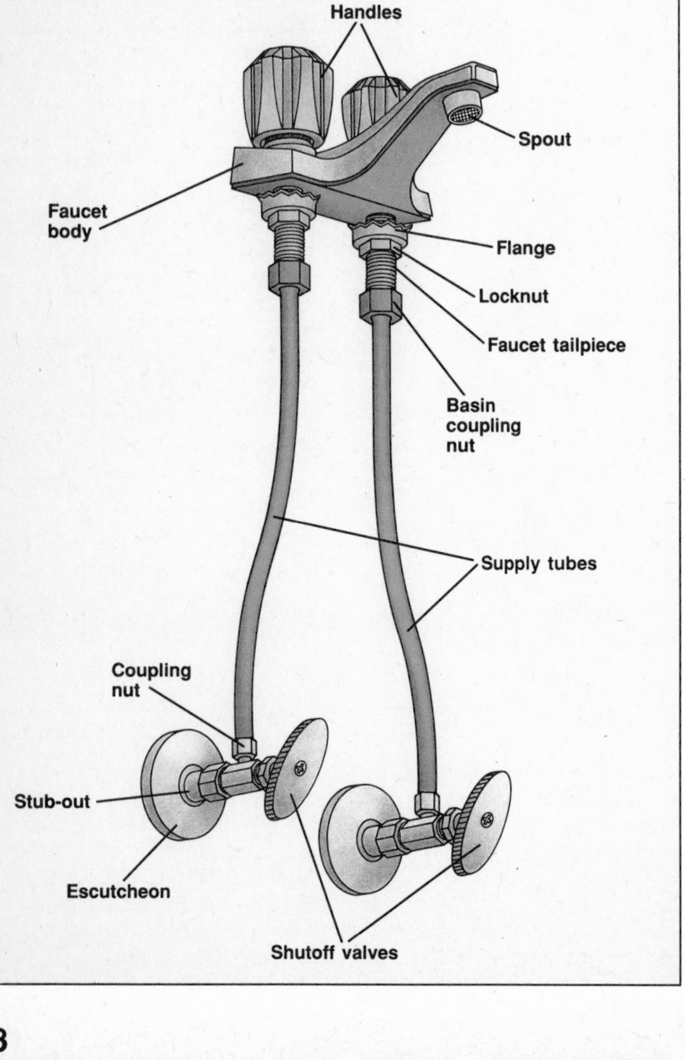

SINGLE-LEVER FAUCET

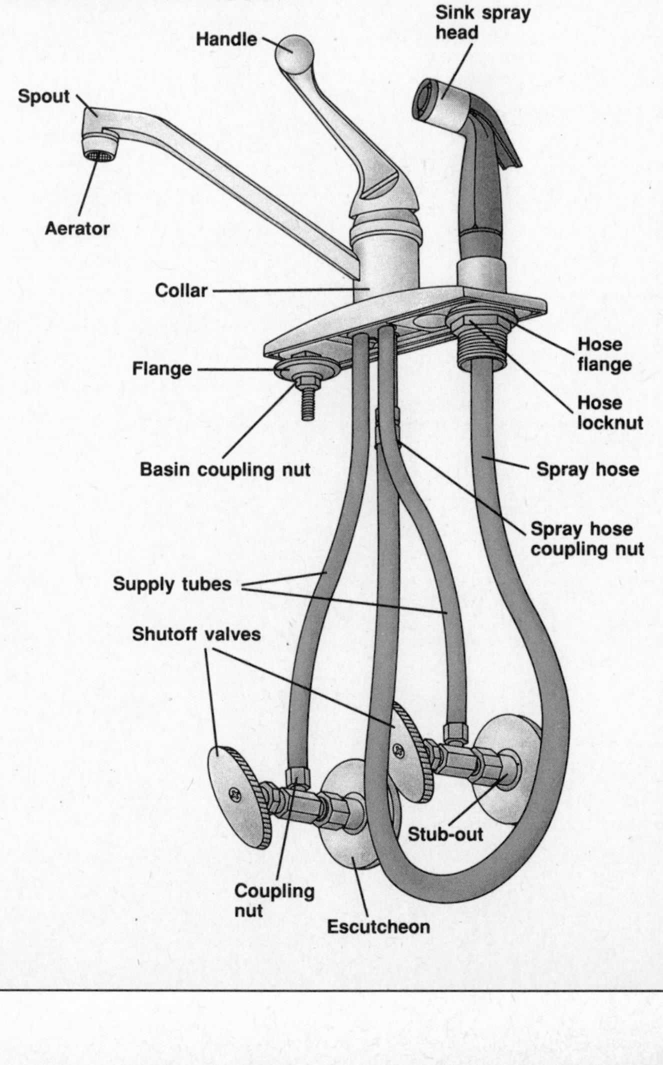

TROUBLESHOOTING GUIDE

SYMPTOM	POSSIBLE CAUSE	PROCEDURE
Double-handle compression faucet drips from spout	Washer worn or damaged	Replace washer *(p. 20)* □○
	Seat pitted or corroded	Newer models: replace or dress seat *(p. 20)* ⬓◒▲; Older models: dress seat *(p. 20)* ⬓◒▲, or replace faucet *(p. 31)* ⬓◒
Double-handle compression faucet leaks from handle	O-rings worn or damaged (newer models)	Replace O-rings *(p. 20)* □○
	Packing nut loose (older models)	Tighten packing nut *(p. 22)* □○
	Stem packing worn (older models)	Replace packing *(p. 22)* □○
	Stem bent (older models)	Straighten or replace stem *(p. 22)* □○ Replace faucet *(p. 31)* ⬓◒
Double-handle reverse-compression faucet drips from spout	Seat washer worn or cracked	Replace seat washer *(p. 23)* □○
Double-handle reverse-compression faucet leaks from handle	Packing washer worn	Replace packing washer *(p. 23)* □○
Double-handle diaphragm faucet leaks from handle	O-ring on sleeve worn	Replace O-ring *(p. 24)* □○
Double-handle diaphragm faucet drips from spout	Diaphragm worn	Replace diaphragm *(p. 24)* □○
Double-handle disc faucet leaks from handle	O-ring on disc assembly worn	Replace O-ring *(p. 25)* □○
Double-handle disc faucet drips from spout	Disc cracked or pitted	Replace disc assembly *(p. 25)* □○
	Seat assembly worn	Replace seat assembly *(p. 25)* □○
Single-lever rotating-ball faucet leaks from handle or drips from spout	Adjusting ring loose	Tighten adjusting ring *(p. 26)* □○
	Cam assembly worn	Replace cam assembly *(p. 26)* □○
	Seat assembly worn	Replace seat assembly *(p. 26)* □○
	Ball cracked or pitted	Replace ball *(p. 26)* □○
Single-lever rotating-ball faucet leaks from collar	Collar O-rings worn	Replace O-rings *(p. 26)* □○
Single-lever cartridge faucet leaks from handle or drips from spout	Cartridge worn, cracked or pitted	Replace cartridge *(p. 28)* □○
Single-lever cartridge faucet leaks around spout collar	Collar O-rings worn	Replace O-rings *(p. 28)* □○
Ceramic disc faucet leaks around base or flow from spout reduced	Aerator blocked to reduce flow	Clean aerator *(p. 34)* □○
	Inlet ports blocked	Clean inlet ports *(p. 30)* □○
	Disc cracked or pitted	Replace disc assembly *(p. 30)* □○
	Inlet seals worn	Replace inlet seals *(p. 30)* □○
Water under sink	Faucet set loose	Tighten locknuts under faucet set *(p. 31)* □○
	Putty dried out or gasket worn	Replace putty or gasket *(p. 31)* ⬓◒
	Faucet body worn	Replace faucet *(p. 31)* ⬓◒
	Supply tubes damaged or kinked	Replace with flexible supply tubes *(p. 33)* ⬓◒
Flow from spout reduced	Aerator blocked	Clean aerator *(p. 34)* □○
Aerator leaks around edge	Washer in aerator worn	Replace washer *(p. 34)* □○
Spray hose leaks	O-ring on diverter valve worn	Replace O-ring on diverter valve *(p. 34)* □○
	Diverter valve bent	Replace diverter valve *(p. 34)* □○
	Washer at base of spray head worn	Replace washer *(p. 35)* □○
Flow of water from spray head reduced	Diverter valve dirty	Clean diverter valve *(p. 34)* □○
	O-ring on diverter valve worn	Replace O-ring *(p. 34)* □○
	Screen in spray head blocked	Clean spray head *(p. 35)* □○
	Handle on spray head broken	Replace spray head *(p. 35)* ⬓◒
	Spray hose cracked or kinked	Replace spray hose *(p. 35)* ⬓◒

DEGREE OF DIFFICULTY: □ **Easy** ⬓ **Moderate** ■ **Complex**
ESTIMATED TIME: ○ **Less than 1 hour** ◒ **1 to 3 hours** ● **Over 3 hours** ▲ **Special tool required**

DOUBLE-HANDLE COMPRESSION FAUCET (Newer models)

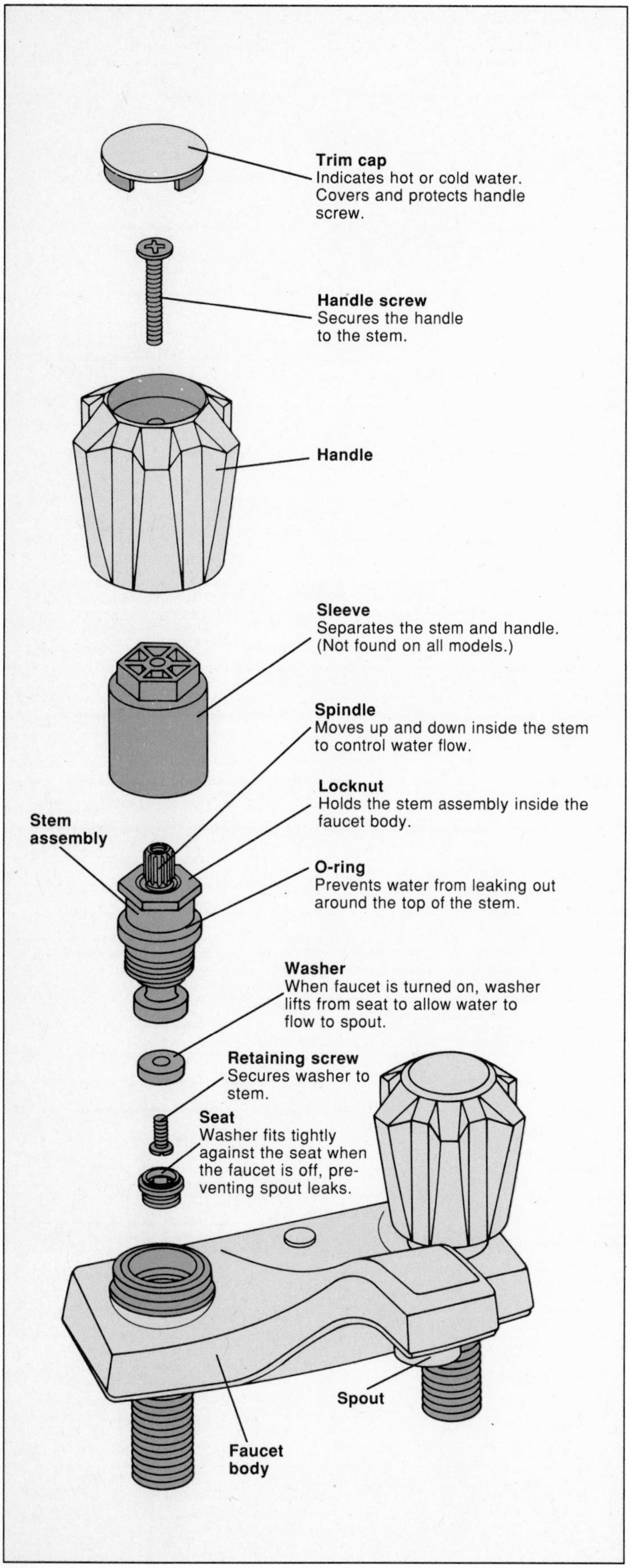
Trim cap
Indicates hot or cold water. Covers and protects handle screw.
Handle screw
Secures the handle to the stem.
Handle
Sleeve
Separates the stem and handle. (Not found on all models.)
Spindle
Moves up and down inside the stem to control water flow.
Locknut
Holds the stem assembly inside the faucet body.
Stem assembly
O-ring
Prevents water from leaking out around the top of the stem.
Washer
When faucet is turned on, washer lifts from seat to allow water to flow to spout.
Retaining screw
Secures washer to stem.
Seat
Washer fits tightly against the seat when the faucet is off, preventing spout leaks.
Spout
Faucet body

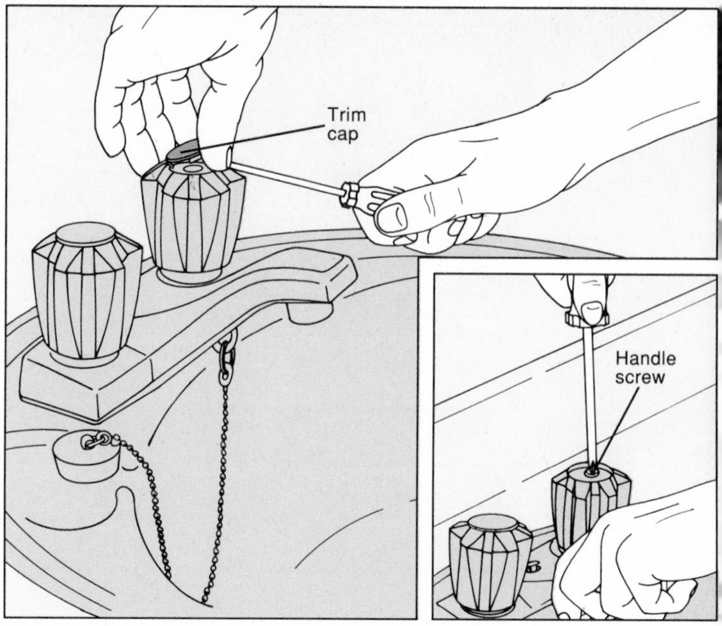
Trim cap
Handle screw

1 Opening up the faucet. To find out which handle needs servicing when the spout drips, turn.off one of the shutoff valves underneath the sink. If the leak stops, the problem is with that handle; if it persists, the other handle is at fault. (When there are no individual shutoff valves for the sink, service both handles.) Turn off the water supply, open the faucet and close the drain to prevent loss of parts. Carefully pry off the trim cap with a knife or small screwdriver *(above)*. Remove the screw that secures the handle to the stem *(inset)*.

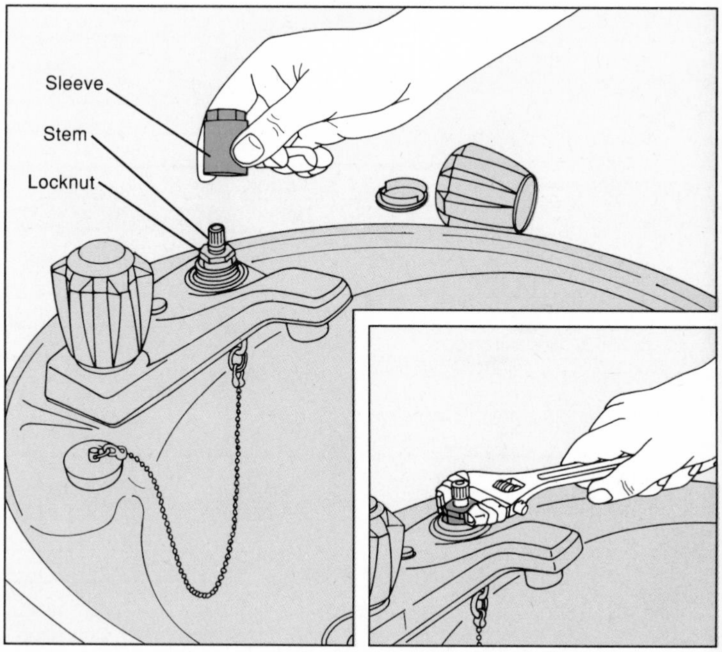
Sleeve
Stem
Locknut

2 Getting at the stem. Lift off the faucet handle and sleeve *(above)*. If it will not budge, apply penetrating oil and wait an hour before trying again. Never strike the handle or sleeve with a hammer; you might damage the soft brass stem. Open the faucet one-half turn, then unscrew the locknut that secures the stem to the faucet body *(inset)*.

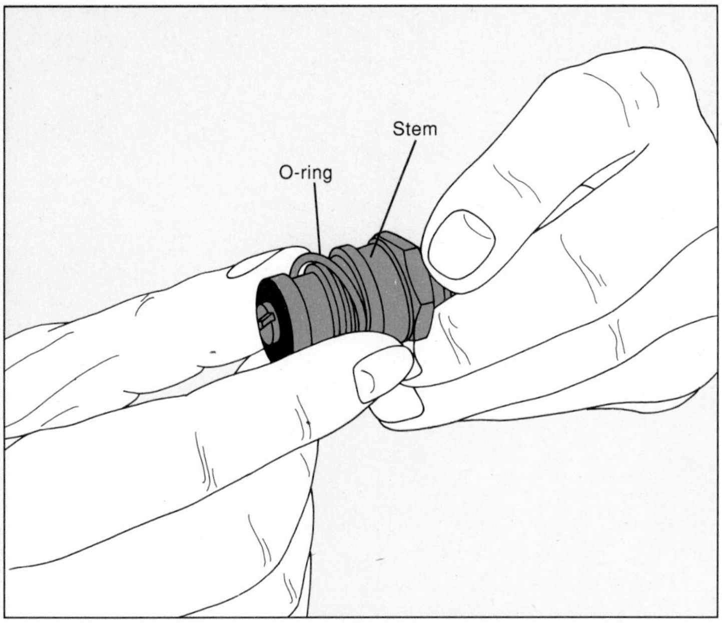

3 Replacing the O-ring. Grasp the stem spindle with taped pliers and lift it out of the faucet body. To stop leaks around the handle, pinch off the O-ring *(above)*, lubricate a new O-ring with petroleum jelly, then roll it onto the stem until it is firmly seated. Reassemble and test the faucet.

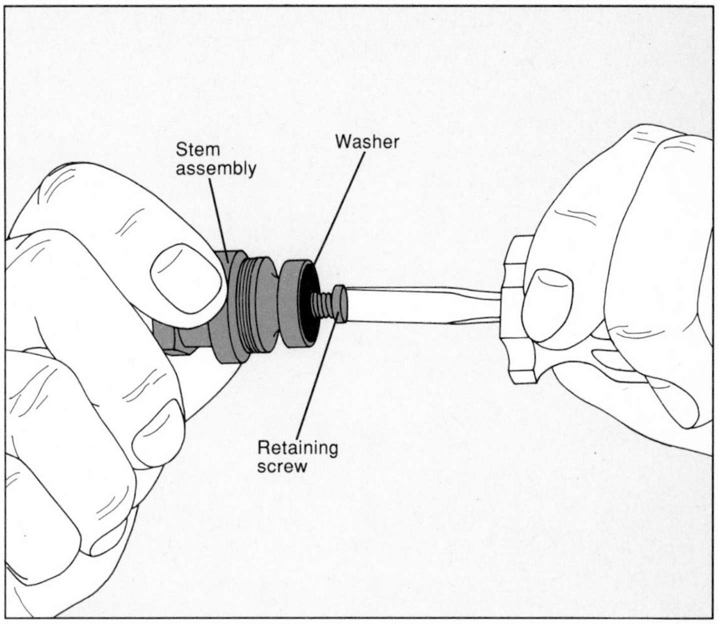

4 Replacing the washer. A dripping spout points to a worn stem washer. Carefully remove the retaining screw *(above)* and pry the washer from the stem with the tip of a screwdriver or knife. (If the screw is tight, reinstall the faucet handle for better leverage.) Replace with an identical washer, its flat side facing the stem. With the washer in place, tighten the retaining screw until it presses the washer squarely into the stem. Reassemble and test the faucet. If the spout still leaks, replace or dress the seat *(step 5)*.

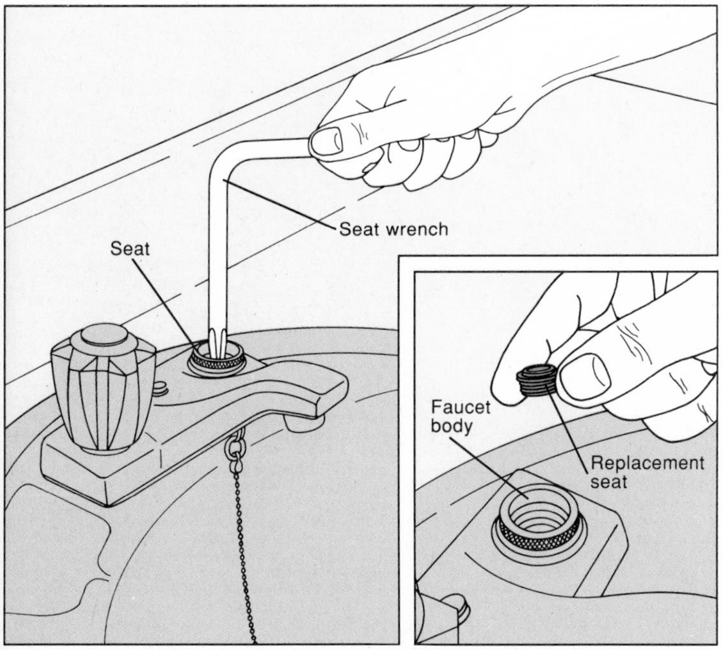

5 Removing the seat. Use a hex wrench or special faucet seat wrench *(above)* to unscrew a leaking seat by turning it counterclockwise. If the seat will not budge, apply penetrating oil, wait overnight and try again. Once the old seat is free, fit an identical replacement seat into the faucet body by hand *(inset)* or with a pair of long-nose pliers. Screw it in tightly with the hex wrench or seat wrench. If the seat cannot be removed—it may be built into the faucet—it must be ground smooth with a valve-seat dresser *(next step)*.

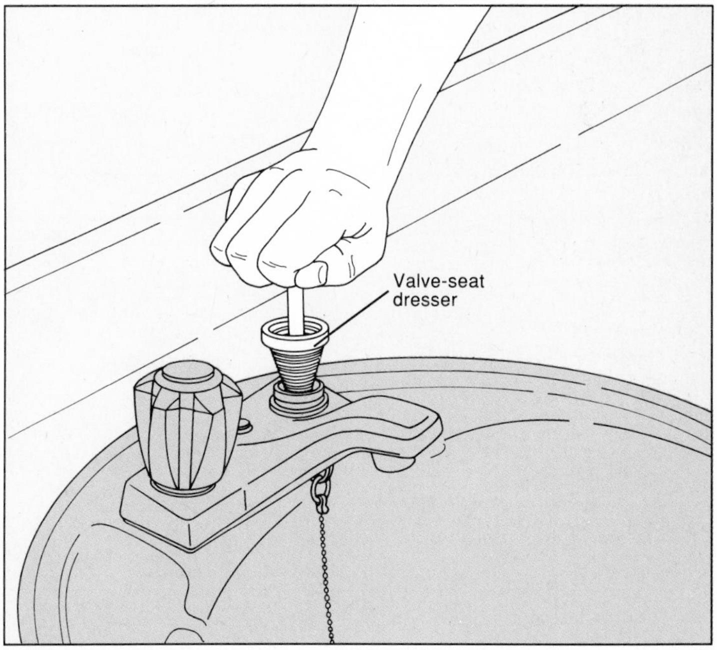

6 Dressing the seat. Buy or rent a valve-seat dresser with the largest cutter that fits the faucet body, and screw on a guide disc that fits the valve seat hole. Slide the cone down snugly into the faucet body. Turn the handle clockwise several times to grind the seat smooth *(above)*. Wipe out the filings with a damp cloth, and reassemble the faucet. If the leak persists, try again; if a second attempt fails, replace the entire faucet set *(page 31)*.

DOUBLE-HANDLE COMPRESSION FAUCETS (Older models)

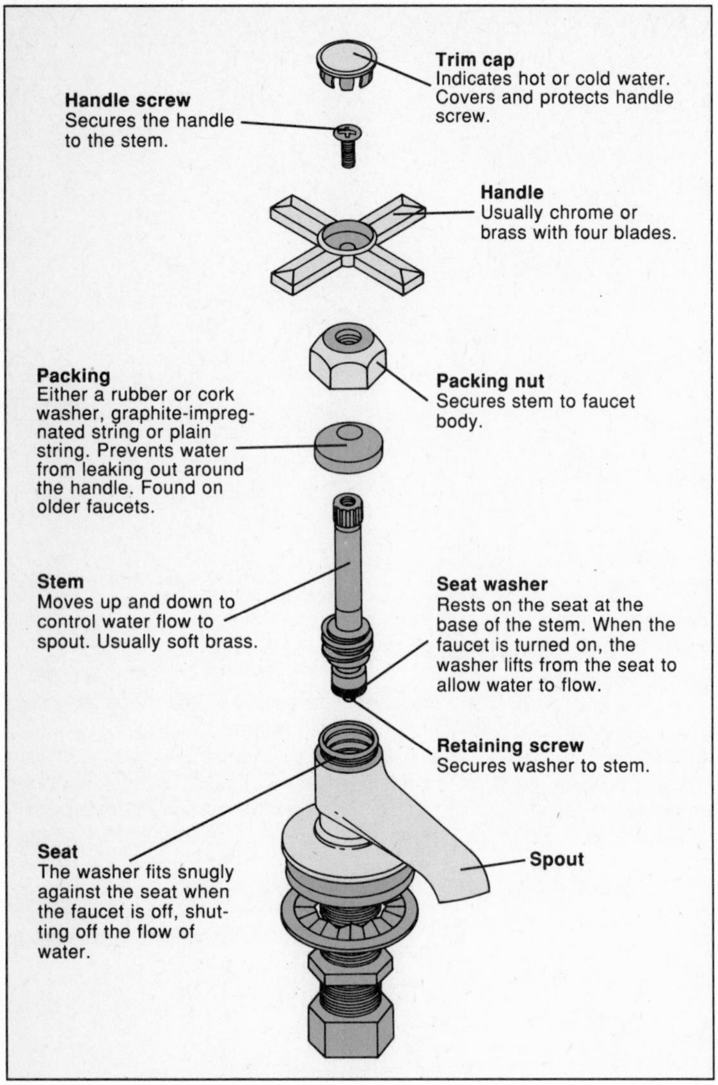
Trim cap
Indicates hot or cold water. Covers and protects handle screw.
Handle screw
Secures the handle to the stem.
Handle
Usually chrome or brass with four blades.
Packing
Either a rubber or cork washer, graphite-impregnated string or plain string. Prevents water from leaking out around the handle. Found on older faucets.
Packing nut
Secures stem to faucet body.
Stem
Moves up and down to control water flow to spout. Usually soft brass.
Seat washer
Rests on the seat at the base of the stem. When the faucet is turned on, the washer lifts from the seat to allow water to flow.
Retaining screw
Secures washer to stem.
Seat
The washer fits snugly against the seat when the faucet is off, shutting off the flow of water.
Spout

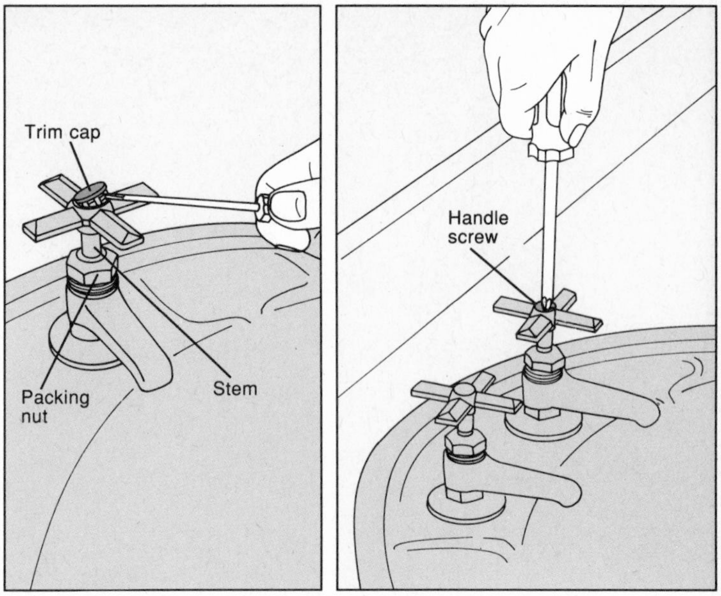
Trim cap
Packing nut
Stem
Handle screw

1 Opening up the faucet. To stop leaks around the handle, try tightening the packing nut by turning it clockwise with an adjustable wrench, its jaws taped to protect the chrome finish. If this has no effect, service the packing washer or string *(step 2)* or the stem and washer *(page 20)*. Turn off the water. Carefully pry off the trim cap with a small screwdriver *(above, left)* or knife. Then unscrew the handle screw *(above, right)* and pull off the handle. If a bent stem appears to be causing the leak, you can try to straighten it with a pair of tape-covered pliers. Should the stem be badly damaged, replace the entire faucet *(page 31)*.

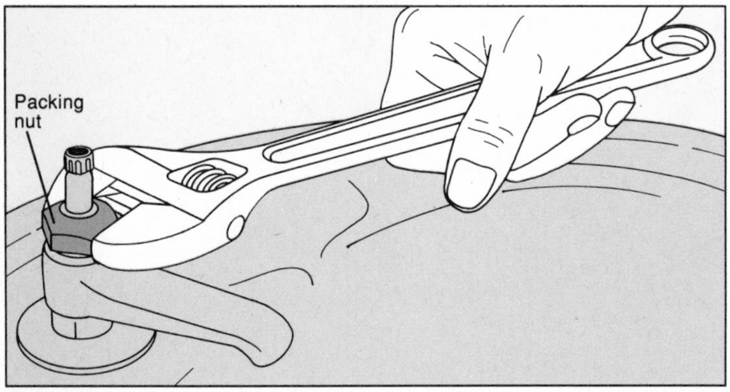
Packing nut

2 Removing the old packing. Use an adjustable wrench to unscrew the packing nut from the faucet body *(above)*. Pry off the old packing washer or unwind old packing string. Scour the base of the stem thoroughly with steel wool to remove mineral deposits, and replace the packing *(step 3)*. If the spout drips, also change the washer at the base of the stem or dress the valve seat *(page 21)*. Kitchen faucets may have additional packing at the base of the spout. Unscrew the spout nut with tape-covered pliers, lift off the spout, remove the packing and replace it as in step 3.

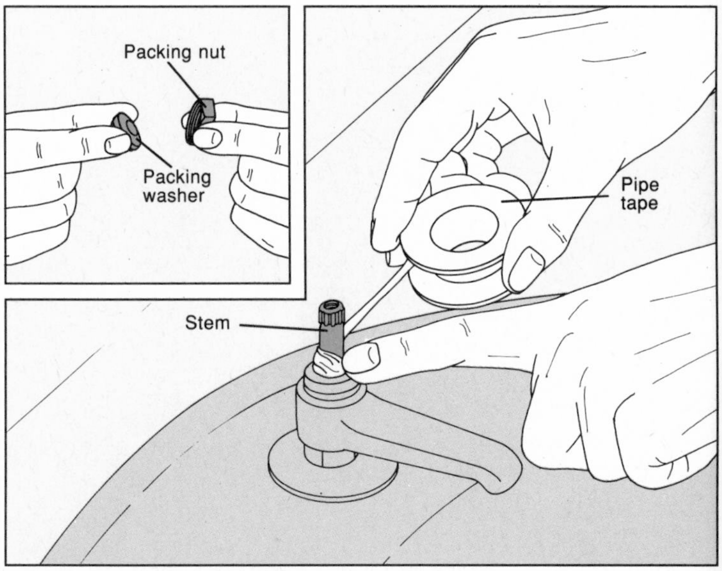
Packing nut
Packing washer
Pipe tape
Stem

3 Changing the packing. Insert a replacement packing washer into the packing nut *(inset)*; replace packing string with pipe tape *(above)* or new packing string. Wrap the tape or string several times around the base of the stem, stretching and pressing it down as you go. Thread the packing nut back on, but do not overtighten. The nut will compress the packing when you screw it down. Reassemble the handle and test the repair.

DOUBLE-HANDLE REVERSE-COMPRESSION FAUCETS

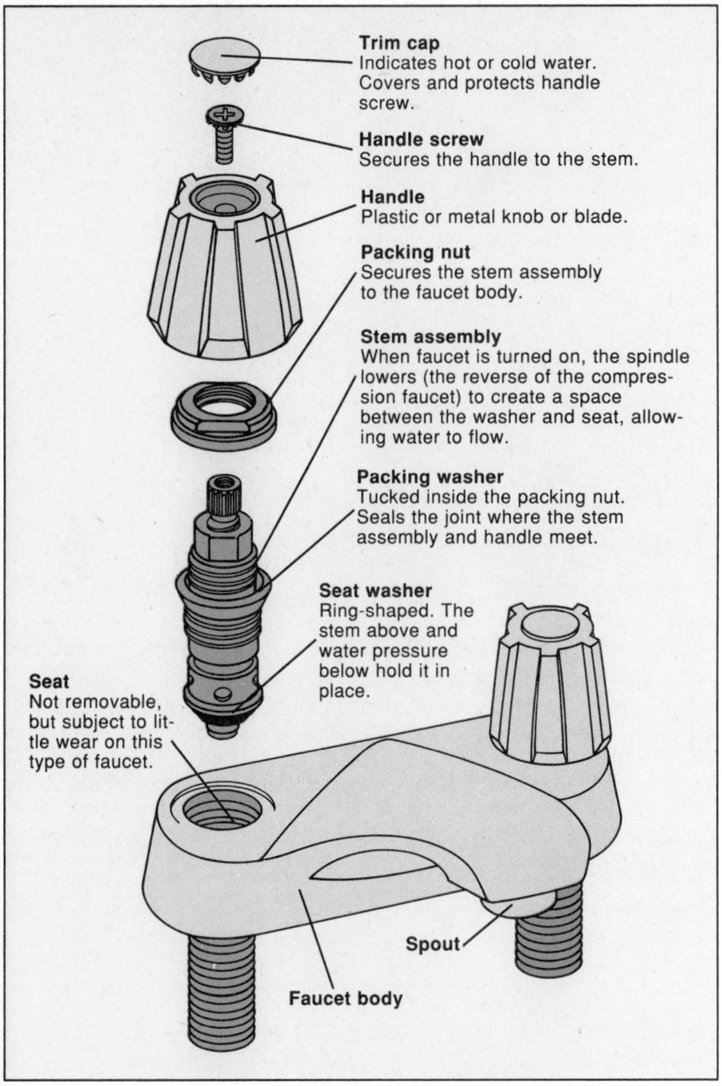
Trim cap
Indicates hot or cold water. Covers and protects handle screw.
Handle screw
Secures the handle to the stem.
Handle
Plastic or metal knob or blade.
Packing nut
Secures the stem assembly to the faucet body.
Stem assembly
When faucet is turned on, the spindle lowers (the reverse of the compression faucet) to create a space between the washer and seat, allowing water to flow.
Packing washer
Tucked inside the packing nut. Seals the joint where the stem assembly and handle meet.
Seat washer
Ring-shaped. The stem above and water pressure below hold it in place.
Seat
Not removable, but subject to little wear on this type of faucet.
Spout
Faucet body

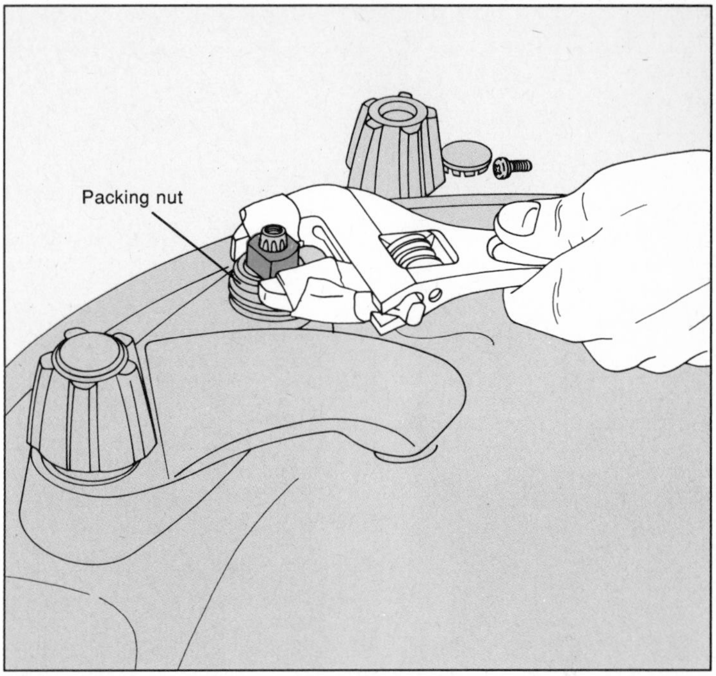
Packing nut

1 Opening up the faucet. Turn off the water supply, open the faucet one-half turn and close the drain to prevent loss of parts. Carefully pry off the trim cap with a small screwdriver or knife, then remove the handle screw and pull off the handle. Use an adjustable wrench to unscrew the packing nut securing the stem assembly to the faucet body *(above)*.

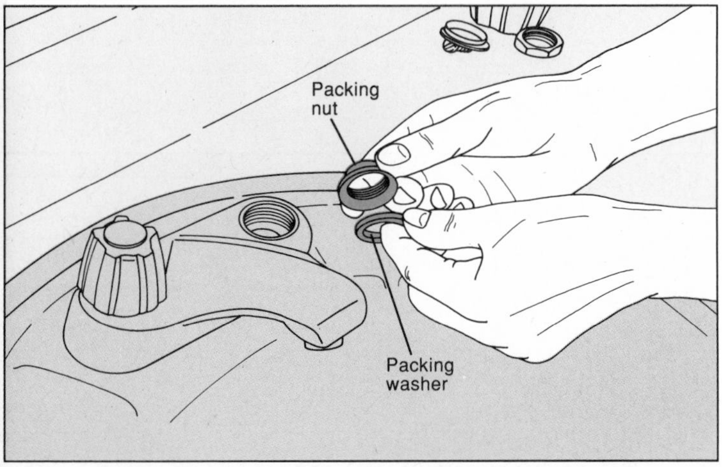
Packing nut
Packing washer

2 Changing the packing washer. To stop water leaking from a handle, pry out the washer inside the packing nut and press a new one in place *(above)*. Reassemble the handle. If water is dripping from the spout, replace the seat washer *(step 3)*. Kitchen faucets have additional packing washers at the base of the spout. Unscrew the spout nut with tape-covered pliers, lift off the spout, and replace the washer.

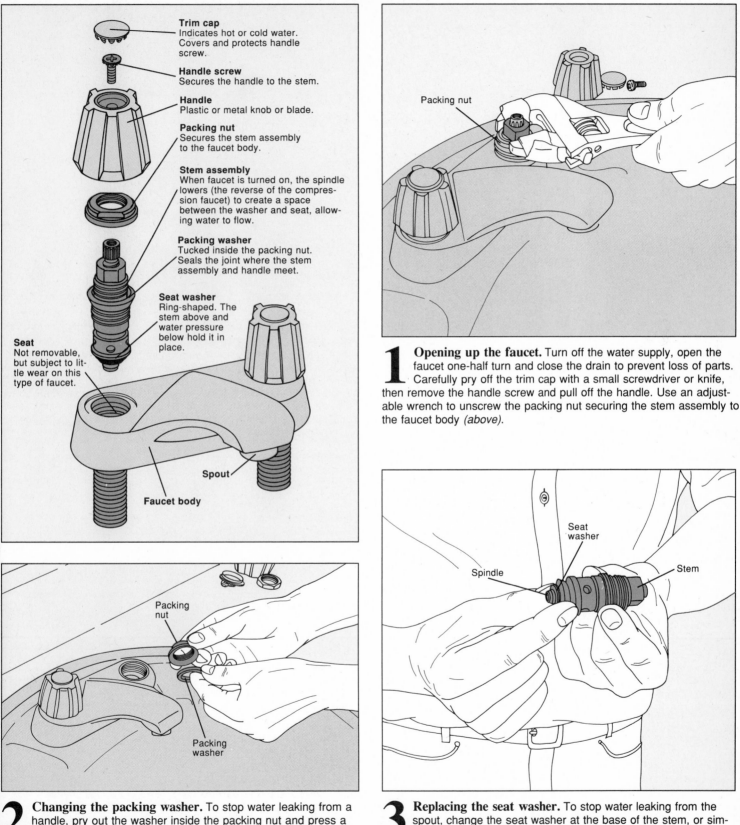

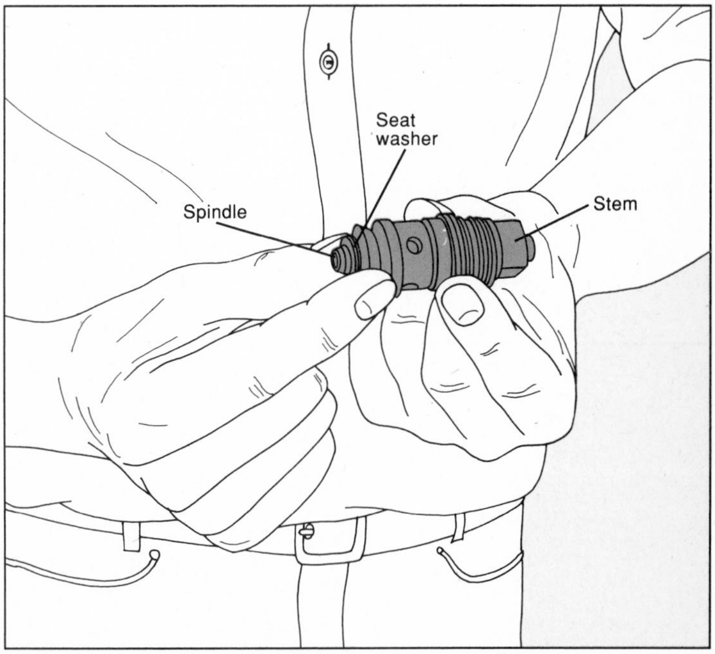
Seat washer
Spindle
Stem

3 Replacing the seat washer. To stop water leaking from the spout, change the seat washer at the base of the stem, or simply replace the entire stem assembly. Turn the spindle by hand (or with tape-covered pliers) to release the cap that holds the washer in place. Pry off the old washer with a pair of long-nose pliers, insert an exact replacement, and rethread the spindle and washer back into the stem *(above)*. Reassemble the faucet.

DOUBLE-HANDLE FAUCETS (Diaphragm type)

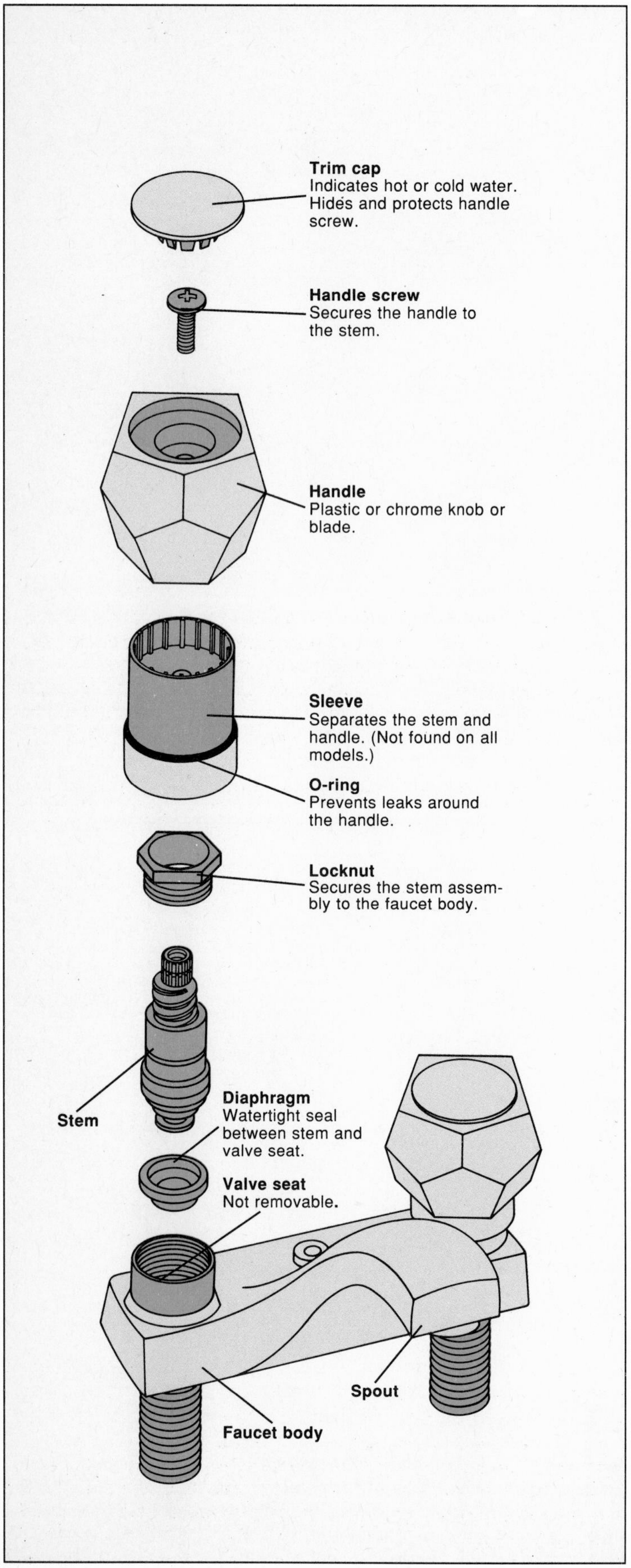
Trim cap
Indicates hot or cold water. Hides and protects handle screw.
Handle screw
Secures the handle to the stem.
Handle
Plastic or chrome knob or blade.
Sleeve
Separates the stem and handle. (Not found on all models.)
O-ring
Prevents leaks around the handle.
Locknut
Secures the stem assembly to the faucet body.
Stem
Diaphragm
Watertight seal between stem and valve seat.
Valve seat
Not removable.
Spout
Faucet body

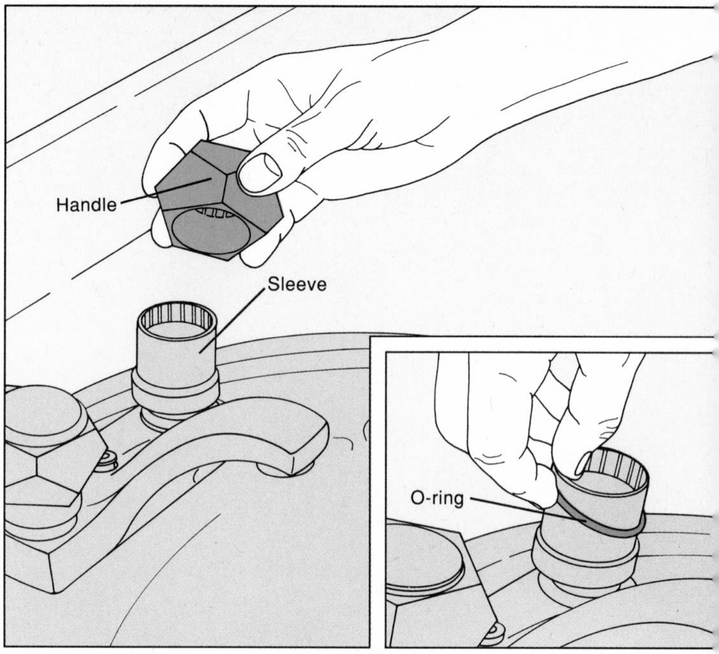
Handle
Sleeve
O-ring

1 **Opening up the faucet.** Turn off the water supply, open the faucet one-half turn and close the drain to prevent loss of parts. Carefully pry off the trim cap with a knife or small screwdriver. Remove the handle screw and pull off the handle *(above)*. To stop water leaking from the handle, replace the O-ring. Roll it off the sleeve *(inset)*, lubricate a new O-ring with petroleum jelly and roll it into place. Reassemble and test the faucet. If the leak persists, replace the diaphragm *(next step)*.

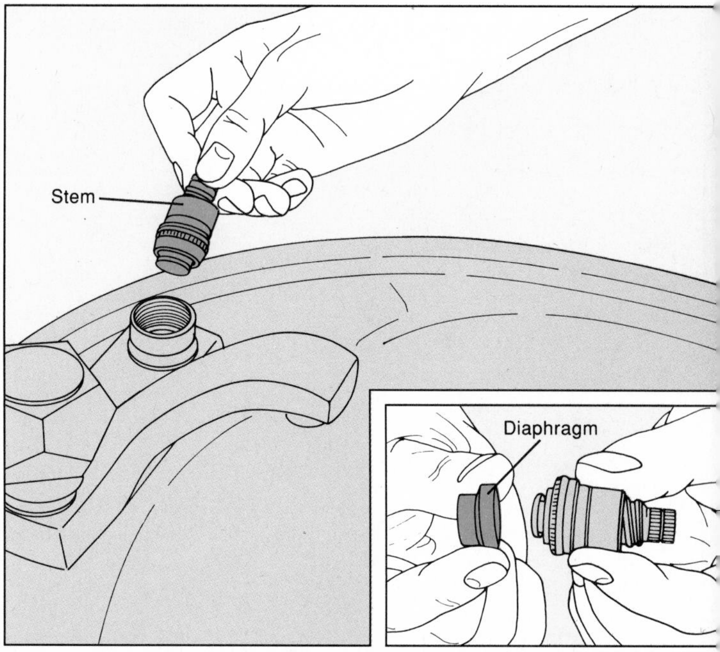
Stem
Diaphragm

2 **Replacing the diaphragm.** To stop leaks from the handle or spout, replace the hat-shaped diaphragm. Lift off the chrome sleeve with tape-covered pliers, exposing the locknut. Unscrew the locknut with a tape-covered adjustable wrench and lift the stem from the faucet body *(above)*. Pry off the diaphragm by hand *(inset)*, and press an exact replacement into place. Reassemble the faucet.

DOUBLE-HANDLE FAUCETS (Disc type)

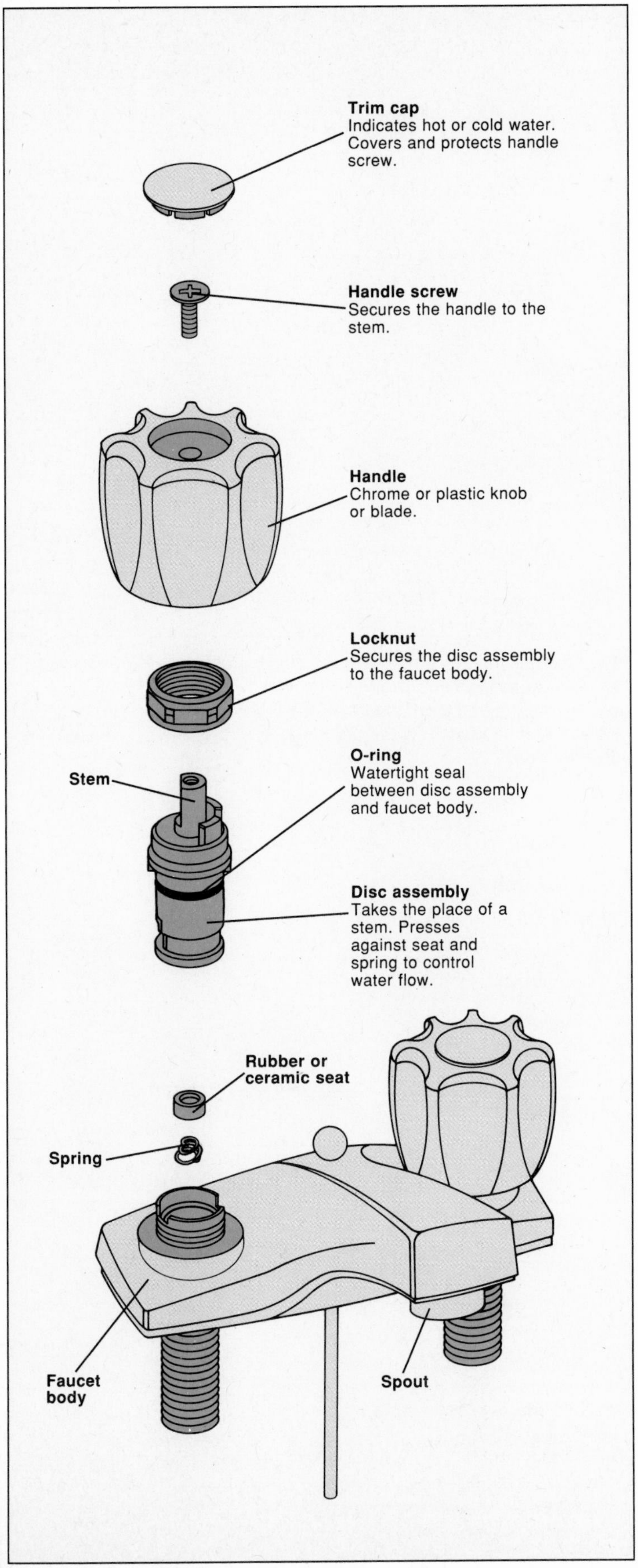
Trim cap
Indicates hot or cold water. Covers and protects handle screw.
Handle screw
Secures the handle to the stem.
Handle
Chrome or plastic knob or blade.
Locknut
Secures the disc assembly to the faucet body.
O-ring
Watertight seal between disc assembly and faucet body.
Stem
Disc assembly
Takes the place of a stem. Presses against seat and spring to control water flow.
Rubber or ceramic seat
Spring
Faucet body
Spout

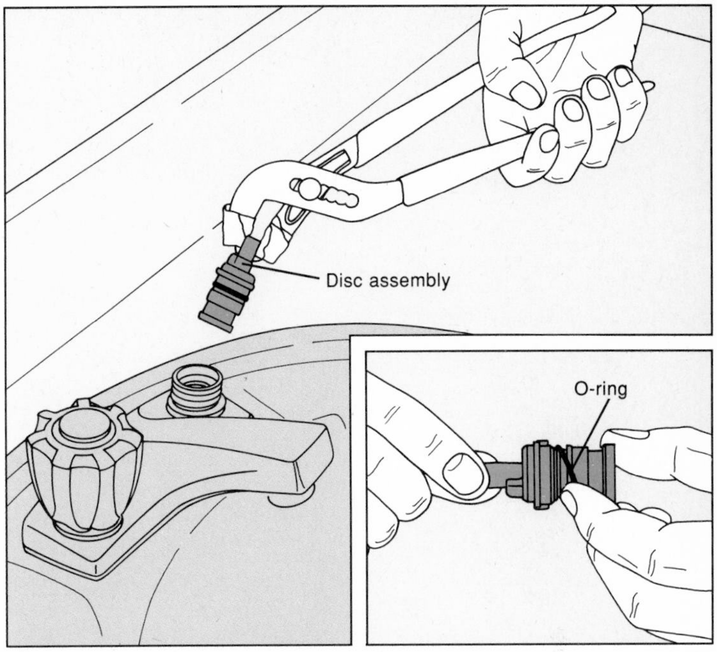
Disc assembly
O-ring

1 Replacing the O-ring. Turn off the water, open the faucet one-half turn and close the drain to prevent loss of parts. Carefully pry off the trim cap with a knife or small screwdriver. Remove the handle screw and pull off the handle. Unscrew the locknut with a taped adjustable wrench. Lift the cartridge from the faucet body *(above)*. Water leaking around the handle may be caused by a cracked or pitted disc assembly. If so, buy a new assembly and O-ring. Insert the assembly, lining up its slots with those on the faucet body. If only the O-ring is worn, pinch it off the assembly *(inset)*, lubricate an exact replacement with petroleum jelly and slip it on. Reassemble the faucet.

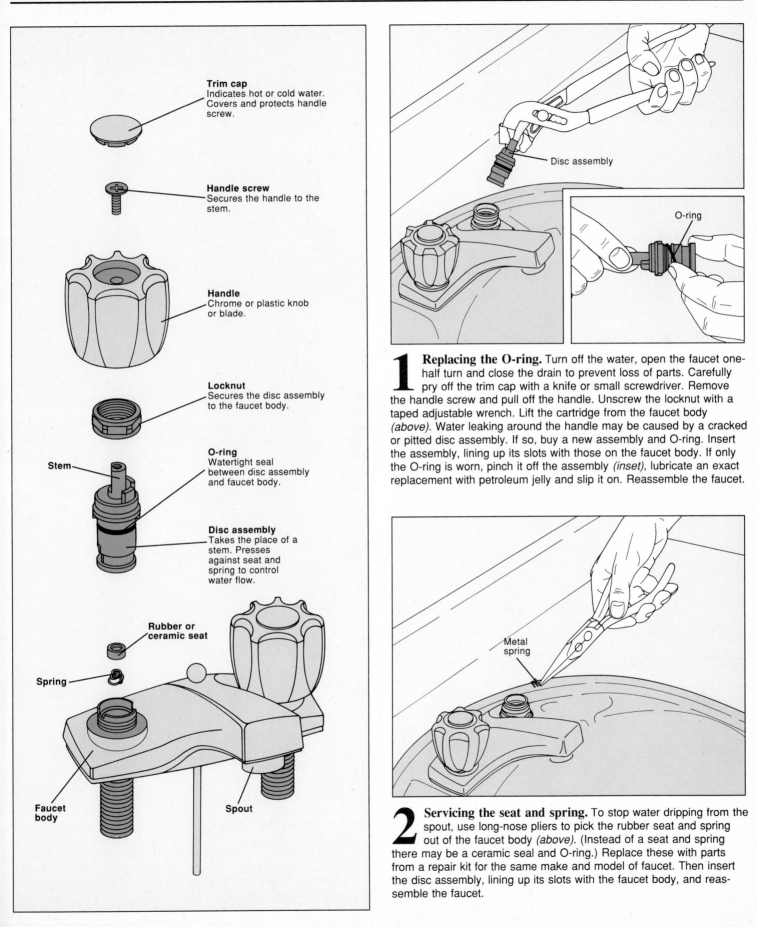

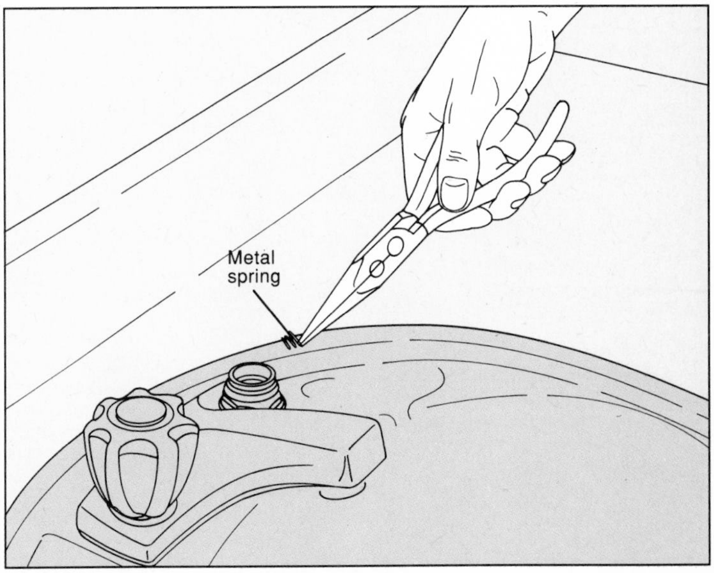
Metal spring

2 Servicing the seat and spring. To stop water dripping from the spout, use long-nose pliers to pick the rubber seat and spring out of the faucet body *(above)*. (Instead of a seat and spring there may be a ceramic seal and O-ring.) Replace these with parts from a repair kit for the same make and model of faucet. Then insert the disc assembly, lining up its slots with the faucet body, and reassemble the faucet.

SINGLE-LEVER FAUCETS (Rotating-ball type)

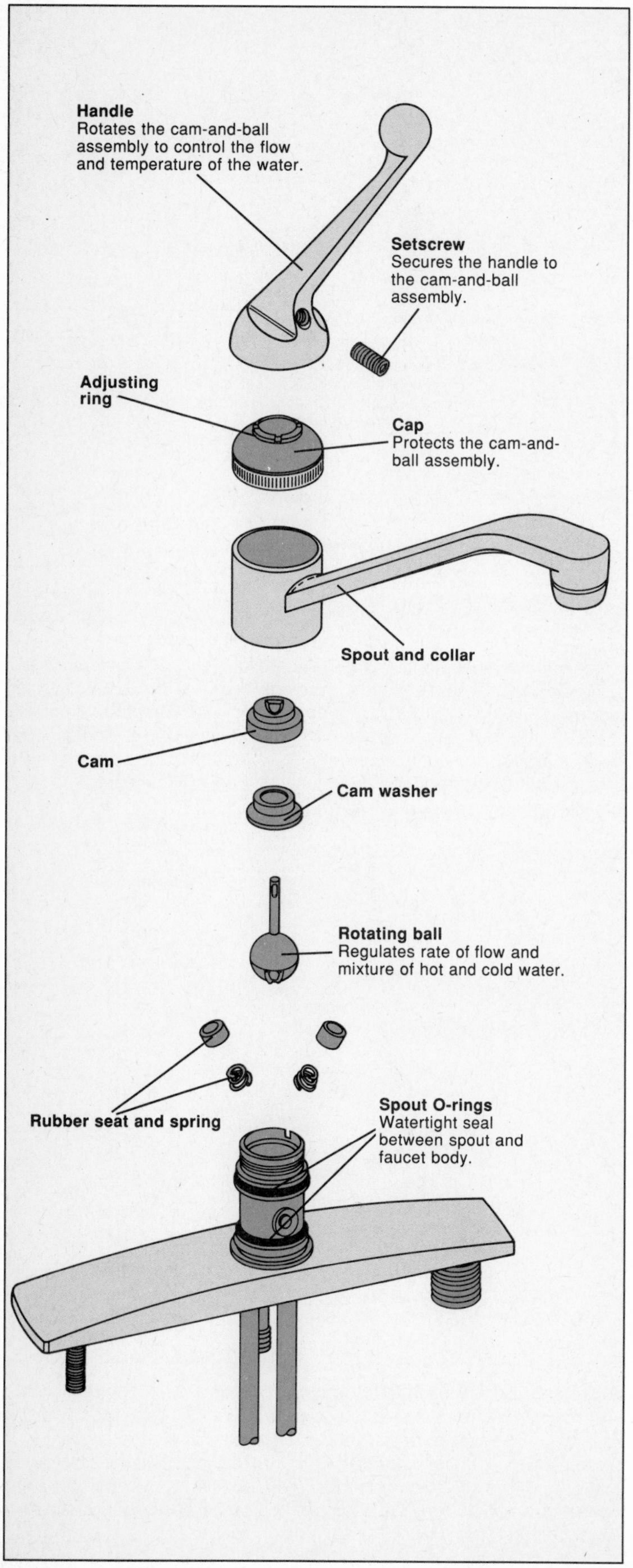
Handle
Rotates the cam-and-ball assembly to control the flow and temperature of the water.
Setscrew
Secures the handle to the cam-and-ball assembly.
Adjusting ring
Cap
Protects the cam-and-ball assembly.
Spout and collar
Cam
Cam washer
Rotating ball
Regulates rate of flow and mixture of hot and cold water.
Rubber seat and spring
Spout O-rings
Watertight seal between spout and faucet body.

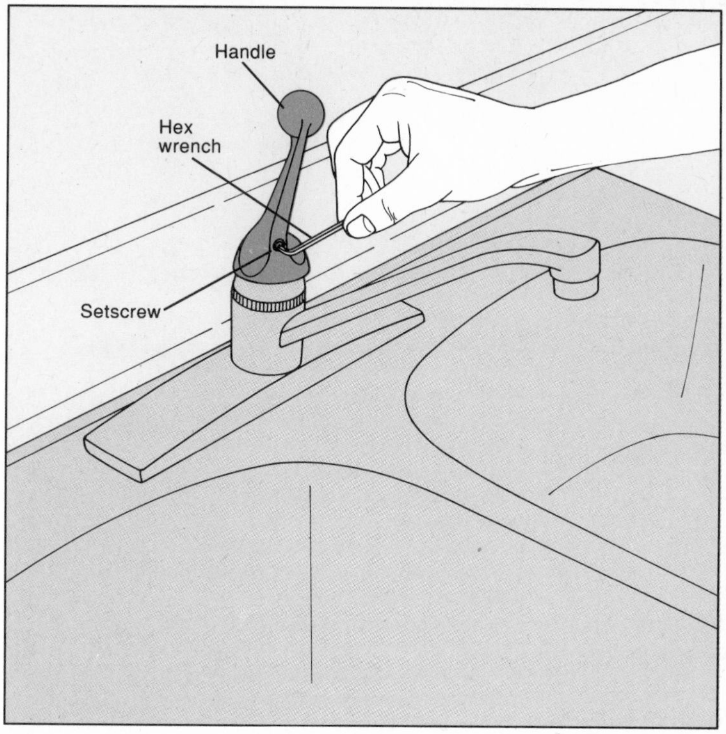
Handle
Hex wrench
Setscrew

1 Removing the handle. The setscrew that secures the handle to the faucet body is underneath the handle. Use a small hex wrench to loosen the screw *(above)*, but leave the screw in the handle since it is small and easily lost. Lift off the handle to expose the adjusting ring.

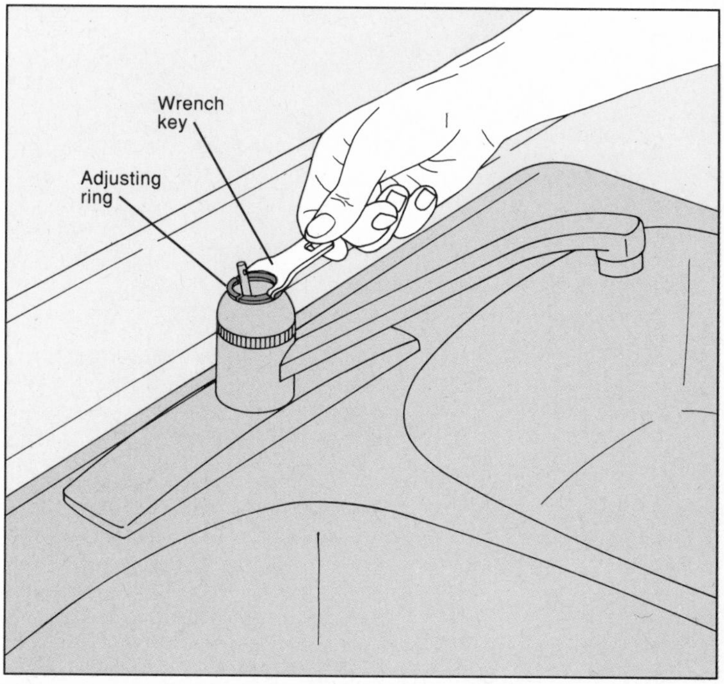
Wrench key
Adjusting ring

2 Tightening the adjusting ring. To stop water leaking from the faucet handle, use the edge of an old dinner knife or a special wrench included in the repair kit to tighten the adjusting ring clockwise, as shown. (The ball should move easily without the handle attached.) Reassemble the handle and test for leaks. Tighten the adjusting ring again, if necessary. If the leak persists, go to step 3.

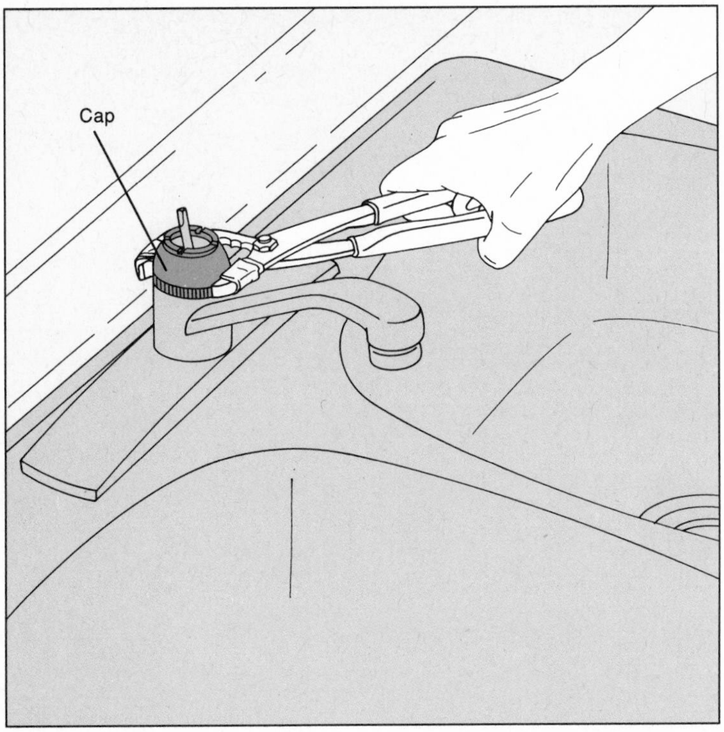

3 Removing the cap. Service the cam-and-ball assembly if the handle continues to leak after you have tightened the adjusting ring, or if the spout drips. First turn off the water supply, open the faucet and close the drain to prevent loss of parts. Unscrew the cap by hand or with a pair of channel-joint pliers taped to protect chrome parts *(above)*.

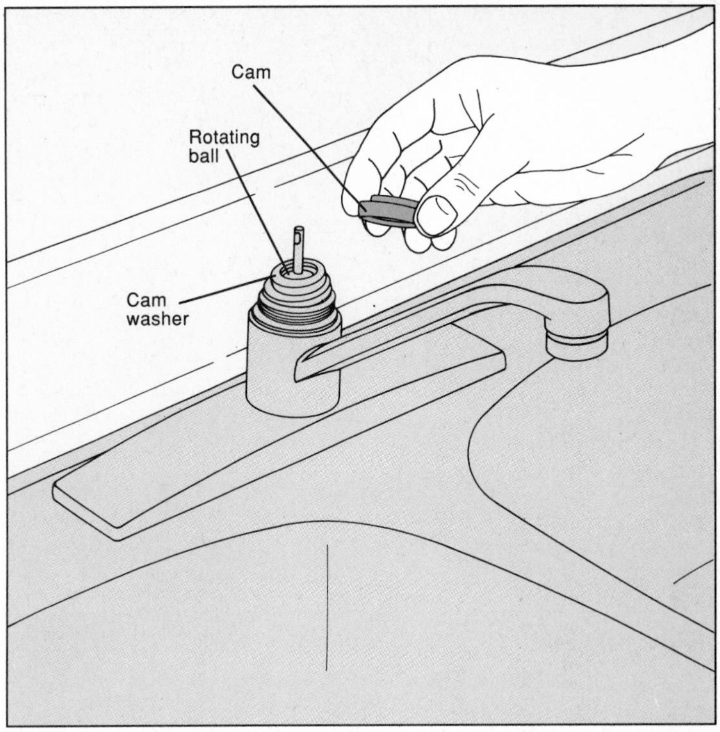

4 Replacing the cam assembly. Lift off the plastic cam *(above)*, exposing the cam washer and rotating ball. Buy a repair kit that includes replacement parts for your make and model of faucet. Service worn or damaged parts individually *(step 5)* or replace all of them while the faucet is disassembled.

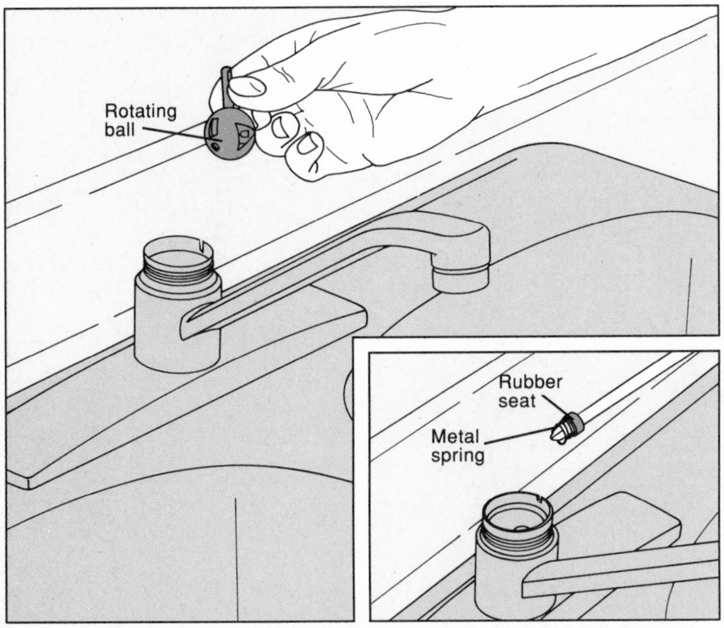

5 Replacing the seats, springs and ball. Lift the rotating ball from the faucet body *(above)*, then reach into the faucet body with long-nose pliers or the end of a screwdriver and remove the two sets of rubber seats and metal springs *(inset)*, or two sets of ceramic seals and O-rings. Replace these parts from the kit, making sure they are properly seated in the faucet body before reassembling the faucet. If the cam-and-ball assembly appears damaged, replace it at the same time.

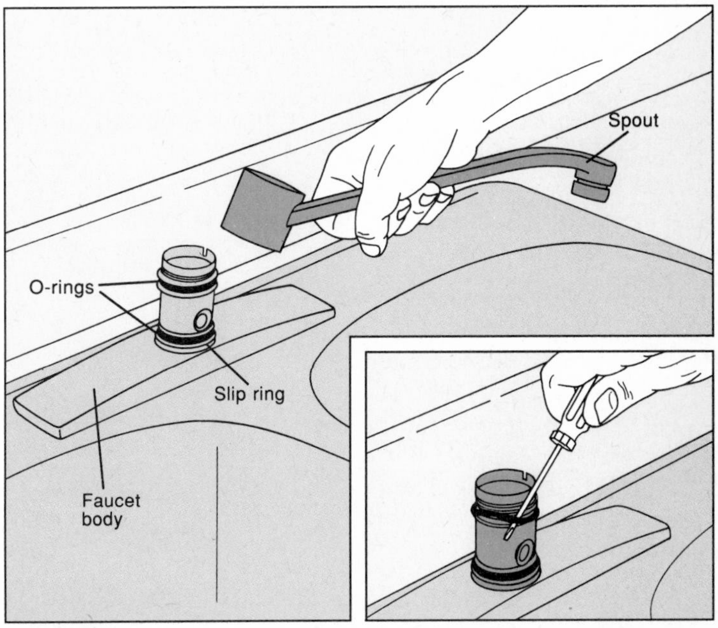

6 Replacing the spout O-rings. If water leaks from the spout collar, twist the spout off *(above)* to expose its O-rings. Slip the end of a small screwdriver under the O-rings to pry them off the faucet body *(inset)*. Lubricate the new O-rings with petroleum jelly, and roll them into place. Lower the spout straight down over the body and rotate it until it rests on the plastic slip ring at the base. Reassemble the faucet.

SINGLE-LEVER FAUCETS (Cartridge type)

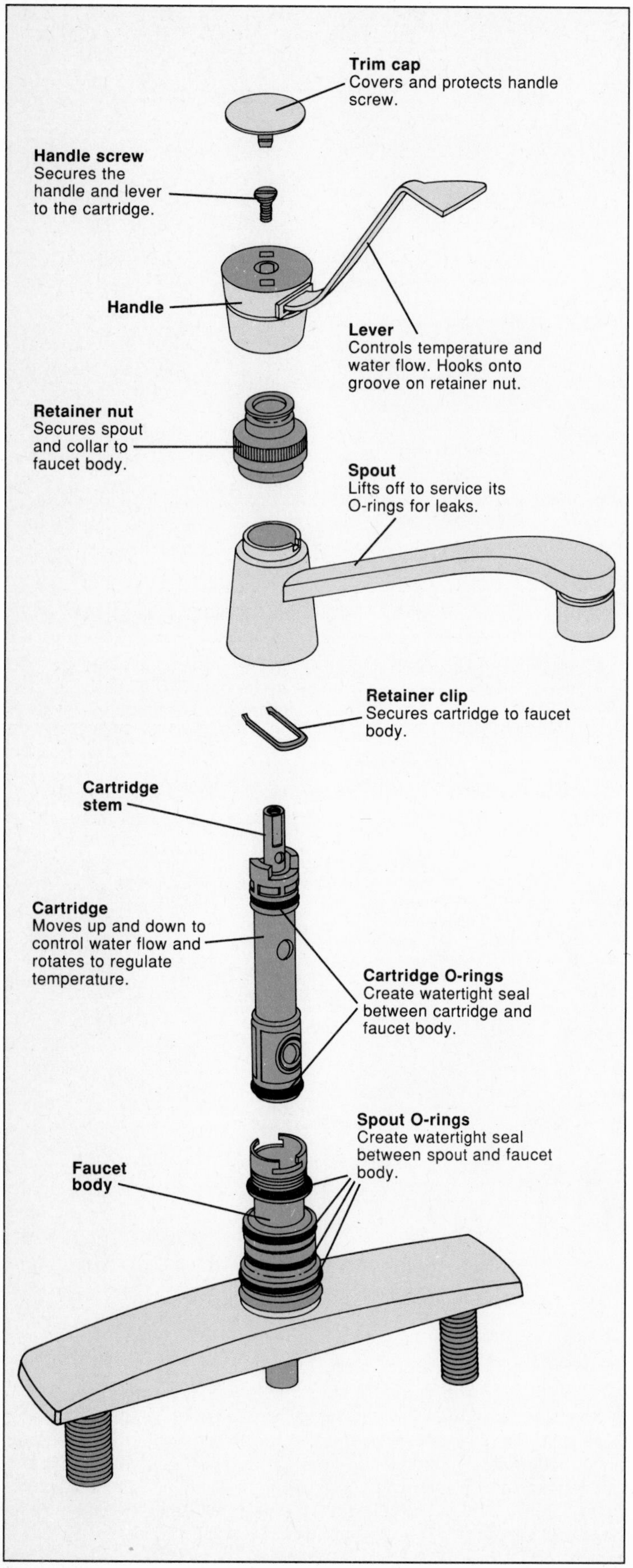
Trim cap
Covers and protects handle screw.
Handle screw
Secures the handle and lever to the cartridge.
Handle
Lever
Controls temperature and water flow. Hooks onto groove on retainer nut.
Retainer nut
Secures spout and collar to faucet body.
Spout
Lifts off to service its O-rings for leaks.
Retainer clip
Secures cartridge to faucet body.
Cartridge stem
Cartridge
Moves up and down to control water flow and rotates to regulate temperature.
Cartridge O-rings
Create watertight seal between cartridge and faucet body.
Spout O-rings
Create watertight seal between spout and faucet body.
Faucet body

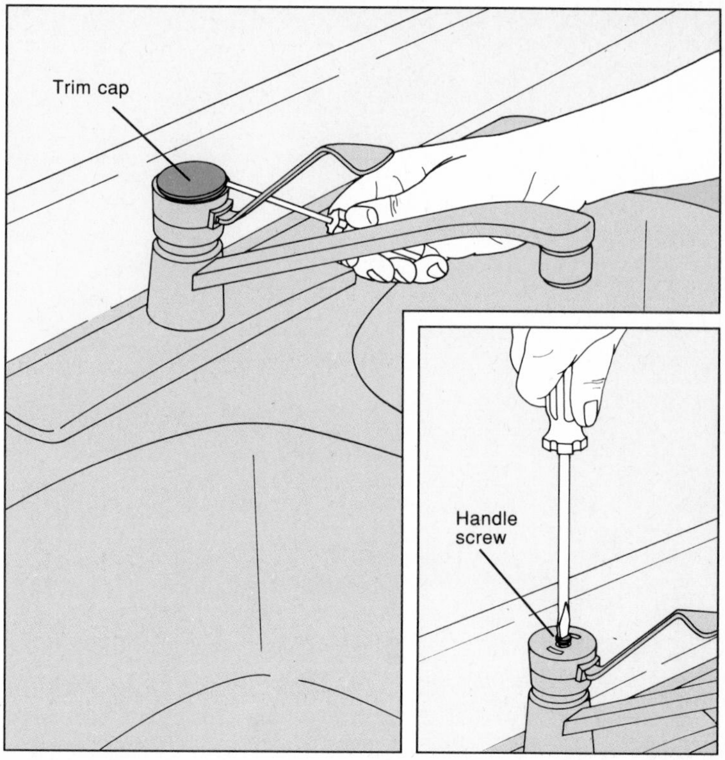
Trim cap
Handle screw

1 Removing the handle. Turn off the water supply, lift the handle several times to drain the faucet, and close the drain to prevent loss of parts. Carefully pry off the trim cap with a small screwdriver *(above)* or knife. Remove the handle screw that secures the handle assembly to the cartridge *(inset)*.

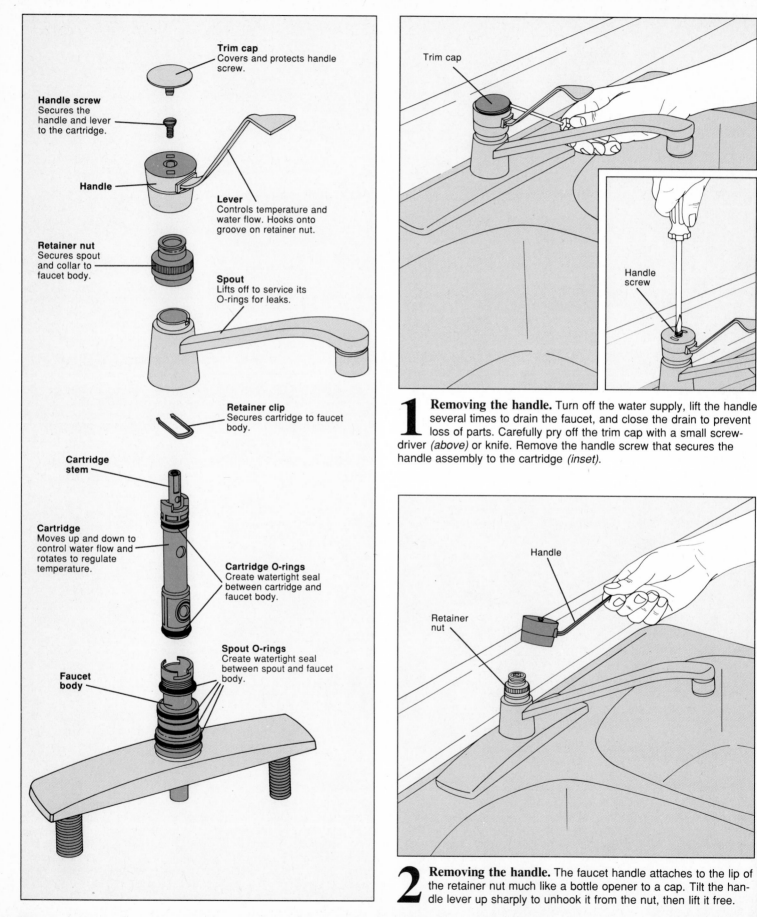

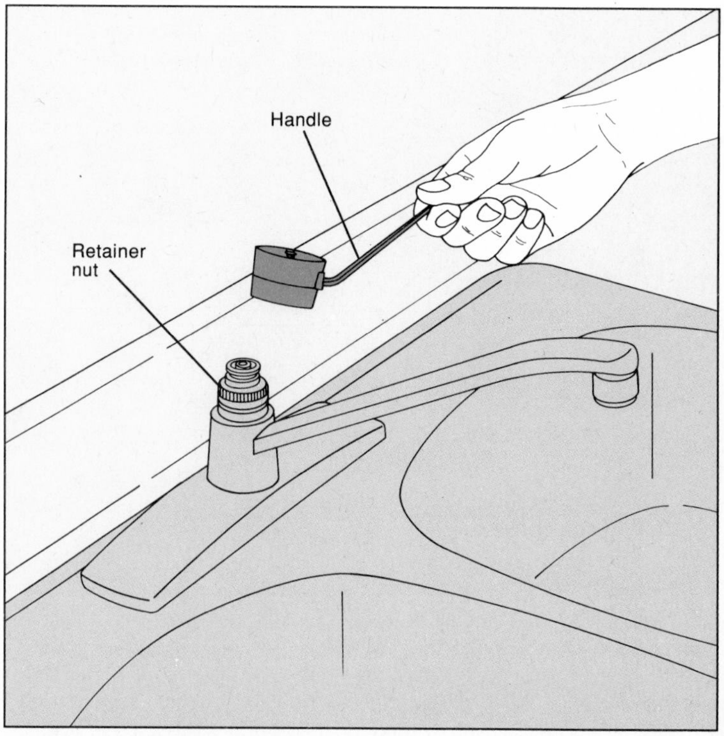
Handle
Retainer nut

2 Removing the handle. The faucet handle attaches to the lip of the retainer nut much like a bottle opener to a cap. Tilt the handle lever up sharply to unhook it from the nut, then lift it free.

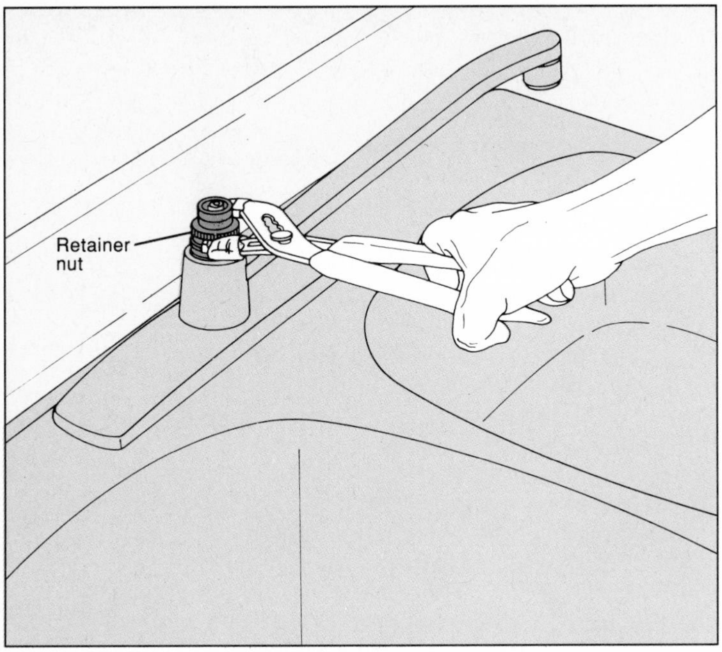

3 Removing the retainer nut. Unscrew the retainer nut with taped channel-joint pliers *(above)* and lift it off the faucet body. To stop a leaking handle or a dripping spout, replace the O-rings or the entire cartridge *(step 4)*. To stop leaks from the spout collar, replace the spout O-rings *(step 6)*.

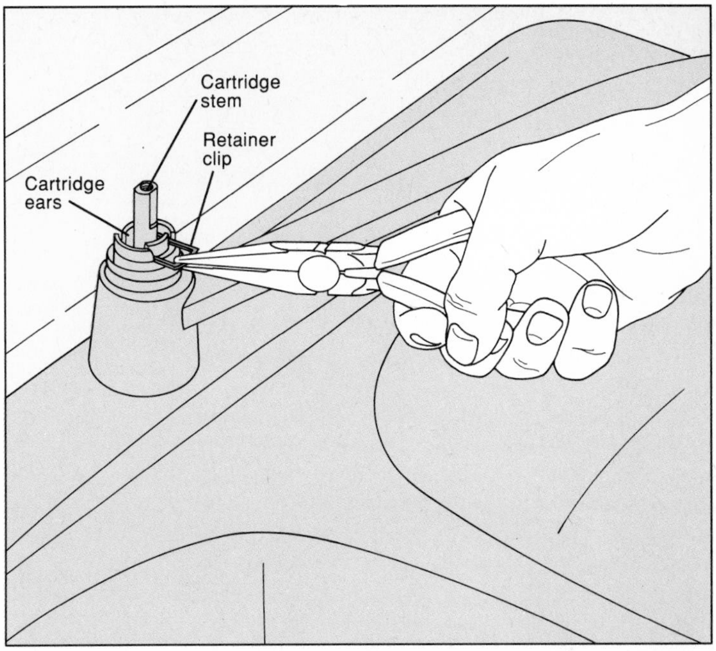

4 Freeing the cartridge. Locate the U-shaped retainer clip that holds the cartridge in place in the faucet body. Using long-nose pliers or tweezers, pull the clip from its slot *(above)*, being careful not to drop it down the drain.

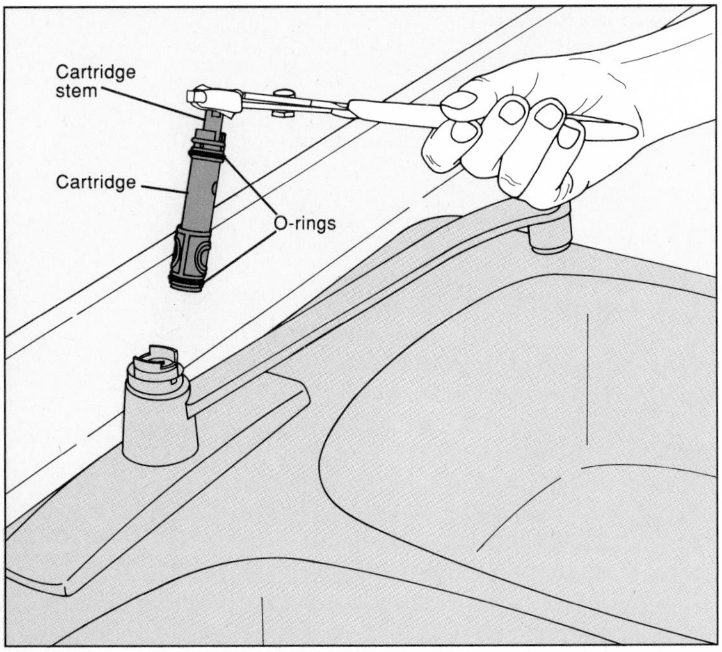

5 Servicing the cartridge. Grasping the cartridge stem with taped pliers, lift the cartridge out of the faucet body. Examine the O-rings and replace them if they are worn or cracked. Pry the old rings off the cartridge with the tip of an awl or other pointed tool. Lubricate the new O-rings with a dab of petroleum jelly and roll them down over the cartridge until they rest in the appropriate grooves. If the cartridge itself is worn or damaged, replace it with a new one. Reinsert the cartridge, align it properly in its seat, and replace the spout and retainer nut. Attach the handle by hooking its inside lever on the lip of the retainer nut. If the hot and cold water are reversed, remove the handle and rotate the cartridge stem one-half turn.

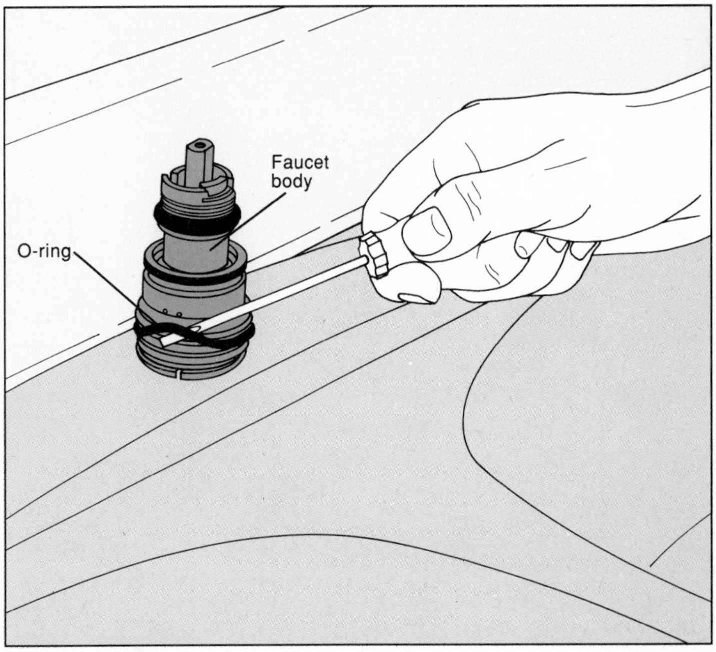

6 Replacing the spout O-rings. If water leaks around the spout collar, replace the spout O-rings. Lift off the spout and pry off the cracked or worn rings *(above)*. Lubricate new O-rings with petroleum jelly and roll them into the grooves on the faucet body. Replace the spout and reassemble the faucet, hooking the inside edge of the handle lever onto the lip of the retainer nut.

SINGLE-LEVER FAUCETS (Ceramic disc type)

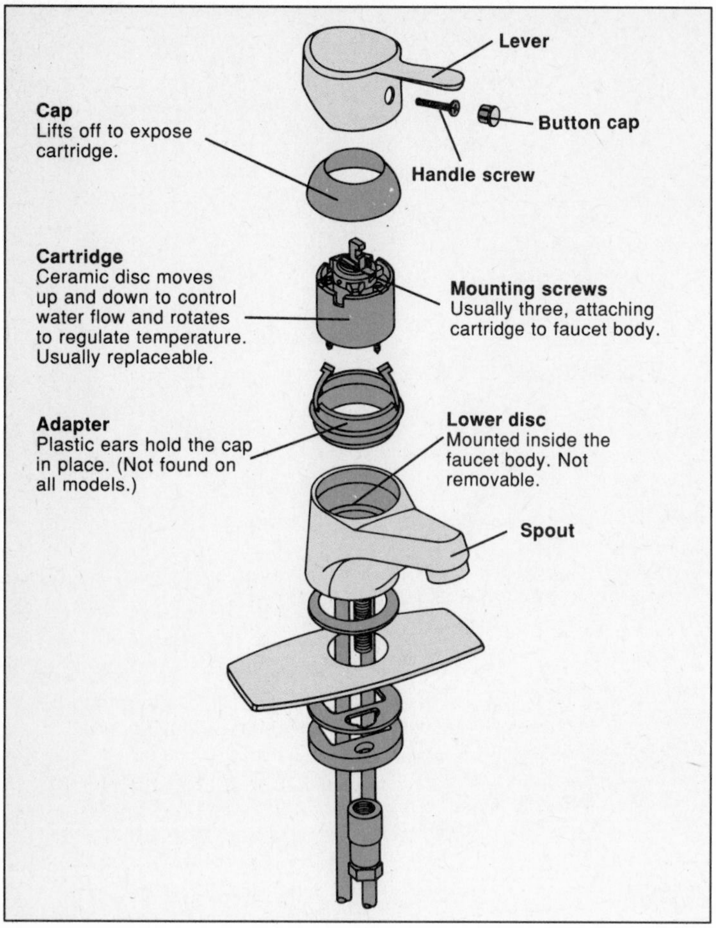
Lever
Cap
Lifts off to expose cartridge.
Button cap
Handle screw
Cartridge
Ceramic disc moves up and down to control water flow and rotates to regulate temperature. Usually replaceable.
Mounting screws
Usually three, attaching cartridge to faucet body.
Adapter
Plastic ears hold the cap in place. (Not found on all models.)
Lower disc
Mounted inside the faucet body. Not removable.
Spout

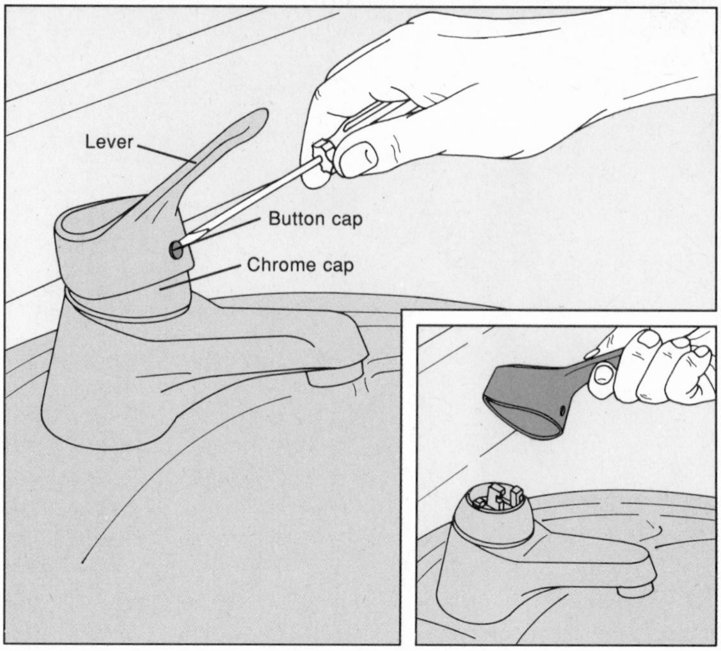
Lever
Button cap
Chrome cap

1 Removing the lever. Turn off the water supply and drain the faucet by lifting the lever to its highest position. Close the drain to prevent loss of parts. Pry off the button cap at the base of the lever with a knife or small screwdriver *(above),* and remove the handle screw. (On some models the screw is underneath the lever body and there is no cap.) Lift off the handle *(inset).*

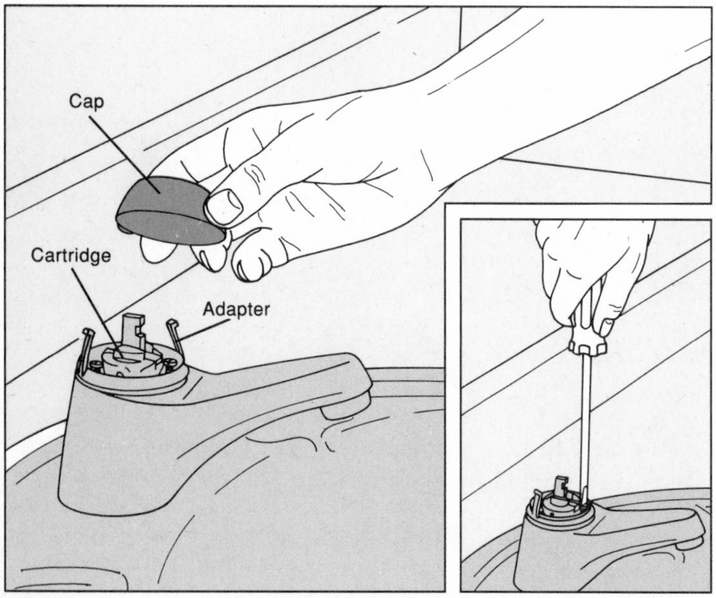
Cap
Cartridge
Adapter

2 Freeing the cartridge. Pry the cap off its plastic adapter *(above)* or, on some faucets, unscrew it from the faucet body. Loosen the two or three brass screws holding the cartridge to the faucet body *(inset),* and lift out the cartridge.

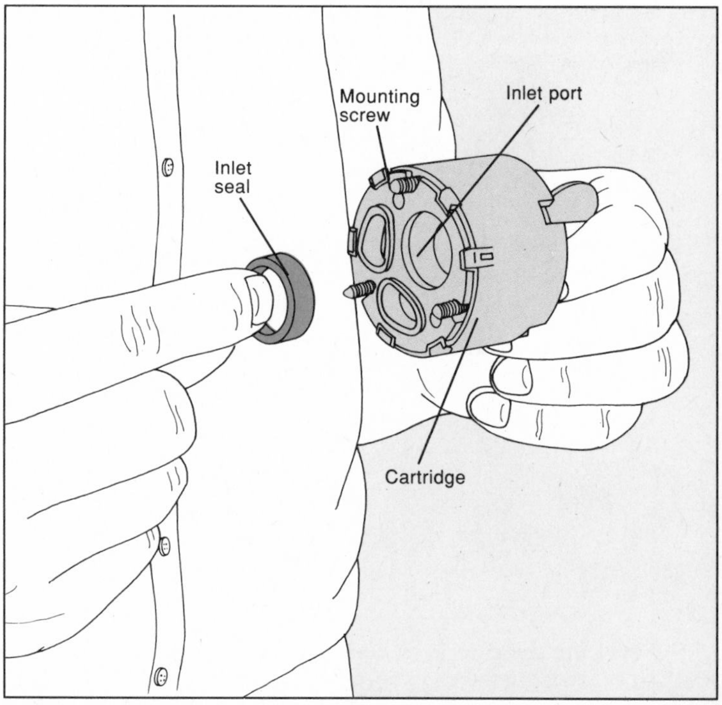
Mounting screw
Inlet port
Inlet seal
Cartridge

3 Servicing the cartridge. First check to be sure that the leak is not caused by a piece of dirt caught between the ceramic discs. Clean the inlet ports and the surface of the bottom disc. If the upper disc is cracked or pitted, buy a replacement cartridge for the same make and model. Insert the new seals in the disc, position the cartridge in the faucet body and screw it in place. Check that the three ports on the bottom of the cartridge align with those of the faucet body.

REPLACING A FAUCET SET

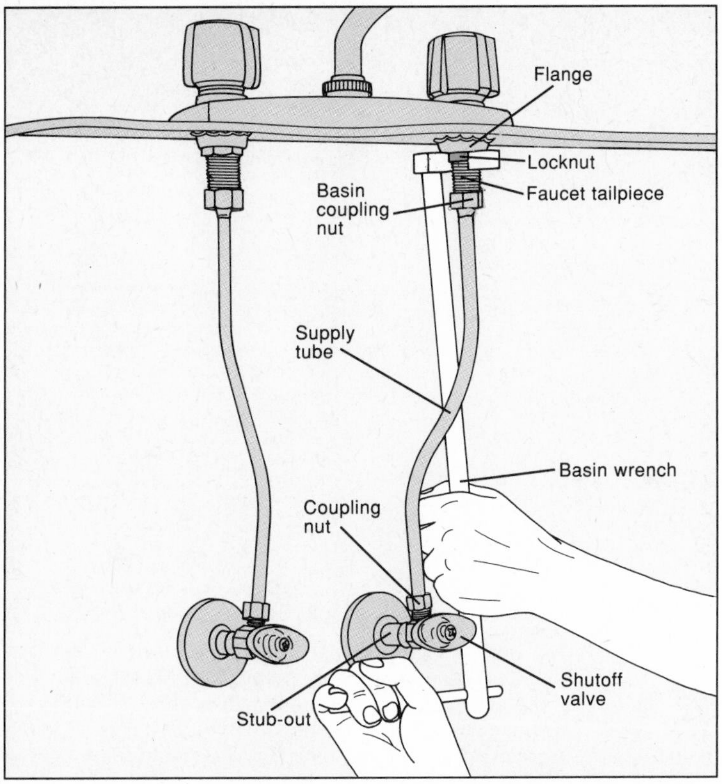

1 Tightening the fittings. To stop water leaking from beneath the faucet, first tighten the locknuts under the sink clockwise with a basin wrench *(above)*. If leaks persist, or the faucet body is badly worn, replace the faucet set. Installing a new faucet not only eliminates chronic drips, but also modernizes your kitchen or bathroom. For stubborn fittings, apply penetrating oil and wait overnight before beginning work. If a poor seal between the faucet body and sink is causing the leak, proceed only through step 4.

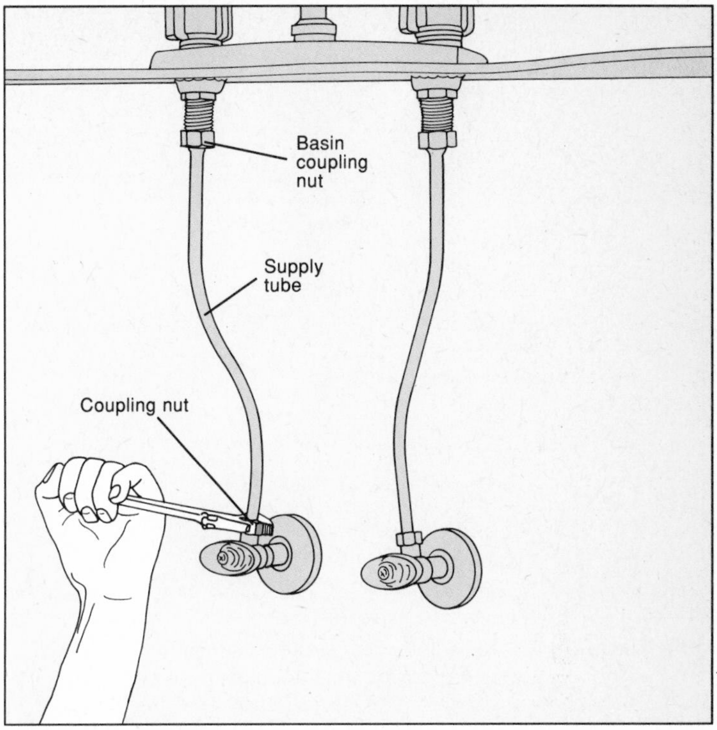

2 Loosening the supply tubes. Close the shutoff valves and open the faucet to relieve pressure. With an adjustable wrench, loosen the coupling nuts at the shutoff valves *(above)*, then unscrew them by hand. Slide the coupling nuts up the supply tubes, and pull the tubes out of the valves. Use a basin wrench to loosen the coupling nuts that secure the supply tubes to the faucet tailpieces. Then loosen the locknuts holding each flange to the bottom of the sink.

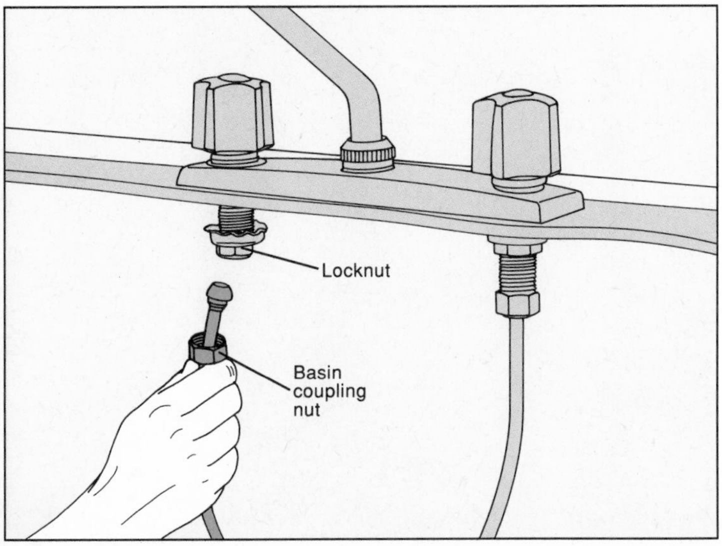

3 Removing the supply tubes. Unscrew the basin coupling nuts under the sink, using a basin wrench if space is cramped, then pull the supply tubes from the faucet. Remove the locknuts from the tailpiece. When you reinstall the old faucet–or install a single-lever faucet–consider replacing the old supply tubes with flexible connectors *(page 33)*.

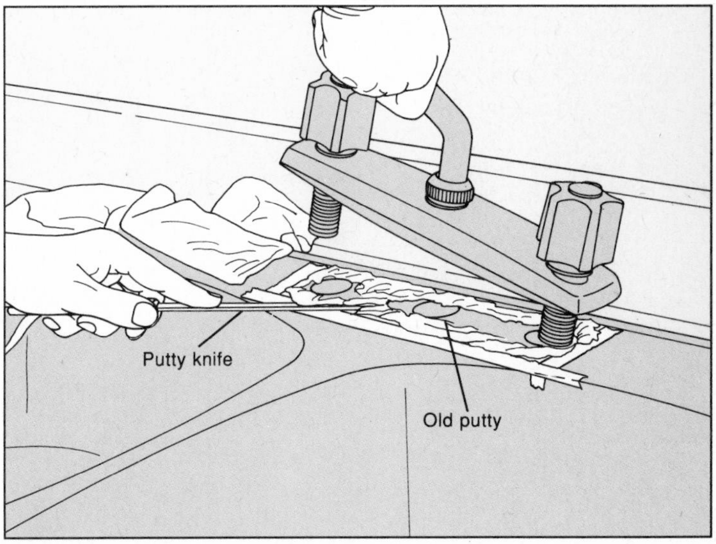

4 Replacing the putty. If the putty underneath the faucet set is dry and cracked, replacing the putty may stop leaks between the sink and faucet body. If the underside of the faucet body is corroded, however, replace the faucet set. Protect the sink with masking tape, then scrape away the old putty with a putty knife *(above)*. Scour the sink deck clean with fine steel wool. Roll some new plumber's putty into a long rope and place it on the deck so that it will lie beneath the new faucet set. Press the faucet down firmly, squeezing out a small amount on all sides. Scrape away the excess putty.

REPLACING A FAUCET SET (continued)

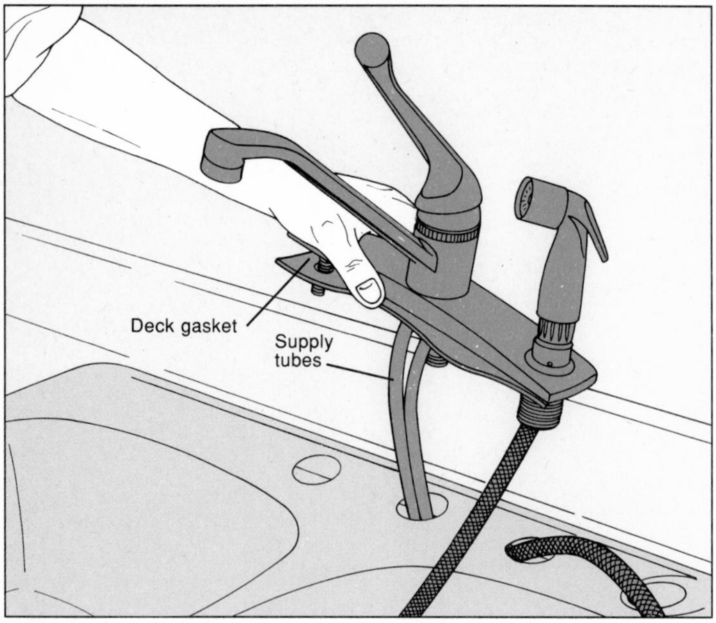

5 **Dropping in the new faucet set.** Single-lever faucets are widely available for three-hole bathroom sinks with a 4-inch space between centers, and four-hole kitchen sinks with 6- or 8-inch separation. Carefully work the faucet's two supply tubes through the center hole of the sink *(above)*. Feed the spray hose, if any, through its separate hole. Press the faucet set down firmly into the putty, squeezing out a small amount on all sides. (Many new faucets come with deck gaskets and do not need putty.)

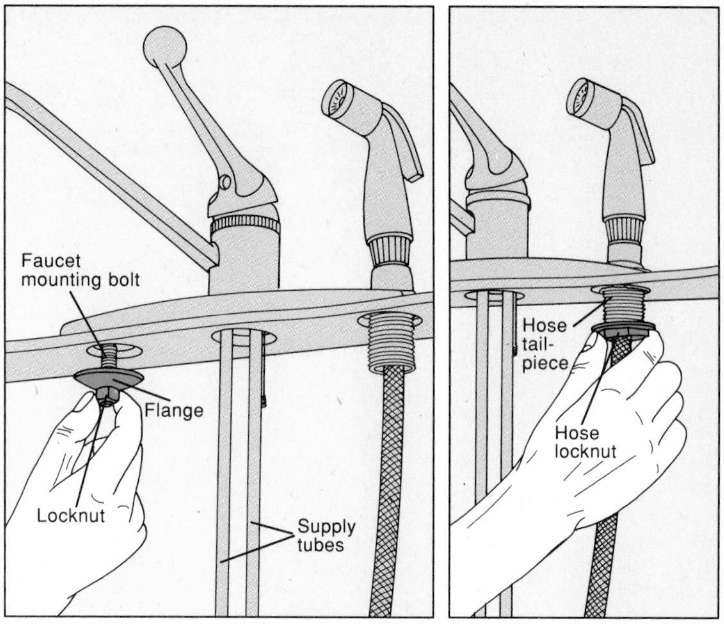

6 **Tightening the sink connections.** Slip the flange onto the faucet mounting bolt beneath the sink *(above, left)*. Thread a locknut onto the bolt and tighten it with a basin wrench (tighten a plastic locknut by hand). Attach a second locknut to the hose tailpiece in the same way *(above, right)*. Tighten with a basin wrench (or by hand for plastic connectors). Scrape away excess putty from around the faucet body.

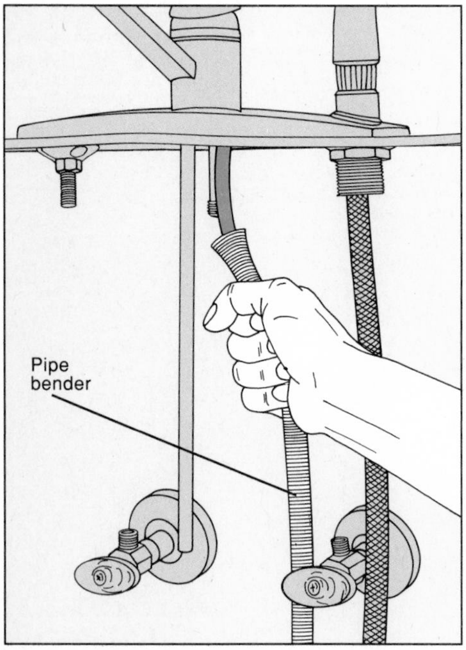

7 **Bending the supply tubes.** Bend the copper supply tubes by hand, or with a coiled pipe bender to prevent kinks *(above)*. If the tubes are not long enough to reach the shutoff valves, add flexible supply tubes *(page 33)*. If there are no shutoff valves, now is a good time to install them *(page 111)*.

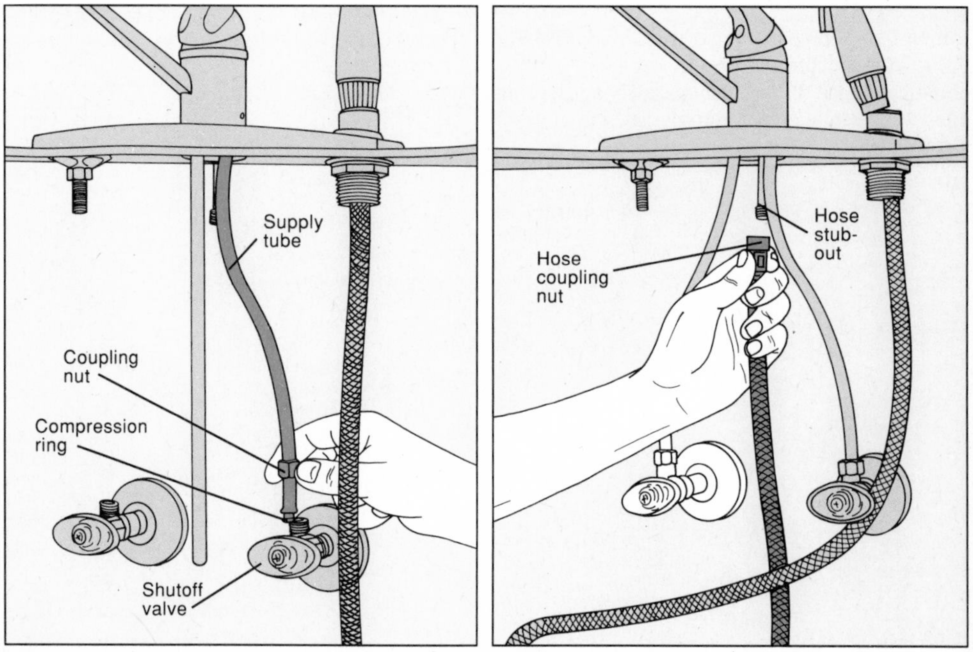

8 **Connecting the tubes and hose.** Fit a coupling nut and compression ring onto the free end of a supply tube *(above, left)*. Push the end of the tube into the shutoff valve as far as it will go, tighten the nut by hand, then give it a quarter-turn with an adjustable wrench. Attach the other flexible connector to its outlet. To attach a spray hose that is part of the faucet set, screw its coupling nut onto the stub-out behind the supply tubes *(above, right)*. Tighten the nut with a basin wrench. If the new faucet set does not have a spray hose, but there is an outlet for one, screw on a cap or plug. Turn on the water, slowly at first. If leaks occur, tighten the coupling nuts another quarter-turn, but take care not to overtighten. Remove the aerator, if any, and run water full force to flush the lines before replacing it.

INSTALLING FLEXIBLE SUPPLY TUBES

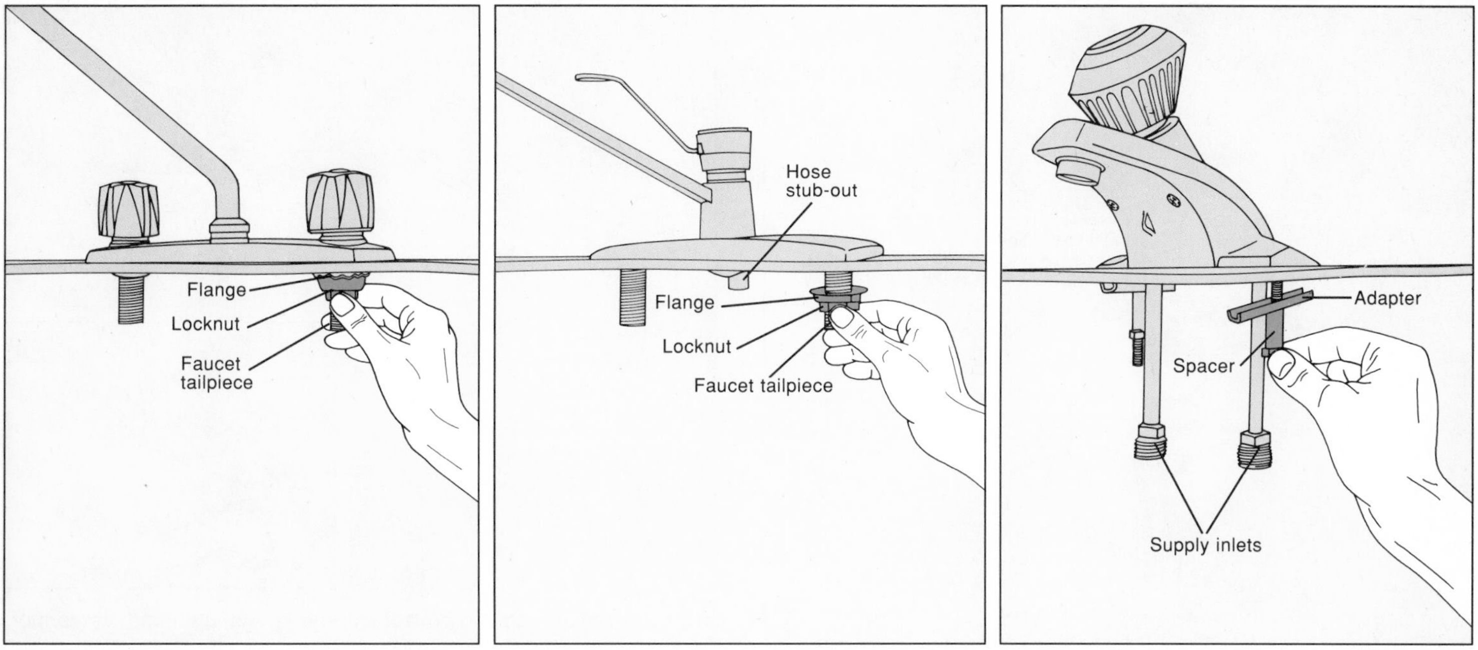

1 Choosing a faucet set. Some single-lever and double-handle faucet sets have no supply tubes attached to the faucet. Instead, the tailpieces, a standard 1/2-inch in diameter, are threaded to connect to any type of flexible supply tube *(above)*. Connectors are made of flexible polybutylene (PB), braided steel or corrugated copper and attach to the faucet and shutoff valves by means of compression nuts and rings. They are flexible, do not kink and are available in 12-to 36-inch lengths. They are also easy for a do-it-yourselfer to install, especially in cramped places, and plastic coupling nuts need only hand-tightening.

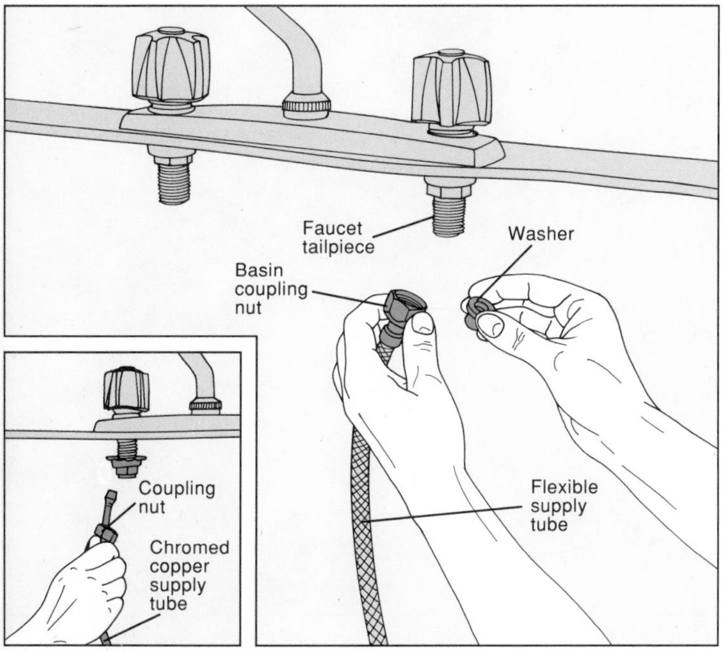

2 Attaching the upper end of the supply tube. Place the washer that comes with the supply tube inside the 1/2-inch coupling nut attached to its upper end *(above)*. Screw the nut onto the faucet tailpiece or threaded adapter and tighten it with a basin wrench. Since chromed copper supply tubes have no washers, simply screw the nut onto the adapter or tailpiece (*inset*) then attach the other end to the shutoff valve.

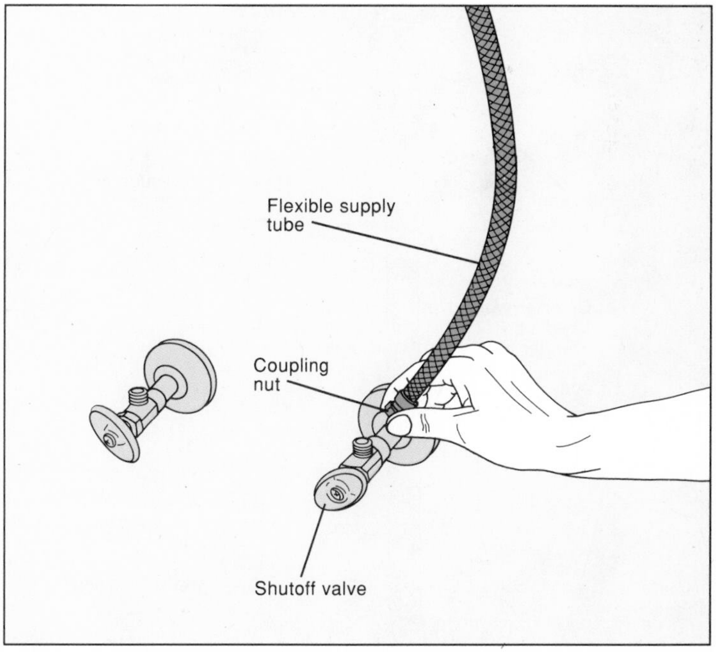

3 Attaching the lower end of the supply tube. Screw the coupling nut at the lower end of the supply tube onto the shutoff valve outlet *(above)*, then give it a quarter-turn with an adjustable wrench. Other tubes push in and hand-tighten. Repeat with the other supply tube, then turn on the shutoff valves. If there are any leaks, tighten the basin coupling nuts above and the compression nuts below another quarter-turn. Remove the aerator and run water, slowly at first, to flush the lines before replacing it.

SINK SPRAYS AND AERATORS

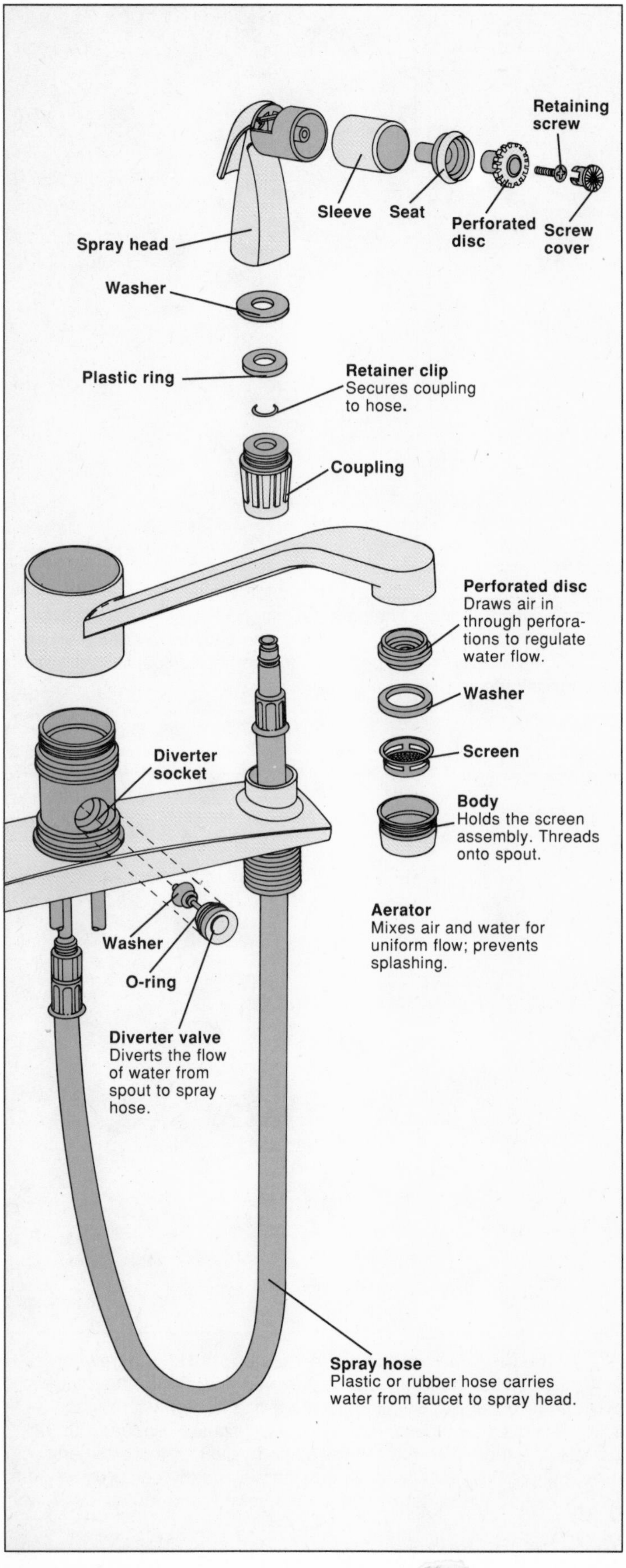
Retaining screw
Sleeve
Seat
Perforated disc
Screw cover
Spray head
Washer
Plastic ring
Retainer clip
Secures coupling to hose.
Coupling
Perforated disc
Draws air in through perfora-tions to regulate water flow.
Washer
Screen
Body
Holds the screen assembly. Threads onto spout.
Aerator
Mixes air and water for uniform flow; prevents splashing.
Diverter socket
Washer
O-ring
Diverter valve
Diverts the flow of water from spout to spray hose.
Spray hose
Plastic or rubber hose carries water from faucet to spray head.

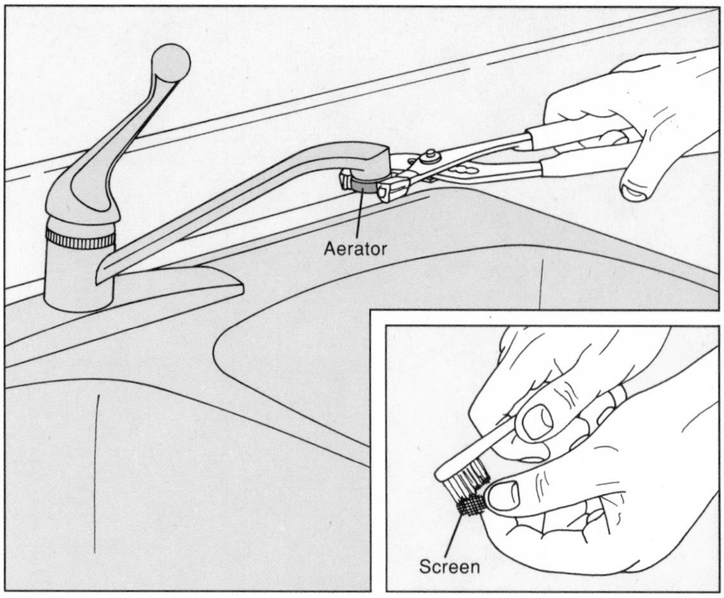
Aerator
Screen

Servicing the aerator. To protect chromed parts, unthread the aerator from the spout end with tape-wrapped channel-joint pliers, as shown. Disassemble the aerator, taking care to place parts in their exact order. Examine the washer and replace it if worn or cracked. Soak the screen in vinegar and scrub with a small brush *(inset)*, then flush with water to remove sediment. Reassemble the aerator, and thread the assembly back on the spout. Tighten a quarter-turn with pliers.

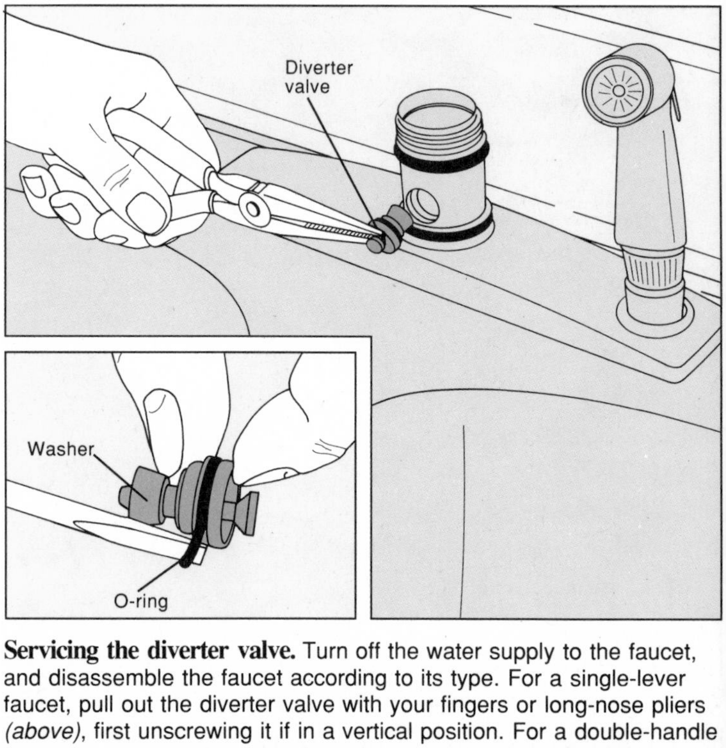
Diverter valve
Washer
O-ring

Servicing the diverter valve. Turn off the water supply to the faucet, and disassemble the faucet according to its type. For a single-lever faucet, pull out the diverter valve with your fingers or long-nose pliers *(above)*, first unscrewing it if in a vertical position. For a double-handle faucet, unscrew and pull out the valve under the spout nut. If the valve is bent, replace it. If in good shape, rinse it off and flush the diverter socket in the faucet body. Pry off the O-ring and roll an exact replacement into the groove on the valve. If the cone-shaped washer is loose, replace the entire diverter assembly. Reassemble the faucet and turn on the water. If leaks persist, clean the spray head *(page 35)*.

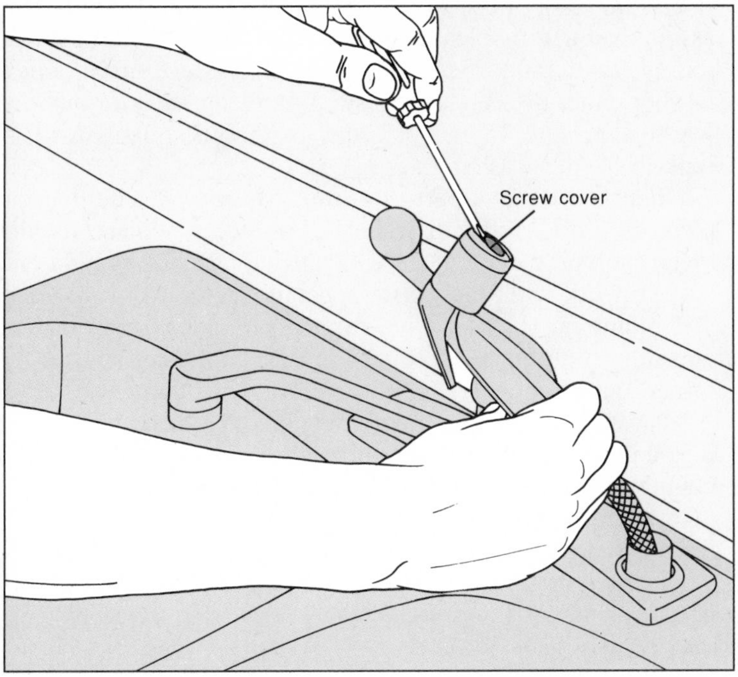

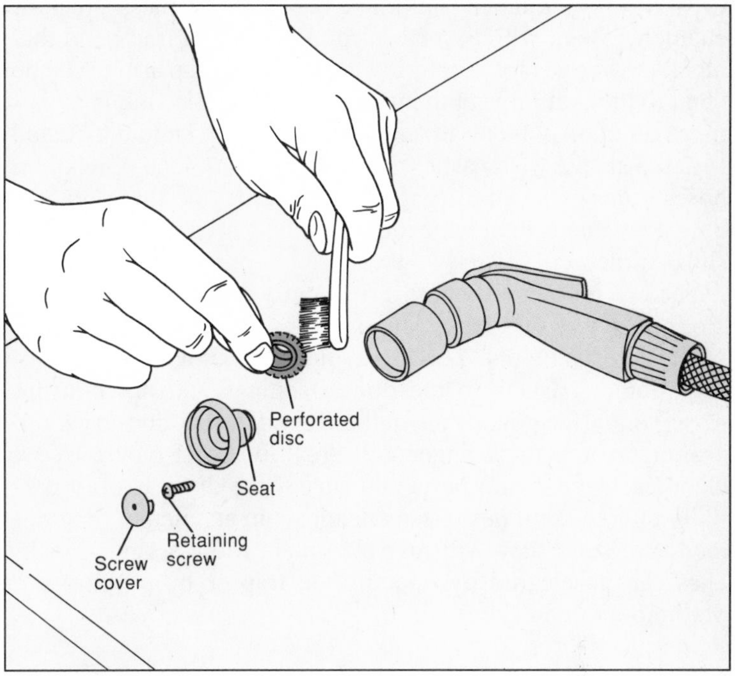

Cleaning the spray head. Pry off the screw cover at the end of the nozzle with a knife or small screwdriver *(above, left)*. Remove the screw to free the perforated disc and seat. Parts vary on different models; remove all parts that can be cleaned and place them aside in exact order of disassembly. If there are washers, replace them. To remove mineral deposits, soak the perforated disc and seat in vinegar and scrub with a small brush *(above, right)*. Reassemble the spray head. If it still leaks, replace the washer in the spray hose *(below)*.

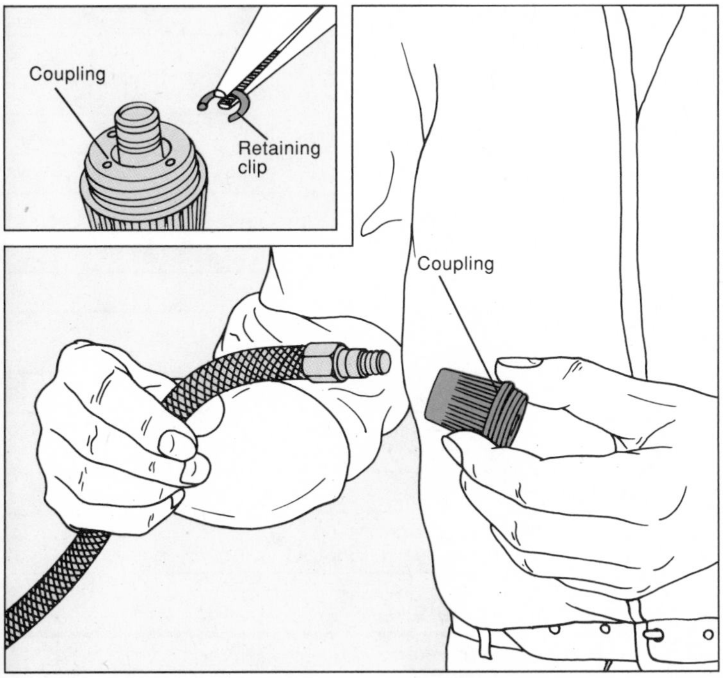

Servicing the spray head. Unscrew the spray head from the coupling. If the washer inside the spray head is worn, replace it. To free the coupling from the hose, snap off the plastic ring and remove the retaining clip with long-nose pliers *(inset)*. If the clip or spray handle is defective, replace the spray head. Replace the hose if it is blocked, kinked or cracked *(next step)*.

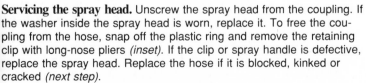

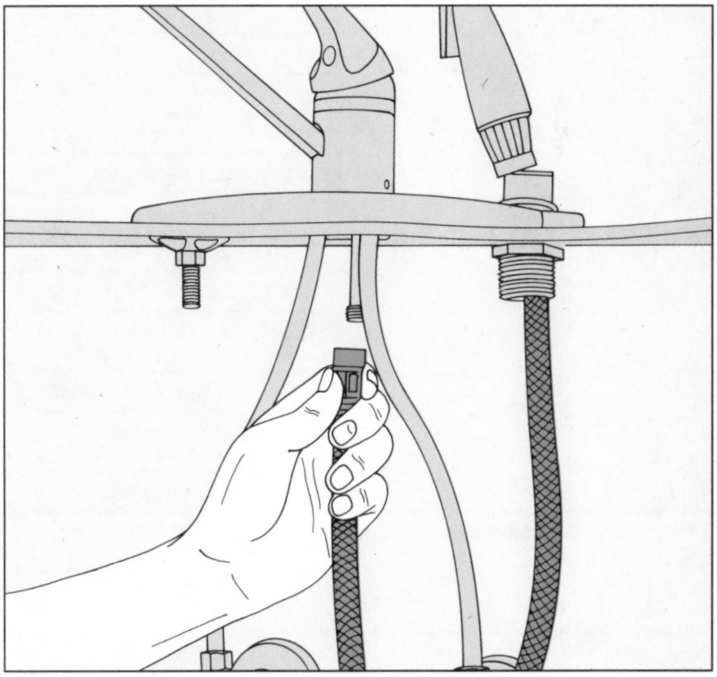

Replacing the spray hose. Use a basin wrench to loosen the coupling nut attaching the hose to the faucet underneath the sink. Then unscrew the hose by hand and pull it free *(above)*. Screw one end of the new hose to the tailpiece. Pull the free end of the hose up through its hole in the faucet set, fit the coupling over the end, and attach the spray head.

KITCHEN SINKS

Most modern kitchen sinks are made of stainless steel or enameled steel, with two basins draining into a trap bend that blocks sewer gas from entering the house. A trap arm joins the bend to the drainpipe at the wall. Under a single sink is a one-piece fixed or swivel trap, consisting of a trap bend connected to a trap arm. A dishwasher fits under any sink; one of its drain hoses attaches to an air gap—a simple device that prevents back-siphonage—and the other to the garbage disposer or sink tailpiece.

The two problems that most frequently plague kitchen sinks, clogged drains and leaky supply pipes, can be handled with basic plumbing tools. You can avoid clogs altogether by placing strainer baskets in the drain openings and not pouring grease or coffee gounds down the drain. If a sink does back up, a plunger or manual auger will break up most clogs. Use a chemical opener in a porcelain sink if the drain is only partially clogged, but never use chemicals in an enamel or stainless-steel sink; they will mar the finish. More serious blockages can be cleared by opening the trap or by probing the drainpipe behind the wall.

You may only need to tighten a loose slip nut on the drain assembly to stop a leak under a sink. If this doesn't work, remove that part of the trap nearest the leak and install a new washer under the connecting slip nut. Keep an assortment of washers on hand; whenever you disassemble a trap it's wise to replace all the washers.

When you remove part of a trap, you may decide that the piece is too corroded to reinstall. You can replace it with metal, polypropylene or PVC plastic. Because it is light and easy to work with, plastic is especially suited for do-it-yourself plumbing (If you leave for the plumbing supply store, close the shutoff valves to ensure that the faucet will not be turned on.)

You will find the job of replacing an old kitchen sink easier if you choose a self-rimming model that you can drop in the countertop. A frame-rimmed sink must be supported when it is installed. Buy a sink the same size as or slightly larger than the old one, so that you do not have to replace the countertop. Connecting most of the supply and waste fittings on the sink before installation will make the job easier.

TROUBLESHOOTING GUIDE

SYMPTOM	POSSIBLE CAUSE	PROCEDURE
Water seeps from sink	Sink basket displaced	Twist basket into place
	Sink basket not sealing	Clean or replace basket
Water under sink	Faucet set loose	Tighten locknuts under faucet set *(p. 31)* □○
	Faucet set worn	Replace faucet set *(p. 31)* ⬓●
	Trap fittings loose	Tighten slip nuts on trap assembly *(p. 40)* □○
	Dishwasher hose clamp loose or worn	Tighten or replace clamp *(p. 42)* ⬓◒
	Dishwasher drain hose worn	Trim or replace hose *(p. 42)* ⬓◒
	Garbage disposer drainpipe washer worn	Replace washer *(p. 42)* ⬓◒
	Trap or drain washers worn	Replace worn parts *(p. 40)* ⬓◒
	Tailpiece or washers worn	Replace worn parts *(p. 40)* ⬓◒
	Leaky joint at sink strainer	Tighten locknut or retainer screws *(p. 41)* □○ Replace plumber's putty or worn parts *(p. 41)* ⬓◒
	Trap assembly worn or damaged	Replace trap assembly *(p. 45)* ⬓◒
	Supply tubes or fittings leak	Tighten coupling nut at shutoff valve □○ Replace supply tubes *(p. 31)* ⬓◒ or shutoff valves *(p. 111)* ⬓◒
	Sink damaged or corroded	Replace sink *(p. 43)* ⬓●
Drain blocked or sluggish	Clog in trap bend or arm	Use boiling water, plunger or hose *(p. 38)* □○ Work through cleanout plug *(p. 38)* □○ or trap *(p. 39)* ⬓◒
	Clog in branch drain	Auger behind the wall *(p. 39)* ⬓◒ Use power auger *(p. 49)* ⬓◒▲
More than one sink clogged	Main drain or vent blocked	Unblock main drain or vent *(p. 84)*

DEGREE OF DIFFICULTY: □ **Easy** ⬓ **Moderate** ■ **Complex**
ESTIMATED TIME: ○ **Less than 1 hour** ◒ **1 to 3 hours** ● **Over 3 hours**

▲ **Special tool required**

The kitchen's three basic plumbing fixtures
For installation and repair purposes, the kitchen sink, garbage disposer and dishwasher are almost a single unit. The disposer is an extension of the sink drain, and all three share a single trap and drainpipe. If leaks are caught at an early stage, most repairs are simply a matter of replacing washers or tightening nuts and screws.

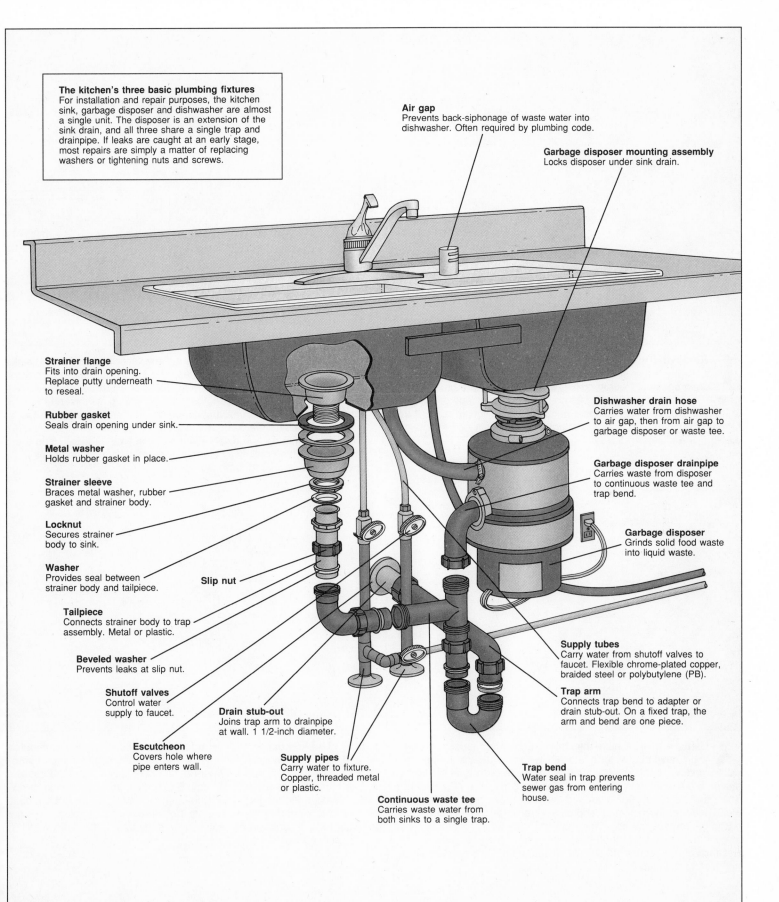

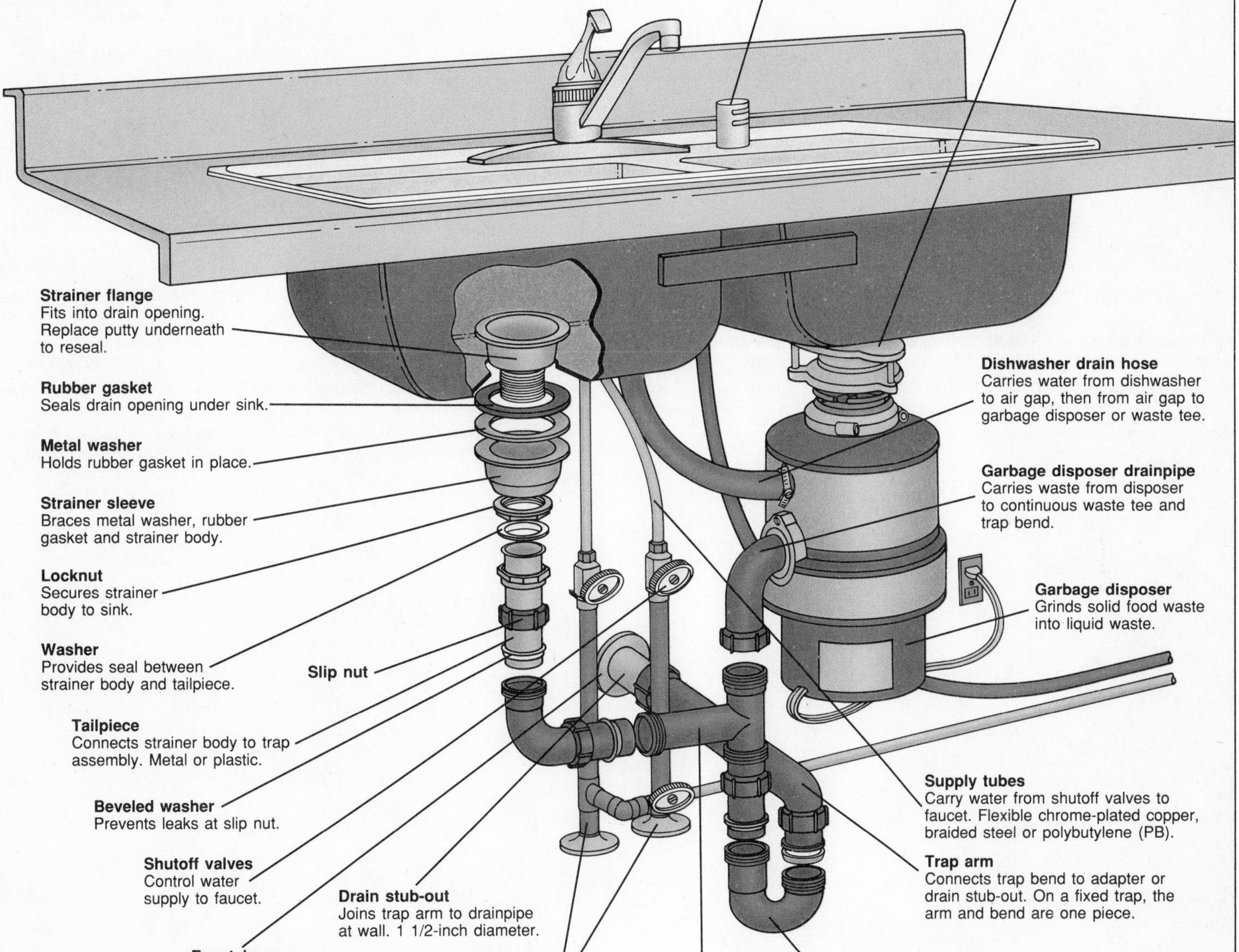
Air gap
Prevents back-siphonage of waste water into dishwasher. Often required by plumbing code.
Garbage disposer mounting assembly
Locks disposer under sink drain.
Strainer flange
Fits into drain opening. Replace putty underneath to reseal.
Rubber gasket
Seals drain opening under sink.
Metal washer
Holds rubber gasket in place.
Strainer sleeve
Braces metal washer, rubber gasket and strainer body.
Locknut
Secures strainer body to sink.
Washer
Provides seal between strainer body and tailpiece.
Slip nut
Tailpiece
Connects strainer body to trap assembly. Metal or plastic.
Beveled washer
Prevents leaks at slip nut.
Shutoff valves
Control water supply to faucet.
Drain stub-out
Joins trap arm to drainpipe at wall. 1 1/2-inch diameter.
Escutcheon
Covers hole where pipe enters wall.
Supply pipes
Carry water to fixture. Copper, threaded metal or plastic.
Continuous waste tee
Carries waste water from both sinks to a single trap.
Dishwasher drain hose
Carries water from dishwasher to air gap, then from air gap to garbage disposer or waste tee.
Garbage disposer drainpipe
Carries waste from disposer to continuous waste tee and trap bend.
Garbage disposer
Grinds solid food waste into liquid waste.
Supply tubes
Carry water from shutoff valves to faucet. Flexible chrome-plated copper, braided steel or polybutylene (PB).
Trap arm
Connects trap bend to adapter or drain stub-out. On a fixed trap, the arm and bend are one piece.
Trap bend
Water seal in trap prevents sewer gas from entering house.

CLEARING A CLOGGED DRAIN

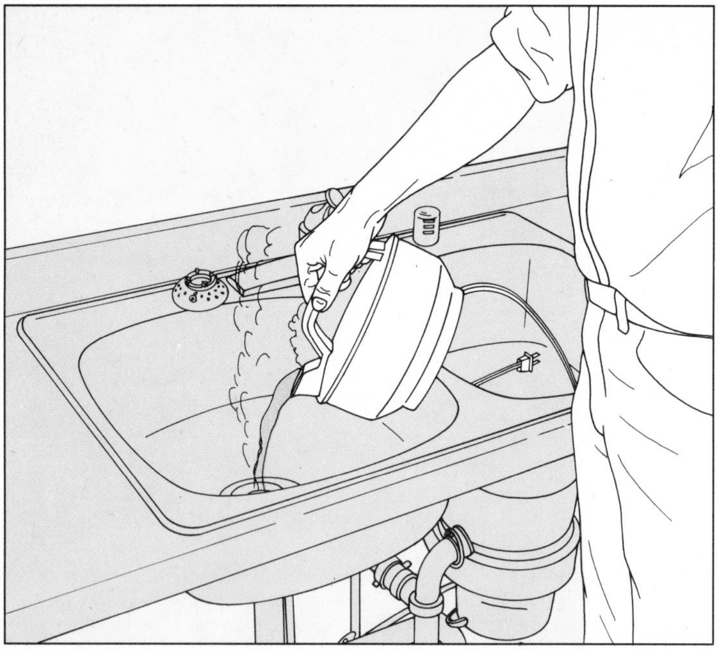

1 Removing grease clogs. Bail out any standing water in the sink, lift out the sink basket and clear any debris caught in the drain opening. Pour in boiling water to break up a grease clog *(above)*, but only if your drain has metal or polypropylene drainpipes, and there is no garbage disposer beneath that particular basin. If the blockage remains, or the sink is fitted with a disposer or PVC drainpipes, use a plunger *(step 2)*.

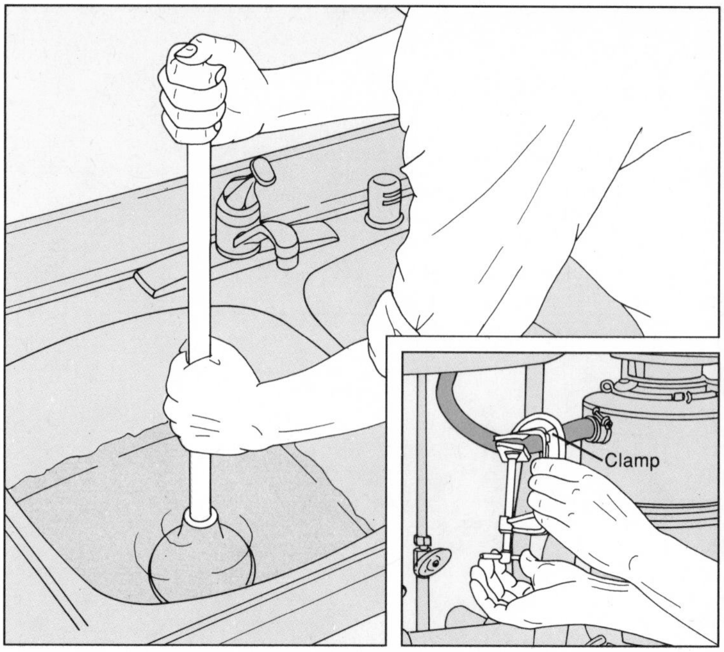

2 Using a plunger. If there is a dishwasher attached to the sink, seal off the drain hose between the air gap and the disposer by tightening a C-clamp over two pieces of wood placed on each side of the hose *(inset)*. Then, for a double sink, pack several rags wrapped in plastic into one drain opening. Run enough water in the other sink to cover the plunger cup, and set the plunger squarely over the drain. Pump up and down about a dozen times without lifting the cup *(above)*, then pull away sharply. Repeat several times. If this is ineffective, use a hose *(step 3)*.

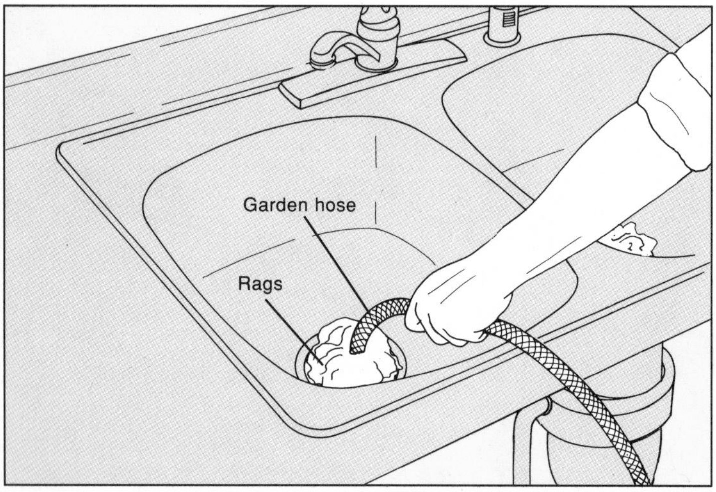

3 Using a hose. A hose attached to an outdoor faucet—or the drain valve of a water heater—may reach the sink through a window. If not, you will need to adapt the hose to fit the kitchen faucet. Block one drain of a double sink, feed the hose down the other drain, and pack rags tightly around the hose. Hold the hose firmly in place and turn the faucet on and off several times *(above)*. If water backs up, remove the hose and rags and work on the cleanout *(next step)*. If there is no cleanout, you must remove the trap *(step 5)*. Because of the danger of back-siphonage, do not leave the hose attached to the faucet.

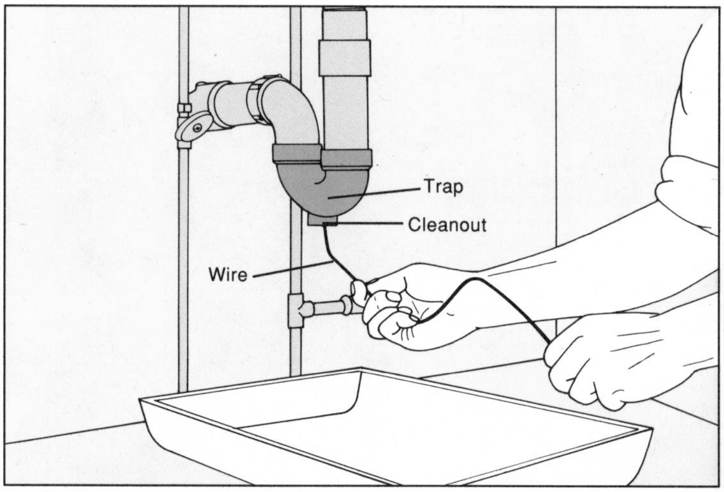

4 Unblocking at the cleanout. Place a bucket or pan beneath the trap. Loosen and remove the cleanout plug by hand or with channel-joint pliers. Water pouring out indicates that the blockage must be elsewhere; go to step 5 to remove the trap. If little or no water emerges, probe through the opening with an auger or bent coat hanger to snag or loosen the clog. Tighten the cleanout plug. Turn the water back on and run hot water to flush the trap. If the drain is still sluggish, remove the trap *(step 5)*.

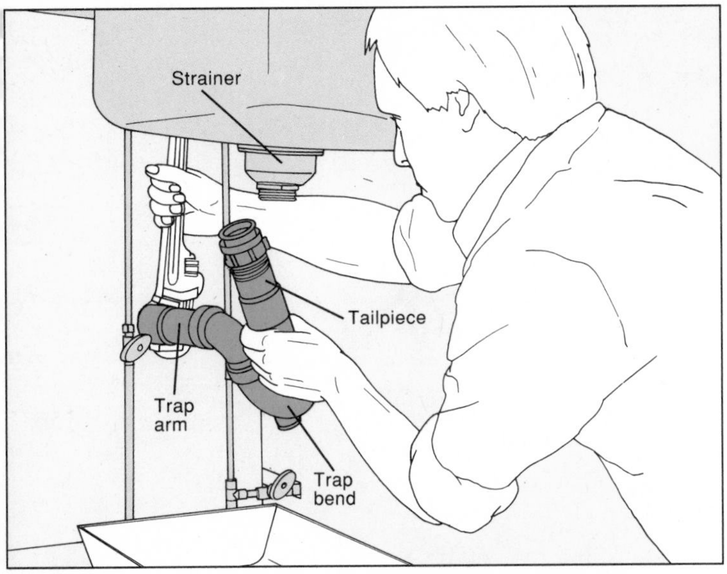

5 Removing the trap under a single sink. With channel-joint pliers, remove the slip nuts connecting one end of the tailpiece to the strainer body, and the other end to the trap bend. For a fixed trap, push the tailpiece down into the trap and loosen the slipnut at the stub-out with a pipe wrench *(above)*. Then remove the trap assembly from the drainpipe by hand. For a swivel trap, loosen the slip nuts, pull the trap bend free and empty it into a container. Scrub the bend with a brush, then rinse. Replace the cleaned trap and test the drain. If you found no clog, clear the drainpipe behind the wall *(step 8)*.

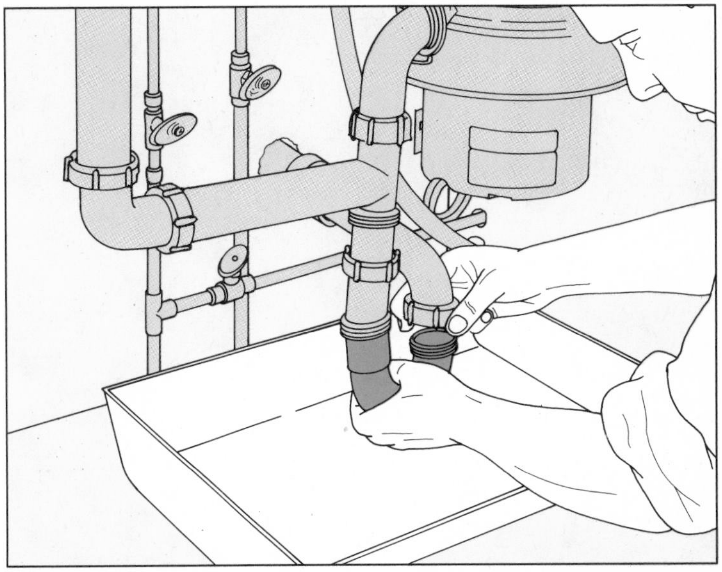

6 Removing the trap under a double sink. Place a pan or pail under the trap. Support the bend with one hand and loosen the slip nuts on each end, using channel-joint pliers if metal. Push the loosened slip nuts and beveled washers onto the pipes above, and pull the trap free *(above)*. Empty the bend into a container, scrub it with a flexible brush—or use an auger—then rinse with water. If you removed the obstruction, replace the trap using new washers and test the drain. Otherwise, remove the trap arm *(step 7)*.

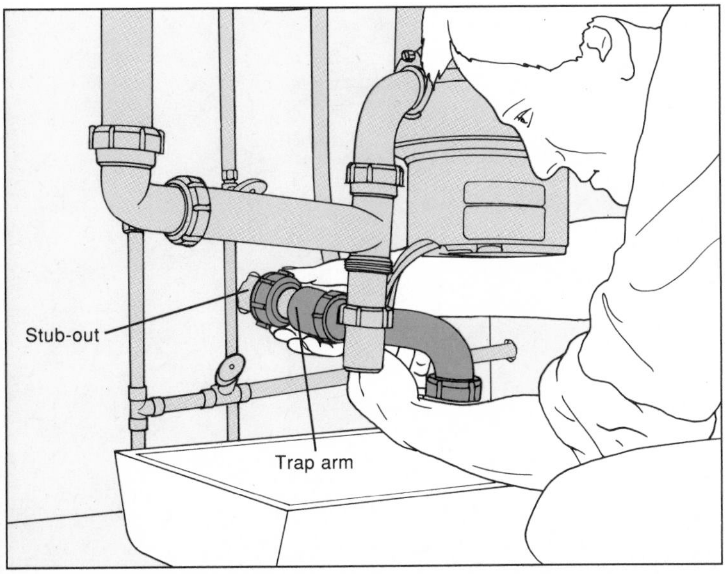

7 Removing the trap arm. Loosen the slip nut that joins the trap arm to the drain stub-out with a pipe wrench—it may be covered by an escutcheon—then unthread the trap assembly by hand *(above)*. Pull out the trap arm and clean it as in step 6. If you removed the clog, reconnect the trap arm and bend, and run water to test the drain. If it remains blocked, work behind the wall to clear the branch drain *(next step)*.

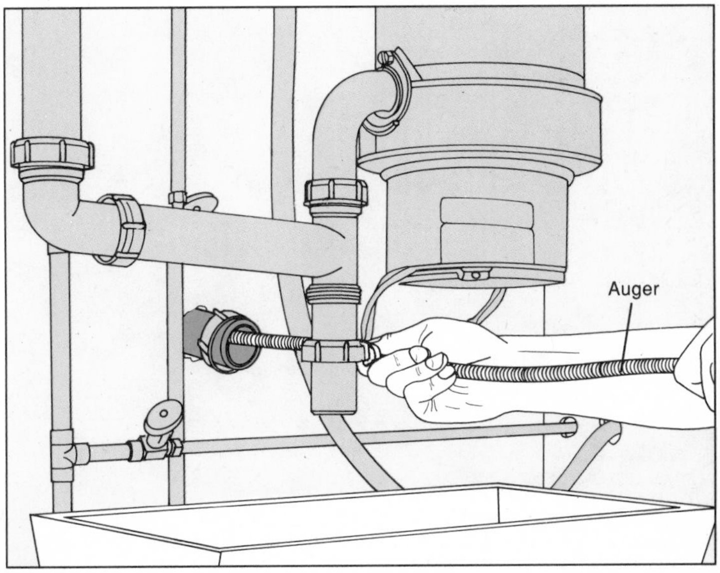

8 Augering into the branch drain. Probe into the branch drain behind the wall with a manual auger *(above)* or use a power auger *(page 49)*. Work carefully with the auger. Ramming it too vigorously can loosen fittings behind the wall and could pierce an old, deteriorated pipe. Reconnect the trap assembly using new washers, and tighten all slip nuts. Test the drain. If it is still clogged, work on the main drain *(page 84)*.

SERVICING THE TRAP AND DRAIN

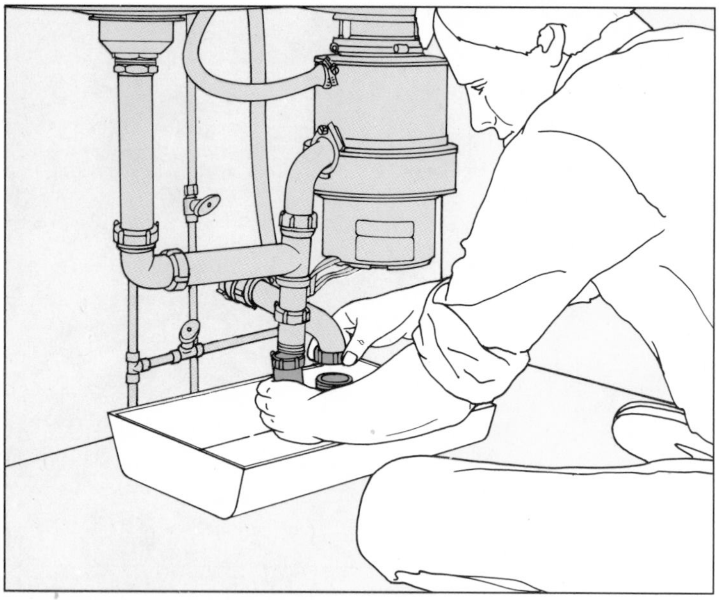

1 Servicing the trap bend. To stop leaks at the trap bend, first try tightening the slip nuts a quarter-turn. If this fails, remove the bend *(page 39)*. Replace if corroded or cracked, and change the washer where the bend meets the trap arm. If you are replacing a metal trap with a plastic one, install a new plastic slip nut on the pipe above. Push the trap bend back onto the pipe above, as shown. If there are leaks from other fittings, go to steps 2, 3 or 4 before reassembling the trap. To reconnect the bend, turn it so that it is directly under the trap arm or stub-out opening and thread the slip nut onto the bend by hand. Tighten metal slip nuts a quarter-turn with channel-joint pliers. Hand-tighten plastic connections.

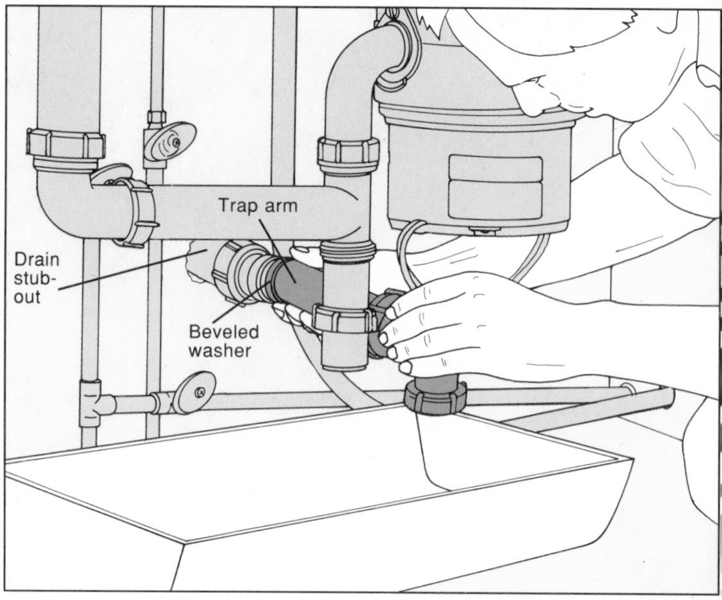

2 Servicing the trap arm. Move the pan under the drain stub-out or place rags below it. Remove the trap bend, if you have not already done so, then the trap arm *(page 39)*. Replace corroded parts, and install new washers. For additional protection against leaks, spread a coat of plumber's putty or silicone sealant inside the slip nuts. Hold the trap arm up to the stub-out and thread the slip nut back on *(above)*. Reconnect the trap bend to the trap arm *(step 1)* and turn on the faucet. If leaks persist, tighten the slip nuts a quarter-turn.

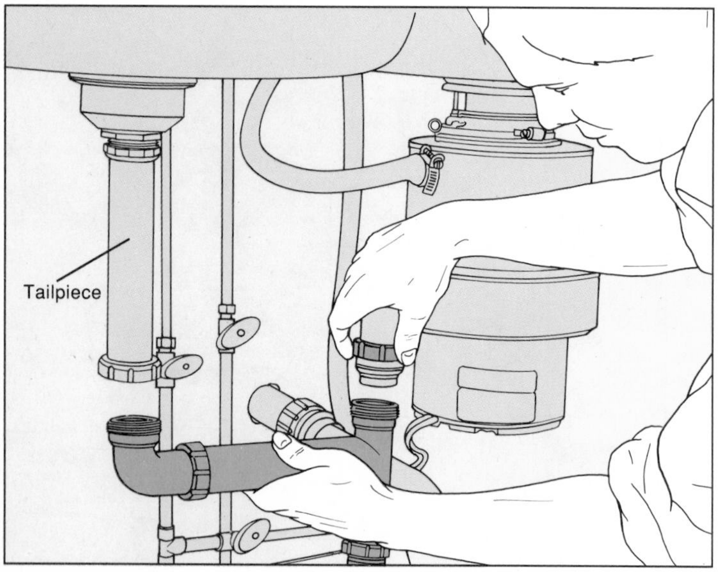

3 Servicing other drain fittings. Place a container under the drain fittings to catch water runoff and debris. One by one, tighten the slip nuts to stop a leak. If this doesn't work, remove the trap bend, then disassemble the fittings and inspect them for damage *(above)*. Replace any part that is cracked or corroded. If there are leaks at the sink strainer and tailpiece, go to step 4. Reassemble the drain and trap using new washers, and coat the inside of the slip nuts with putty or silicone sealant before tightening the connections.

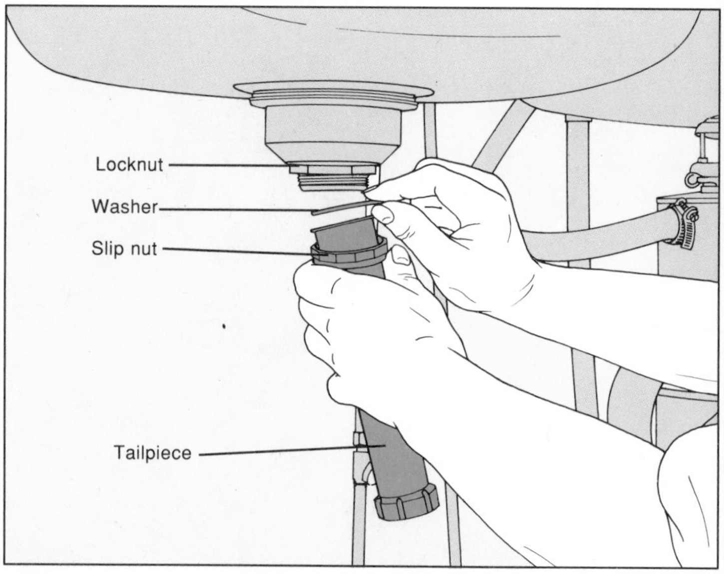

4 Servicing the tailpiece. With channel-joint pliers, loosen the slip nut that connects the tailpiece to the sink. If you have not already removed the drain fittings, loosen one or more other slip nuts as needed to lower the tailpiece an inch or so from the sink. Slip the tailpiece washer out *(above)*, insert a new washer, and reconnect the tailpiece. Reconnect the drain fittings as in step 3, or if water leaks from the sink strainer, reseal or replace it *(page 41)*.

SERVICING THE STRAINER

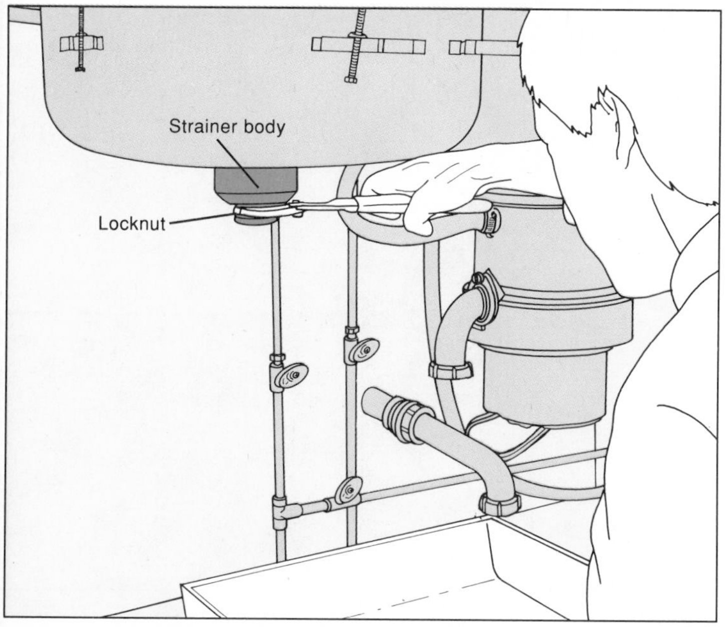

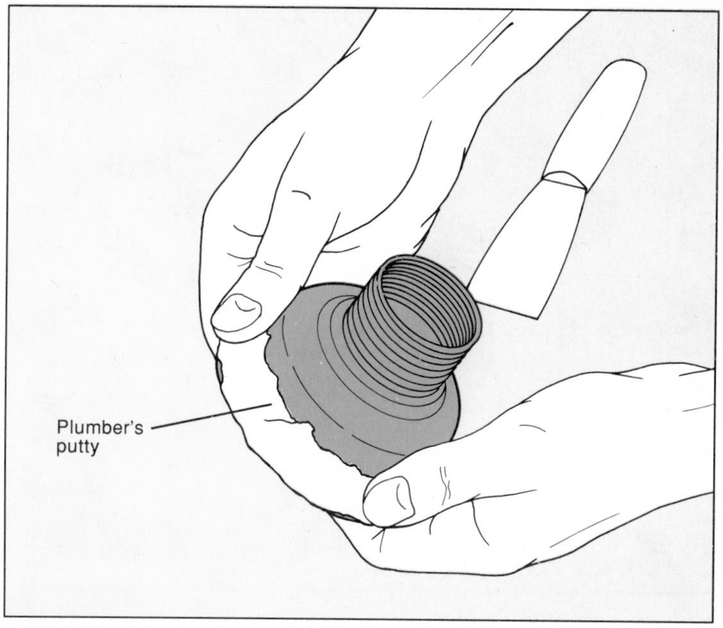

1 Freeing the strainer. Leaks arise here when the strainer body wears through or the putty that seals it to the sink dries out or erodes. To gain access to the strainer, first loosen and remove the faucet tailpiece *(page 40)*. Next, unscrew the locknut that secures the strainer body to the sink *(above)*. If the strainer body starts to turn in the sink while you loosen the locknut, wedge a screwdriver into the drain and hold it steady with your free hand. Another type of strainer is held in place by a plastic retainer and three screws; remove them and twist the retainer a quarter-turn to unlock the strainer body.

2 Stopping leaks around the strainer. If water leaks from around the sink opening, replace the putty and, if necessary, the strainer. Push the strainer body out of the sink from underneath. Scrape the putty from the drain hole, and from the old strainer if you plan to reuse it. Apply a 1/2-inch strip of plumber's putty under the lip of the strainer, as shown. Note that some strainers come with adhesive-coated rubber gaskets, and need no putty.

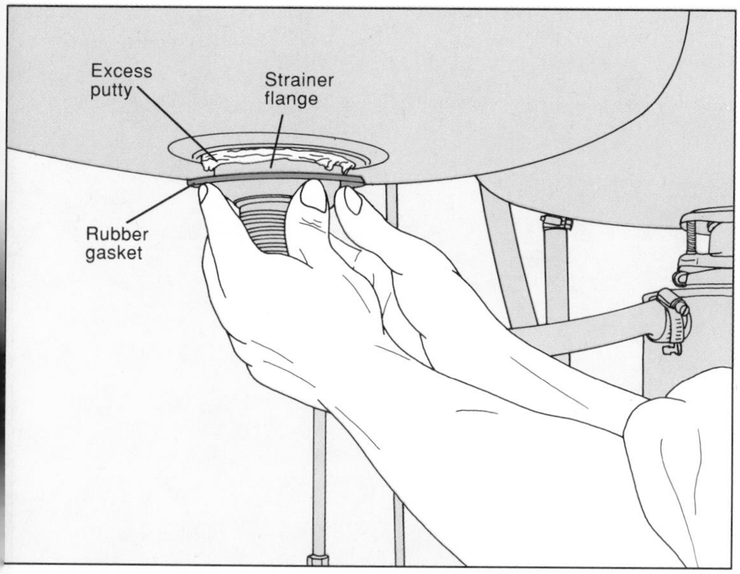

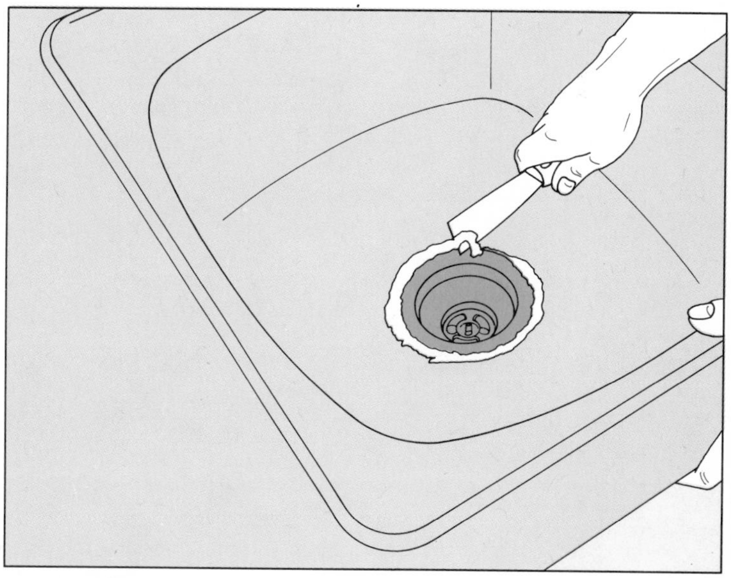

3 Installing the strainer. Lower the strainer body into the sink opening from above. From underneath the sink, slip the rubber and metal washers over the neck of the strainer *(above, left)* then secure the locknut or retainer and screws. Raise the tailpiece into place and tighten the coupling nuts. Scrape away any excess putty around the sink opening with a putty knife *(above, right)*, being careful not to scratch the surface.

SERVICING APPLIANCE CONNECTIONS

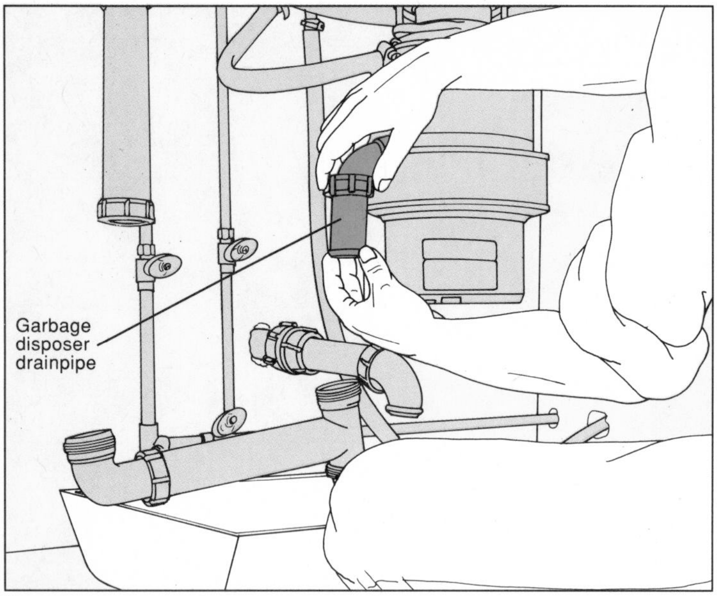

Servicing the garbage disposer drainpipe. Turn off the water and unplug the garbage disposer. Remove the trap bend *(page 39)* and disassemble the garbage disposer drainpipe *(above)*. Remove and replace the beveled washer in the disposer and reassemble the fittings in reverse order. While the drain is apart, replace all slip nut washers and seal the slip nuts with plumber's putty.

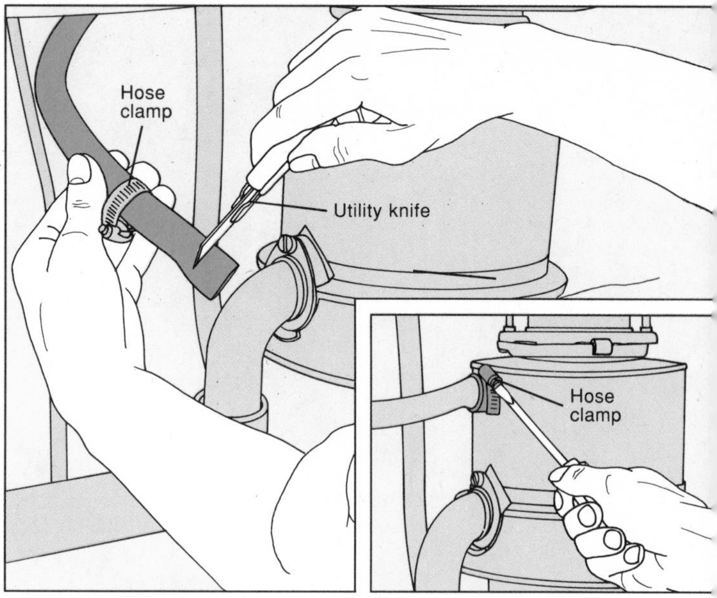

Repairing the dishwasher drain hose. If water is leaking from the dishwasher drain hose between the air gap and the disposer, tighten the fittings or replace the hose. Turn off power to the dishwasher and disposer, and close the dishwasher shutoff valve. Loosen the hose clamps *(inset)* and pull the hose off the disposer, then off the air gap. If only the ends are damaged, cut away an inch or two of the hose wit a utility knife *(above)*. To reconnect the hose or install a new one, slid new clamps onto the hose, push the ends onto the connections and tighten the clamps.

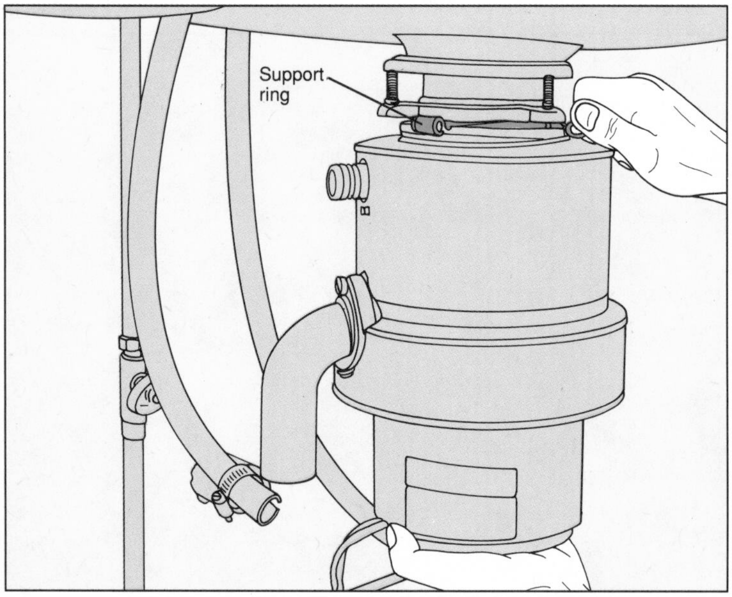

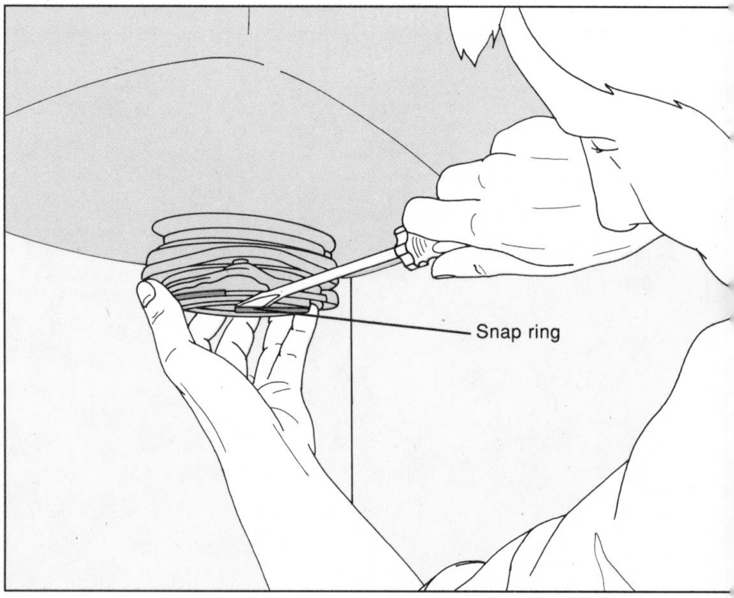

Servicing the disposer mounting assembly. Turn off power to the garbage disposer and place a container under it. Remove the trap bend *(page 40)*, disposer drainpipe, and the dishwasher drain hose. Supporting the disposer with one hand, turn the lower support ring to unlock it from the mounting assembly, then lower the unit from the sink. If the support ring will not turn, use a screwdriver for better leverage. Place the disposer aside. Remove the screws on the mounting assembly and push up the mounting flange at the base of the assembly. Hold it with one hand while you pop the snap ring out of the groove on the strainer flange using a flat-tipped screwdriver *(above, right)*. The mounting flange and its gasket will come off with the snap ring. Push the sink strainer up through the drain hole. Replace dried and cracked putty, as described on page 41. Lower the strainer back into the drain hole, then push a new rubber gasket onto the strainer body from under the sink. Reinstall the mounting flange and snap ring. Screw in the retaining screws. Remount the garbage disposer.

REPLACING A KITCHEN SINK

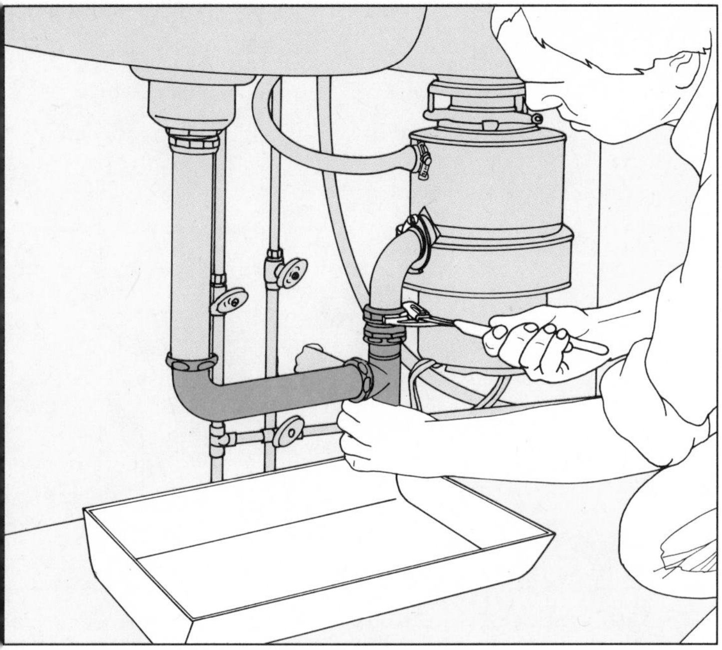

1 Removing the trap bend and drain assembly. Buy a sink the same size as or larger than the old one, and replace the trap bend and drain assembly at the same time. Choose plastic pipes if local codes permit, because they are durable and easy to work with. Turn off all power and water to the sink and its appliances, and have a container at hand to catch water runoff. Use channel-joint pliers to disconnect the trap, arm, tailpiece and drainpipes *(above)*.

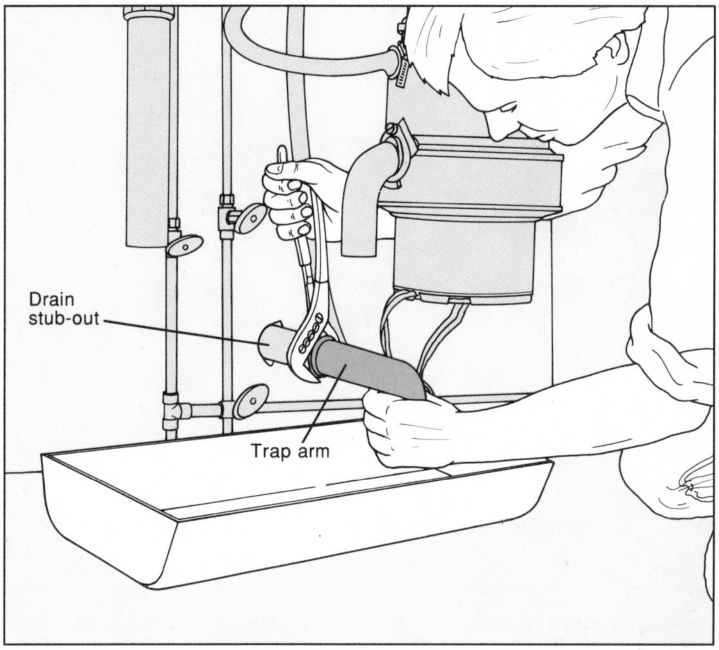

2 Removing the trap arm. If an exposed coupling joins the trap arm to the drain stub-out, loosen the slip nut with channel-joint pliers *(above)*, then unscrew the trap assembly from the drainpipe by hand. If the coupling is hidden, pry the escutcheon loose with a screwdriver, push it onto the trap arm, then loosen the slip nut with the pliers. If there is no dishwasher, disconnect the disposer *(page 42)* and go to step 4.

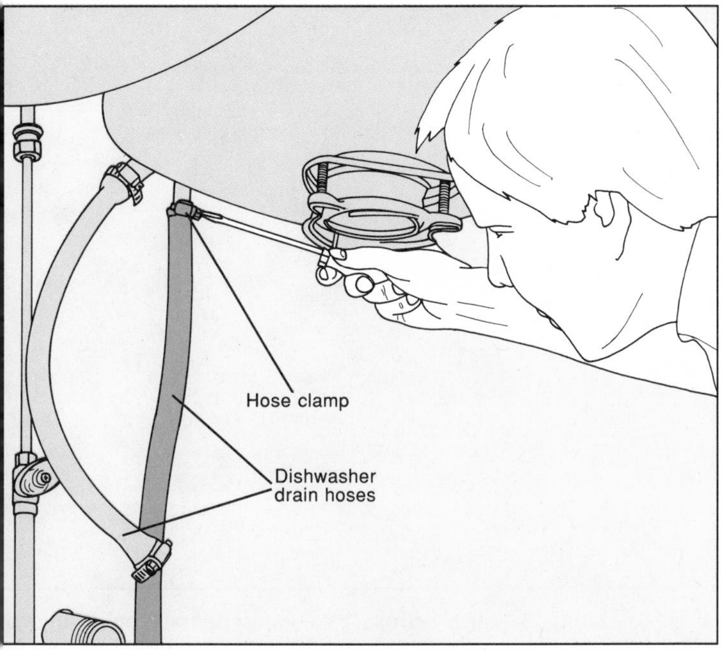

3 Removing the dishwasher drain hoses. Loosen the hose clamp connecting the dishwasher drain hose to the garbage disposer. Pull off the hose and let it hang from the air gap. Dismount the disposer as described on page 42 and set it aside. You now have access to the air gap behind the sink. With a screwdriver, loosen the two hose clamps securing the dishwasher drain hoses *(above)* and pull them free.

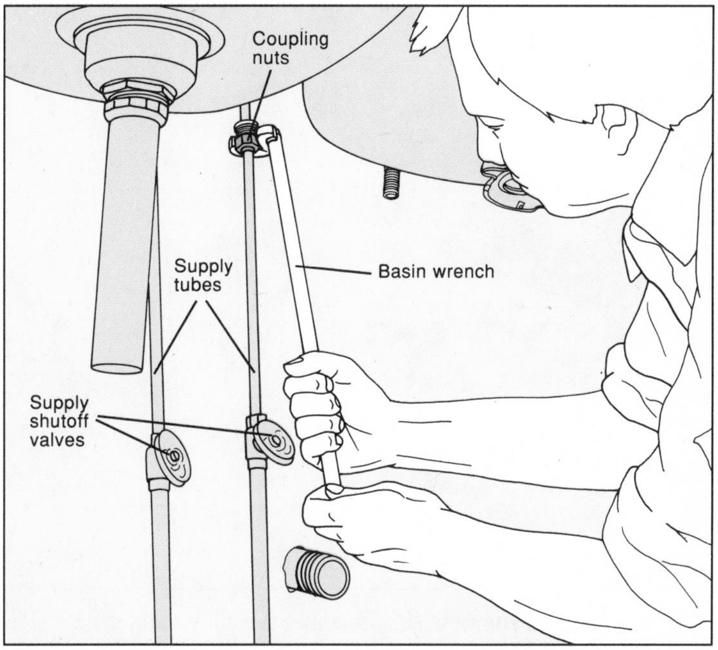

4 Disconnecting the supply tubes. Make sure that the shutoff valves are closed, then slip a basin wrench onto one of the supply tube coupling nuts and turn the wrench counterclockwise to loosen the nut *(above)*. Repeat with the other coupling nut. Remove the nuts by hand and slide them down the supply tubes. If the tubes are in the way of the sink, gently push them aside or remove them.

REPLACING A KITCHEN SINK (continued)

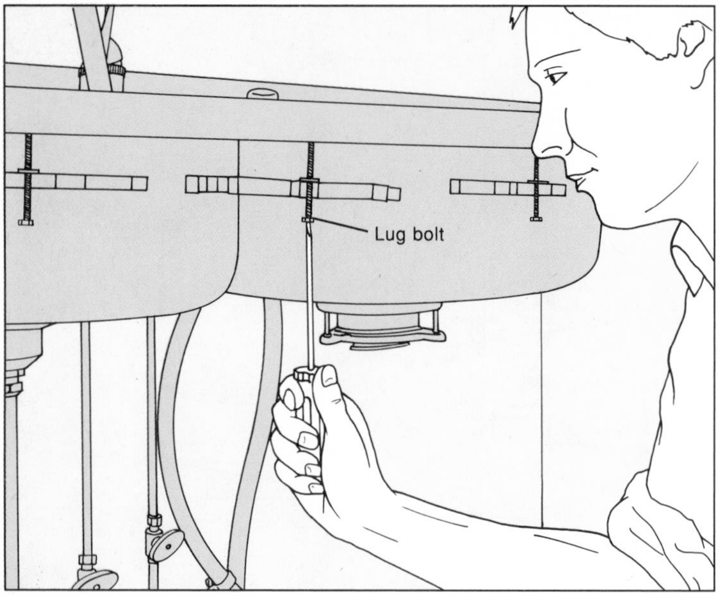

5 Freeing the sink. Once the supply and drain fixtures have been disconnected, you should be able to free a drop-in sink by pushing it up from underneath. First loosen and remove any bolts anchoring the sink underneath the countertop, as shown. If the bolts are stubborn, apply penetrating oil and wait an hour before trying again, or cut them off with a hacksaw.

Frame-rimmed sinks require that you first remove the lugs holding them in place. Have a helper support the sink, by grasping the drain-holes from above while you separate the sink and frame.

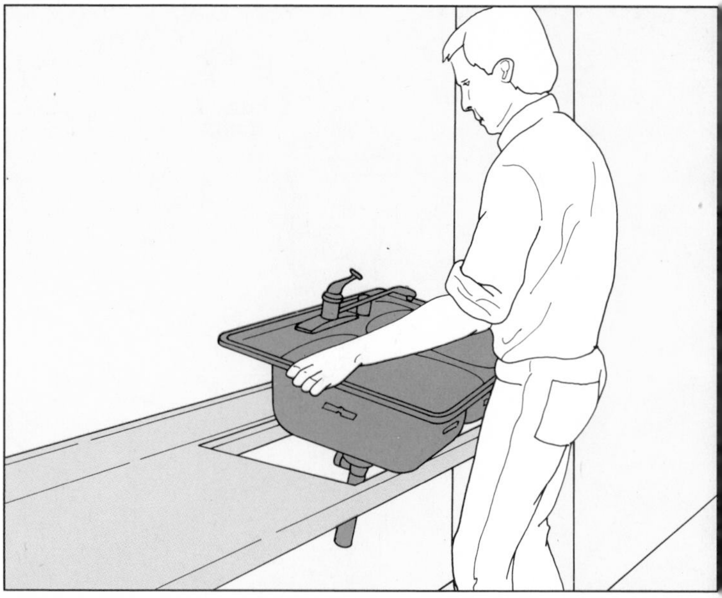

6 Lifting out the old sink. To remove a drop-in sink, slip the tip of a screwdriver between the sink edge and countertop, then twist one side of the sink from the countertop. Have someone hold that side, or wedge a piece of wood beneath it and pry up the other side of the sink. Pull the sink straight up and away from the countertop, as shown. To remove a frame-rimmed sink, grasp the drains in both hands and lift the sink (or lower it, on some models) from the countertop. You will also need a helper if you are removing an old cast-iron sink.

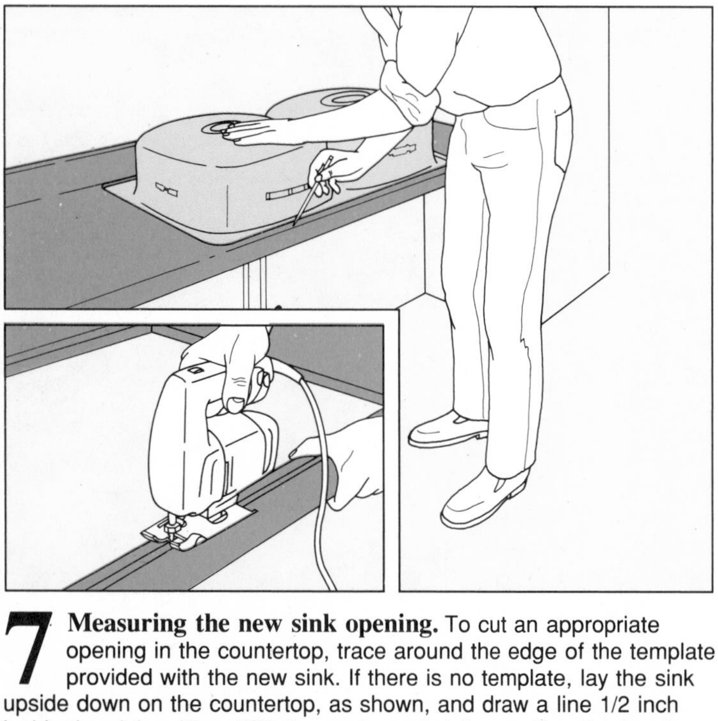

7 Measuring the new sink opening. To cut an appropriate opening in the countertop, trace around the edge of the template provided with the new sink. If there is no template, lay the sink upside down on the countertop, as shown, and draw a line 1/2 inch inside the sink outline. With frame-rimmed sinks, position the frame squarely on the countertop and trace around it. Use a saber saw to cut out or enlarge the opening, taking care not to touch any metal plumbing parts. Test-fit the new sink in the countertop.

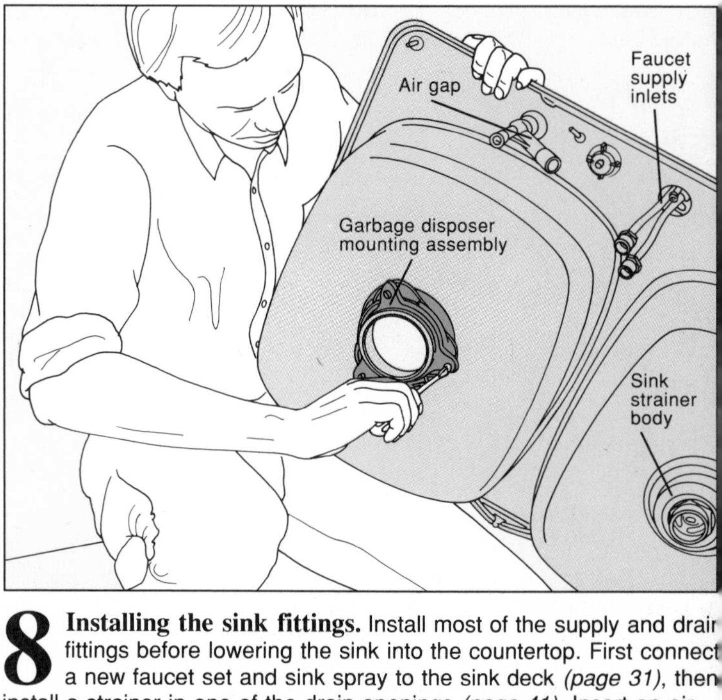

8 Installing the sink fittings. Install most of the supply and drain fittings before lowering the sink into the countertop. First connect a new faucet set and sink spray to the sink deck *(page 31)*, then install a strainer in one of the drain openings *(page 41)*. Insert an air gap into the sink from below, secure it with a locknut and connect the cap and washer from above. Remove the garbage disposer mounting assembly from the old sink (or buy a new one) and transfer it to the drain, securing it with a screwdriver, as shown.

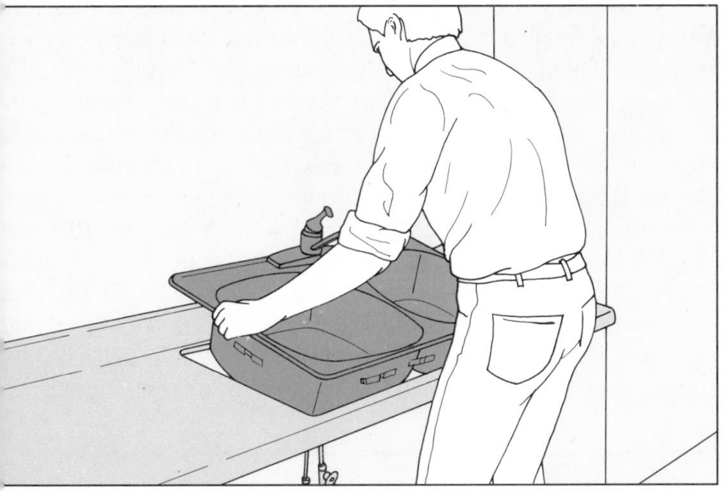

9 Dropping in the new sink. Apply a narrow strip of plumber's putty or silicone sealant around the hole in the countertop. For frame-rimmed sinks, also apply putty or sealant around the top edge of the sink where it will meet the rim. Lift the sink with both hands and lower it into the countertop hole *(above)*. Tighten the lug bolts with a screwdriver until the tops of the bolts loosely grip the underside of the countertop. Tighten the corner bolts first, then the middle bolts, until there are no gaps between the sink edge and the countertop.

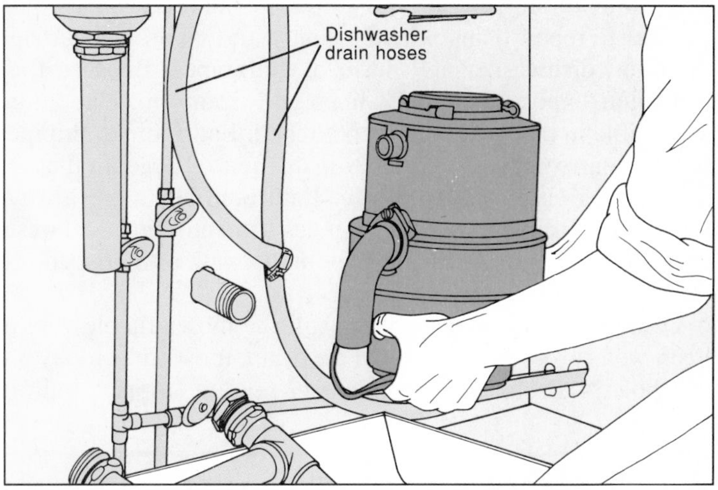

10 Mounting the garbage disposer. Reconnect the tailpiece to the sink strainer body *(page 40)*, then attach the supply tubes to the faucet set and tighten with a basin wrench. Attach the dishwasher drain hoses to the air gap and mount the disposer, as shown *(page 42)*. Connect the free end of the dishwasher drain hose to the disposer and tighten the hose clamp with a screwdriver (but do not overtighten).

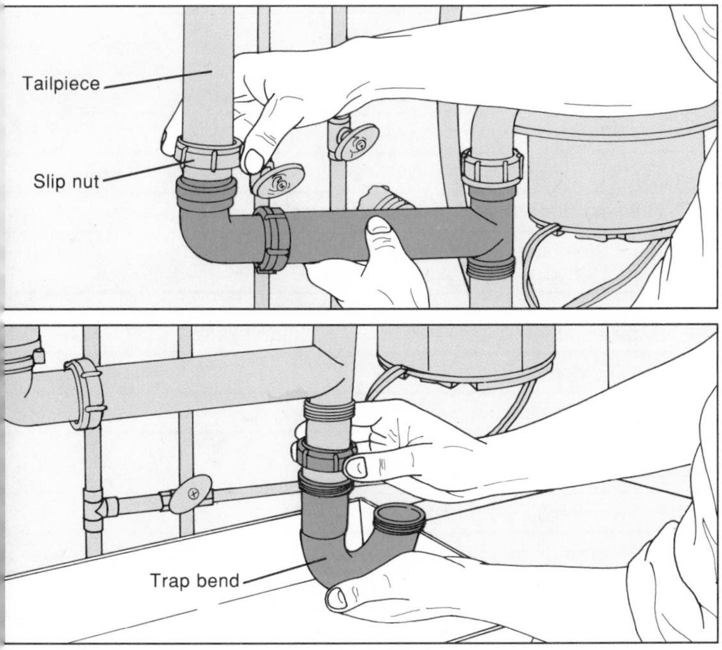

11 Replacing the drain assembly. Buy a packaged drain assembly that telescopes to fit the distance between sink basins. In assembling the drain, work your way from the sink tailpiece to the garbage disposer *(above, top)*, then install the trap *(above, bottom)*. Connect, but do not tighten the fittings until the entire drain is assembled and properly aligned. Test-fit each piece and trim if necessary, using a hacksaw and miter box. Make sure that all washers are properly seated in their slip nuts. For extra protection against leaks, apply plumber's putty or silicone sealant inside the slip nuts before tightening the fittings.

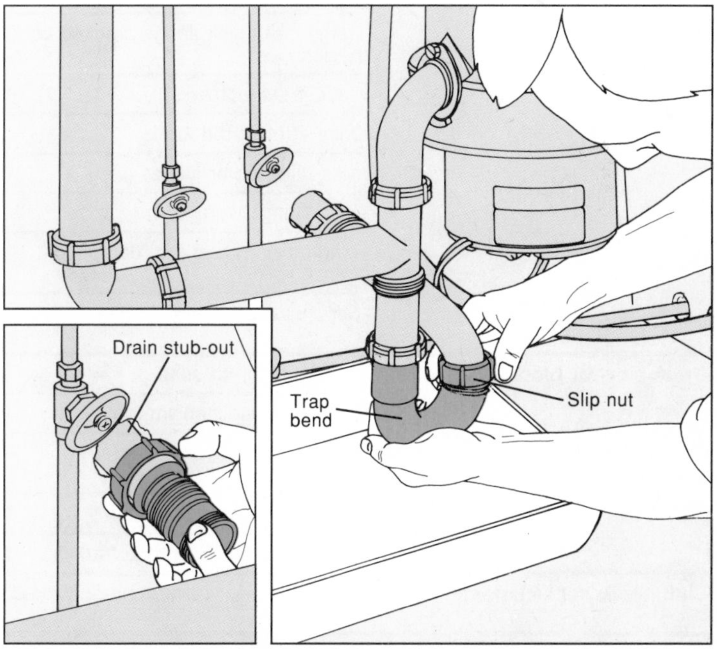

12 Completing the installation. Connect the trap arm to the stub-out in the wall *(inset)* and hand-tighten its slip nut and washer. Finally, fit the trap arm in place and screw on its slip nuts *(above)*. Recheck and tighten each clamp, slip nut and fitting. Open the shutoff valves and turn on the power to the dishwasher and disposer. Run water to test for leaks and tighten any connections, if necessary. If leaks persist, turn off the water and power and reassemble the fittings.

BATHROOM SINKS

With only a single trap and drain, a bathroom sink is far simpler to repair than a kitchen sink with two drains, garbage disposer, dishwasher and maze of drainpipes. Because they are often exposed, suppply lines and drains may be more accessible in the bathroom than kitchen. Bathroom drainpipes are also narrower—1 1/4 inches in diameter—because they do not carry grease or food waste. Bathroom sinks are always fitted with a trap to prevent sewer gas from entering the house: P-shaped if it joins a branch drain in the wall, or S-shaped if it drains through the floor.

While a simple rubber plug is still the most reliable way to keep water from seeping down the drain, most sinks today are equipped with the more convenient pop-up stopper. Pulling one end of a lift rod between the faucet handles causes the stopper to drop into the drain, sealing it. If water seeps away while the basin is full, or drains too slowly when the lift rod is lowered, the stopper or its lift mechanism may be at fault.

The two most common sink problems are leaks and clogs. Water under the sink usually points to worn faucet parts *(page 18)* or loose supply or drain fittings. Replacing dried putty, tightening a slip nut or changing its washer may be all that's needed. Replacing a faucet set *(page 31)*, or the sink itself *(page 54)*, can be tackled by a do-it-yourselfer with a patient and methodical approach. If you decide to change a damaged or outdated sink, the simplest replacement is a self-rimmed model that overlaps the edge of a hole in the countertop.

TROUBLESHOOTING GUIDE

SYMPTOM	POSSIBLE CAUSE	PROCEDURE
Water seeps from sink	Accumulation of soap or hair on stopper	Clean stopper *(p. 50)* □○
	Stopper O-ring worn	Replace O-ring *(p. 50)* □○
	Stopper worn	Replace stopper *(p. 50)* □○
	Lift mechanism too long	Adjust or replace lift mechanism *(p. 51)* ⬓◒
Pop-up stopper does not open or close properly	Lift mechanism disconnected	Reconnect below sink *(p. 51)* □○
	Lift mechanism broken	Service lift mechanism *(p. 51)* □○ Replace drain body and pop-up assembly *(p. 52)* ⬓◒
Water under sink	Faucet set loose	Tighten locknuts under faucet set *(p. 31)* □○
	Putty under sink flange cracked or gasket worn	Replace putty or gasket *(p. 53)* ⬓◒
	Faucet set worn	Replace faucet set *(p. 31)* ⬓●
	Joint at pivot rod loose	Tighten retaining nut or replace washer or gasket *(p. 53)* □○
	Sink tailpiece or fittings worn	Tighten connections; replace washer or tailpiece *(p. 53)* ⬓◒
	Trap fittings worn	Tighten connections; replace worn washers *(p. 53)* ⬓◒
	Trap bend or trap arm damaged	Repair temporarily with tape *(p. 11)* □○ Replace trap bend or arm *(p. 53)* ⬓◒
	Supply tube leaks	Tighten coupling nut at shutoff valve □○ Replace supply tubes *(p. 31)* ⬓◒ or shutoff valves *(p. 111)* ⬓◒
Drain slow or blocked	Accumulation of soap or hair on stopper	Clean stopper *(p. 50)* □○
	Pop-up lift mechanism too short	Adjust or replace lift mechanism *(p. 51)* □○
	Clog in trap bend or arm	Use plunger or auger *(p. 48)* □○ Clean trap *(p. 48)* ⬓◒
	Object dropped down drain	Retrieve object from trap *(p. 48)* ⬓◒
	Clog in drainpipe beyond trap	Auger through drain *(p. 48)* ⬓◒ or use power auger *(p. 49)* ⬓◒▲
Sink needs replacement	Basin chipped, cracked or discolored	Touch up or refinish sink *(p. 130)* ⬓◒ Replace with drop-in sink *(p. 54)* or wall-mount sink *(p. 55)* ⬓●

DEGREE OF DIFFICULTY: □ Easy ⬓ Moderate ■ Complex
ESTIMATED TIME: ○ Less than 1 hour ◒ 1 to 3 hours ● Over 3 hours

▲ Special tool required

However, the new sink must be the same size as or larger than the old one. Otherwise, you will have to replace the countertop at the same time. Another option is replacing a wall-hung sink with a one-piece sink and countertop.

Because the bathroom sink trap is narrow, it tends to clog as soap and hair accumulate in the tailpiece, bend or arm. Start by removing the stopper and clearing the drain opening with a bent coat hanger, then look beneath the sink for a cleanout plug on the trap. If the clog remains, fill the sink halfway and use a plunger, which is far safer for both you and your pipes than a caustic chemical drain opener. If this fails, use a manual or power auger, or disassemble the trap and clean its various sections one by one.

Before working on faucets or supply lines, close the shutoff valves beneath the sink (or the main shutoff valve if the fixture has none) and cover the drain to prevent loss of parts. Have a bucket and rags at hand to catch water runoff from trap-and-drain repairs. Protect chromed pipes and fittings by using smooth-jawed wrenches or by wrapping the jaws of wrenches and pliers with tape. If you must use an electric drill or power auger, be sure that it is double insulated (or used in a GFCI-protected outlet) and do not touch any metal parts while the tool is in use. If the sink is set in a narrow vanity, half of the repair may involve getting at the plumbing. A basin wrench, with its long shaft and spring-loaded jaws, is especially useful in such cramped quarters.

Keeping drains clog-free
Bathroom drains will almost never clog if you follow one simple rule: Do not use them for anything but waste water. Cleaning the stopper and drain regularly will help prevent the accumulation of soap and hair, which can block the trap or drain farther down the system. If the drain does back up, first determine the extent of the blockage by checking other drains in the house. If more than one is clogged, your problem is probably in the main drain rather than a single fixture.

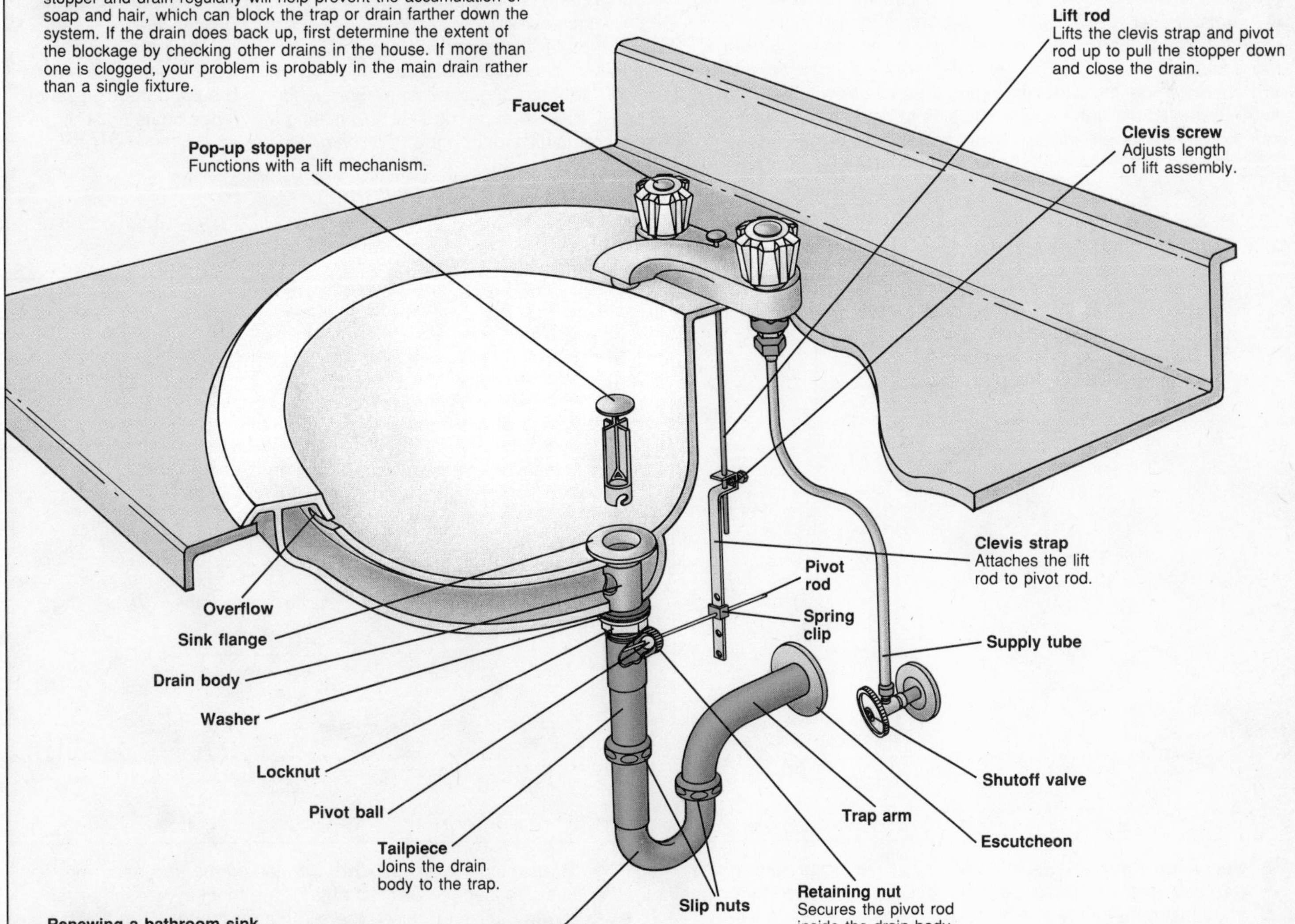
Lift rod
Lifts the clevis strap and pivot rod up to pull the stopper down and close the drain.
Faucet
Clevis screw
Adjusts length of lift assembly.
Pop-up stopper
Functions with a lift mechanism.
Clevis strap
Attaches the lift rod to pivot rod.
Pivot rod
Overflow
Spring clip
Sink flange
Supply tube
Drain body
Washer
Shutoff valve
Locknut
Pivot ball
Trap arm
Escutcheon
Tailpiece
Joins the drain body to the trap.
Slip nuts
Retaining nut
Secures the pivot rod inside the drain body.
Trap
May be P-shaped, as shown, or S-shaped, which is attached to a drainpipe through the floor.

Renewing a bathroom sink
A bathroom sink will usually outlast its faucets and drain. If you are renovating a bathroom and the basin is in good condition, you can save time and money by discarding the old fittings and replacing them with a modern faucet equipped with a pop-up drain.

CLEARING A CLOGGED SINK DRAIN

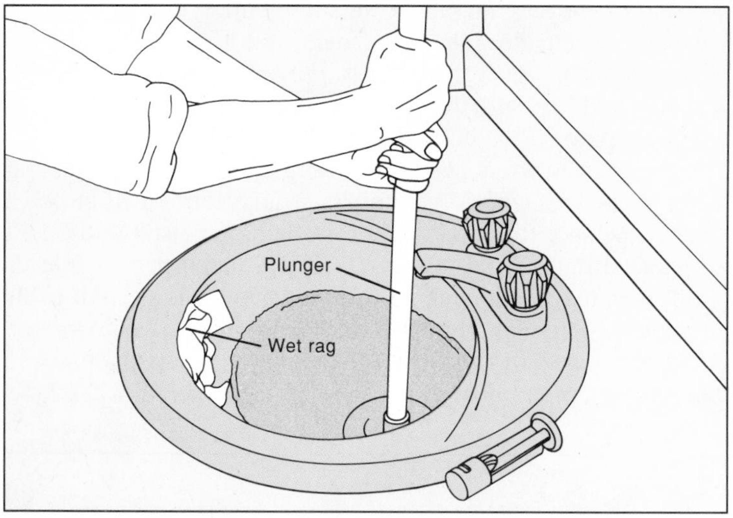

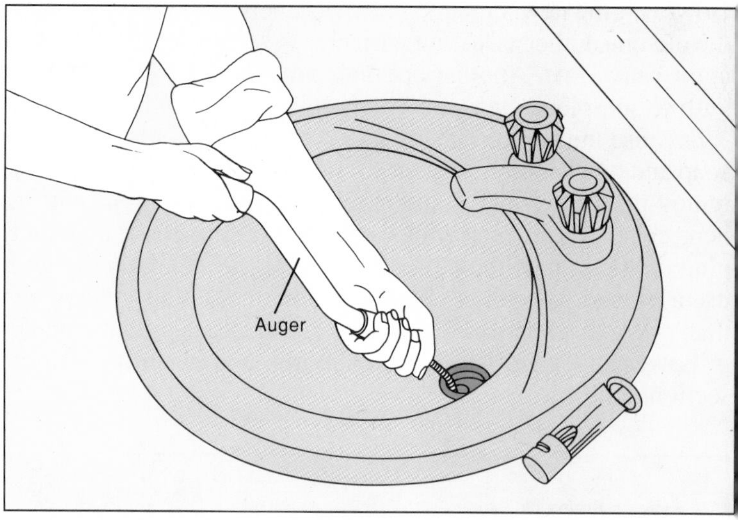

1 Using a plunger or auger. The safest way to unclog a sink drain is with a plunger or auger. Less safe is a caustic chemical drain opener *(page 13)*. Whichever method you choose, begin by removing and cleaning the sink stopper, if any *(page 50)*. Before using a plunger, pack a wet rag (or a rag wrapped in plastic) into the sink's overflow opening. Coat the rim of the plunger with petroleum jelly, and run just enough water in the sink to cover the plunger cup. Insert the plunger at an angle to avoid trapping air in the cup.

Plunge down and up forcefully *(above, left)*, keeping the plunger upright and the cup sealed over the drain. Be patient and repeat the process several times.

If the plunger is ineffective, try using a 1/4-inch auger. Feed the auger into the drain as far as the bend of the trap. Tighten the handle and rotate the auger, twisting and pushing to break up the clog *(above, right)*. Pull the auger up slowly, catching the debris on its hook. If the clog remains, work on the trap *(below)*.

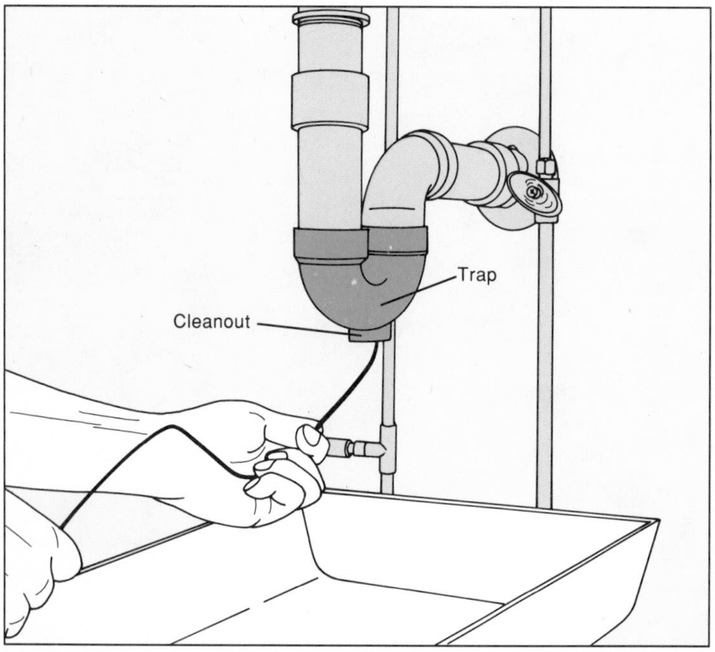

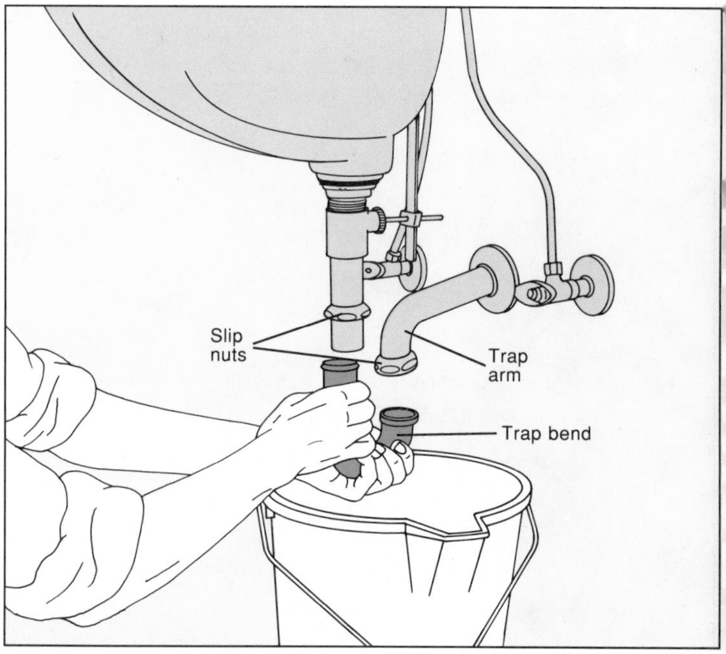

2 Servicing the trap. Place a pan or pail under the bend of the trap and have some rags at hand. If there is no cleanout plug, you must remove the entire trap *(next step)*. If there is a cleanout plug, loosen it with an adjustable wrench, then remove it by hand, catching water and debris in the pail. With a bent coat hanger *(above)*, probe into the opening and try to snag and remove any debris. Replace the cleanout plug and test the drain. If the clog remains, remove the trap bend *(next step)*.

3 Removing the trap bend. Supporting the trap bend with one hand, loosen the two slip nuts with a monkey wrench or tape-covered pliers. (Loosen the slip nut on the trap arm first.) Unscrew the slip nuts by hand and slide them away from the connections. To loosen corroded nuts, spray on penetrating oil and wait 15 minutes before trying again. If the slip nuts will not budge, saw through them vertically with a hacksaw, taking care not to cut into the trap, and pull them off. Pull the trap bend down and off *(inset)* and empty the water into the pail.

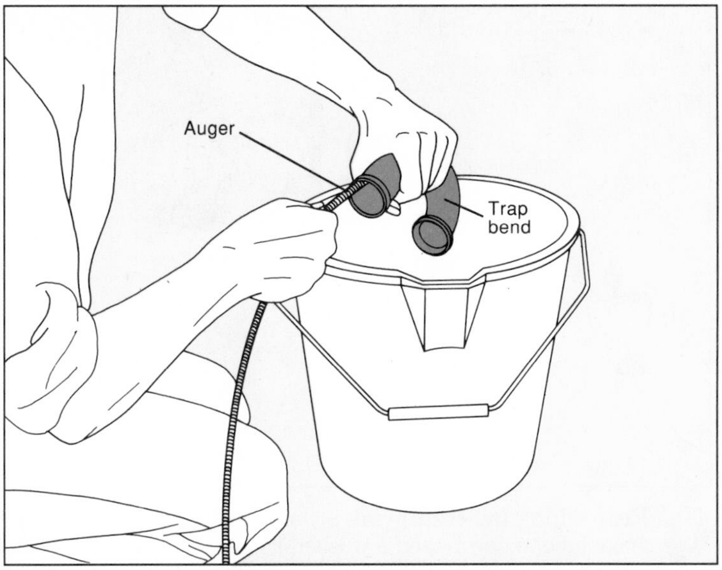

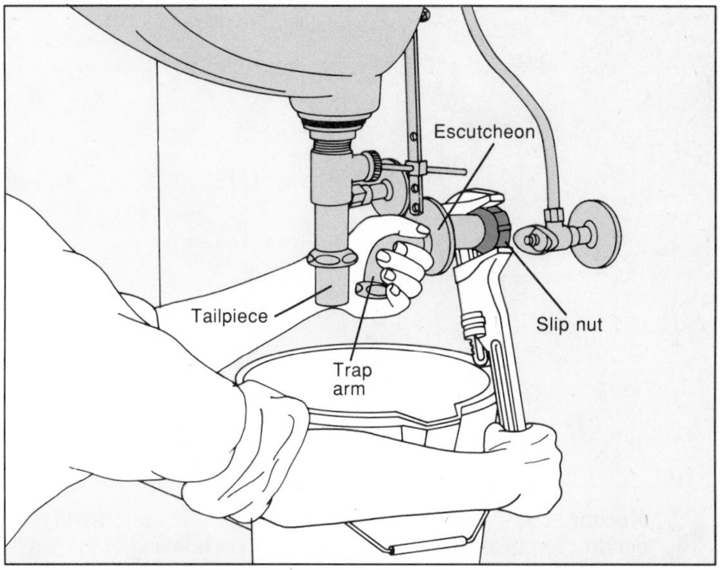

4 Cleaning the trap. Scrub the bend with a bottle brush or auger through it *(above)*, then rinse it in another sink. Inspect a metal trap bend for pinholes and corrosion; check a plastic bend for heat distortion. If debris is dislodged, reassemble the cleaned trap. Push the long end of the bend onto the tailpiece and align the other end with the trap arm. Slide the slip nuts and washers over their fittings and tighten the nuts. Run water into the sink. If the drain is still clogged, remove the bend and work on the trap arm *(next step)*.

5 Removing the trap arm. Using a screwdriver or putty knife, pry the escutcheon away from the wall or floor. Use a monkey wrench to loosen the slip nut that attaches the trap arm to the drainpipe adapter or stub-out at the wall, as shown, or to the drainpipe in the floor if it is an S-trap. Unscrew the slip nut by hand and slide the nut and washer along the trap arm. Twist the trap arm free, and clean it as in step 4. It may also be necessary to remove the tailpiece *(page 53)* to provide enough room to auger behind the wall.

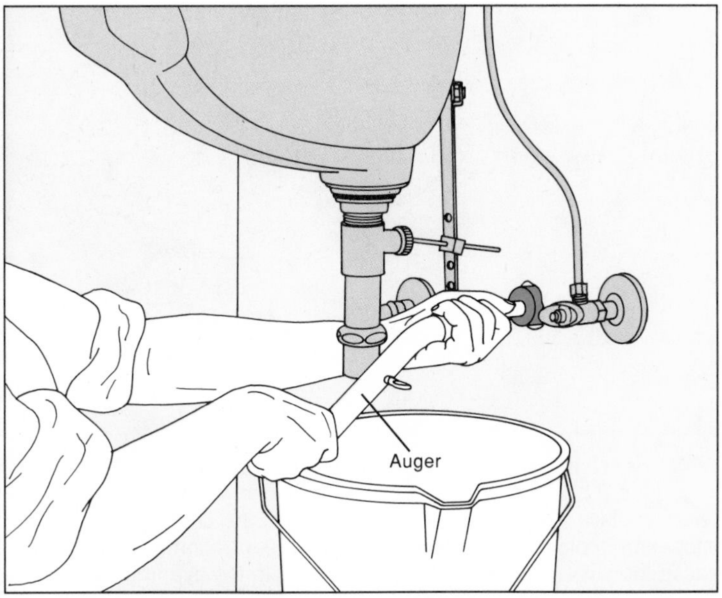

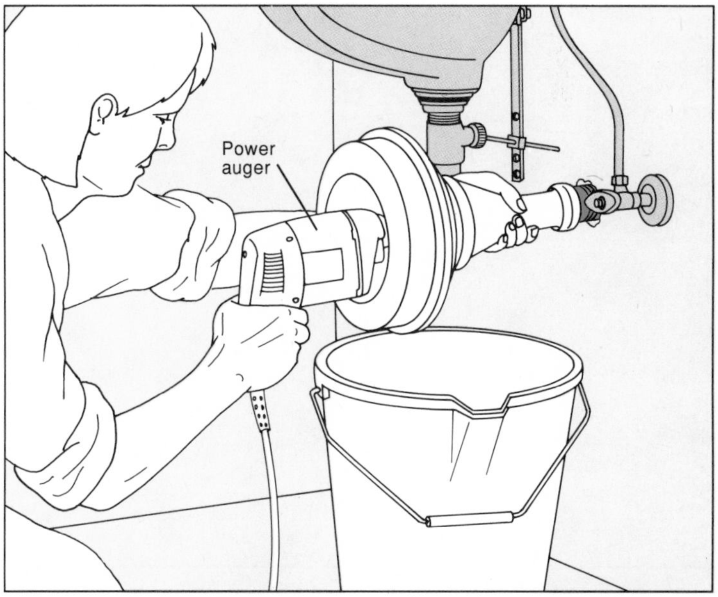

6 Using a manual auger. Have a pail and rags at hand. Feed the auger into the drainpipe, as shown, forcing the coil into the pipe as far as it will go. When you reach the blockage, lock the handle and rotate the auger clockwise. As it breaks up the clog, feed in more coil and continue rotating. Remove the auger slowly, catching any debris in the pail. Once the auger moves easily in the pipe, reassemble the trap *(step 8)*. If the blockage remains, use a hose and expansion nozzle *(page 72)* or a power auger *(next step)*.

7 Using a power auger. If you are confident in working with power tools, consider renting a power auger with a 5/16-inch cable. Insert the auger head into the drainpipe, feed out about 2 feet of cable and tighten the cable lock. Hold the handle with one hand and support the nozzle with the other, as shown. Adjust the pressure on the trigger to control the speed. Pull out the coil carefully, catching any debris in the pail. When you can flush the toilet and hear the sound of running water in the drainpipe, the blockage has been cleared.

CLEARING A CLOGGED SINK DRAIN (continued)

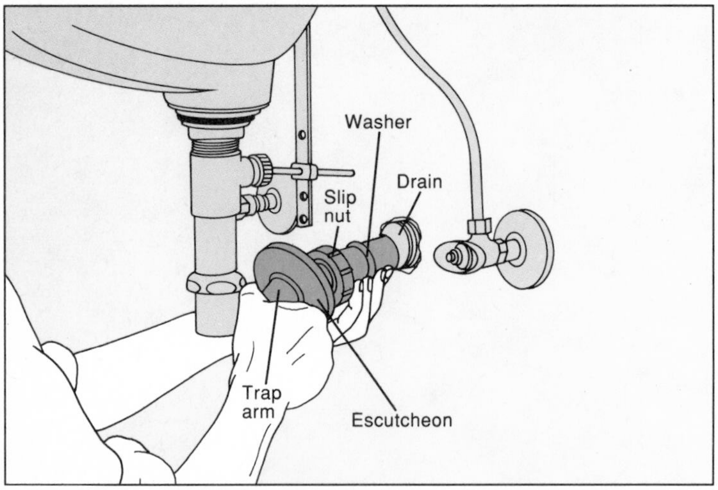

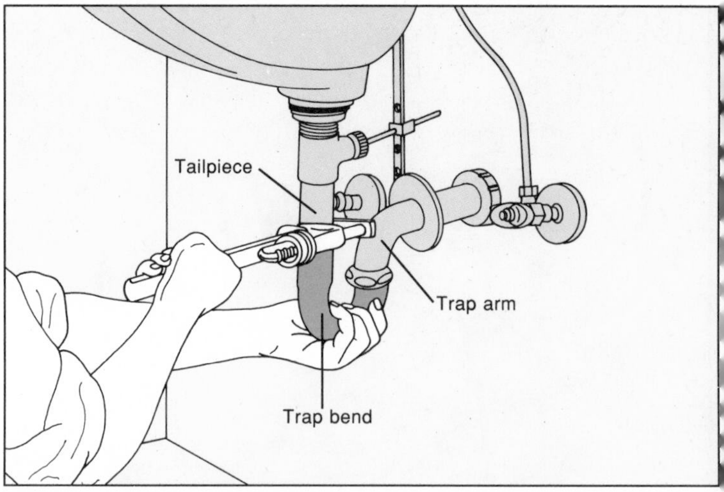

8 Reconnecting the trap arm. Replace a badly corroded trap bend or arm. Clean away any debris at the opening in the wall, then slide the fittings onto the trap arm in the following order: slip nut (threaded end first), escutcheon, another slip nut (threaded end last) and a slip washer. Now push the trap arm about 1 1/2 inches into the drain, as shown. Slide the washer onto the adapter or stub-out and tighten the slip nut over it by hand.

9 Reinstalling the trap bend. Slide a slip nut onto the tailpiece, threaded end down, then a washer, its beveled side down. Push the long end of the bend onto the tailpiece and slide it up until it is aligned with the trap arm. Then slip down the washer and tighten the slip nut over the bend by hand. Push the trap arm deeper into the wall or floor, or pull it out, until it aligns with the bend. Tighten all the nuts with a monkey wrench, as shown, but do not overtighten. (Coating the slip nuts with plumber's putty or silicone sealant before assembly will help prevent leaks.) If any connections leak, tighten them another quarter-turn. Finally, push the escutcheon firmly against the wall or floor.

REMOVING AND CLEANING THE STOPPER

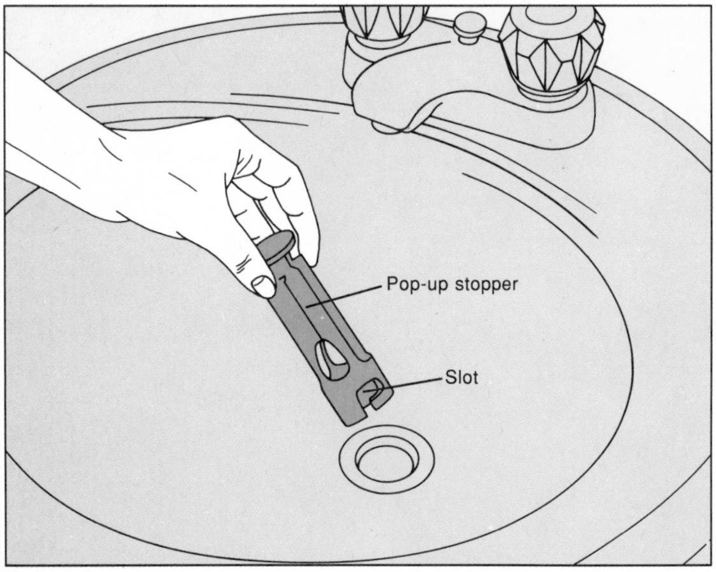

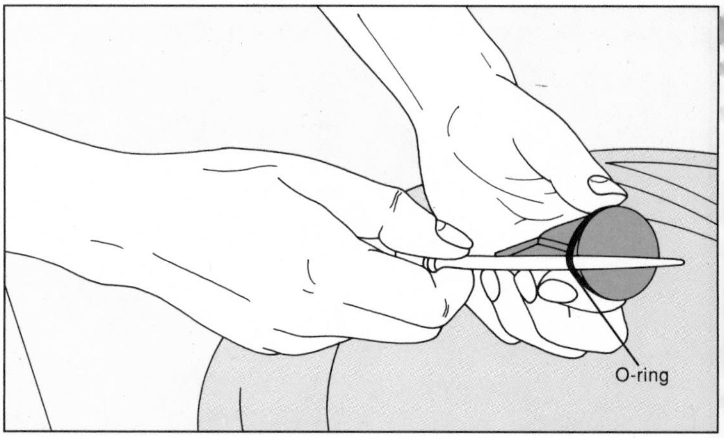

1 Removing a pop-up stopper. Raise the stopper to its open position and pull it out of the drain. If the stopper does not lift out, turn it counterclockwise to free it from the pivot rod, then pull the stopper out as shown. Do not force the stopper if it does not lift out easily. On some models, you must unscrew a retaining nut below the sink. Pull the rod out of the drain body and lift out the stopper from above the sink. Service the stopper *(step 2)* or replace a badly worn stopper. If unavailable, install a new drain body and pop-up assembly *(page 52)*.

2 Servicing the stopper. Clean the stopper with fine steel wool or a stiff brush and soap. Pry off the O-ring *(above)*, if any, and replace it. Lower the stopper back into the drain. Rotate a slotted stopper into place. If the stopper has an eye, position the eye so that it faces the pivot hole. Thread the pivot rod into the eye from beneath the sink and tighten the retaining nut. Run water through the drain. If any water leaks from the retaining nut, or if the stopper does not open fully or close tightly, adjust or replace the lift mechanism *(page 51)*.

REPLACING THE POP-UP LIFT MECHANISM

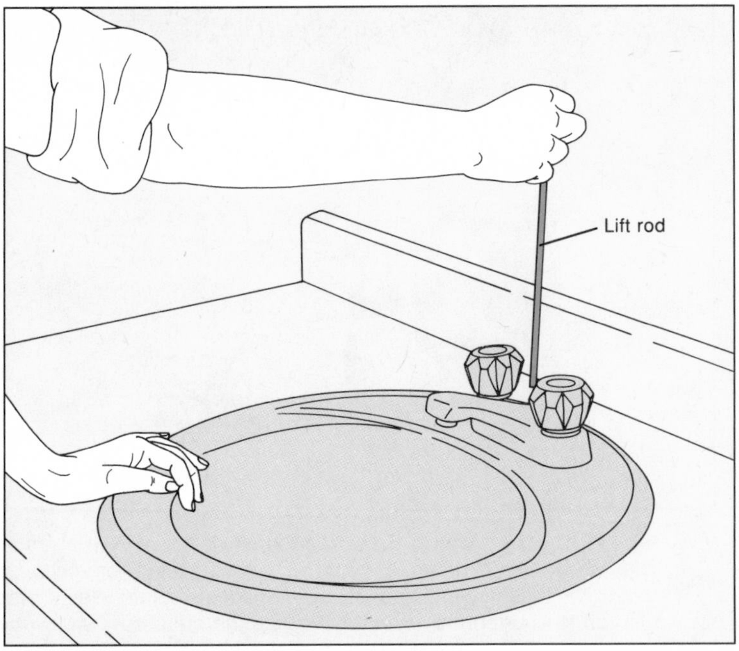

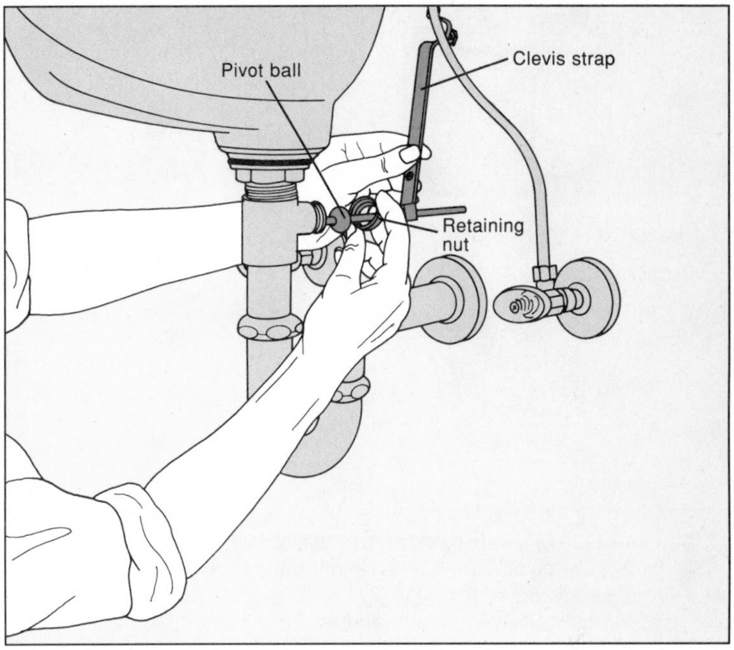

1 Removing the lift rod and pivot. If adjusting the pop-up lift mechanism does not stop water from seeping down the drain, or the stopper does not open properly, replace the mechanism. Remove the stopper *(page 50)* then loosen the clevis screw under the sink. Pull the lift rod up and out from the faucet body or sink *(above, left)*. Loosen the retaining nut and pull the pivot rod and ball out of the drain body *(above, right)*. Replace a damaged lift mechanism if parts are available for your make and model *(below)* or install a new drain body and pop-up assembly *(page 52)*.

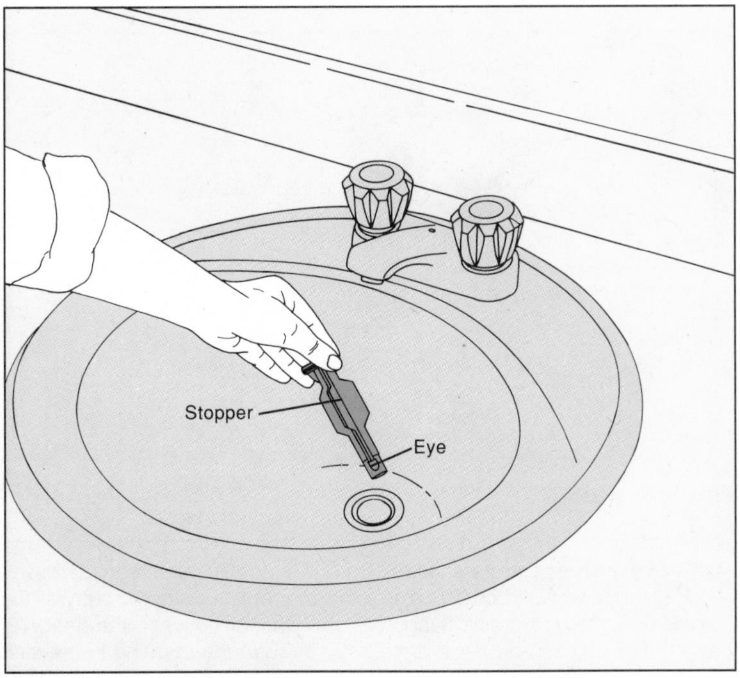

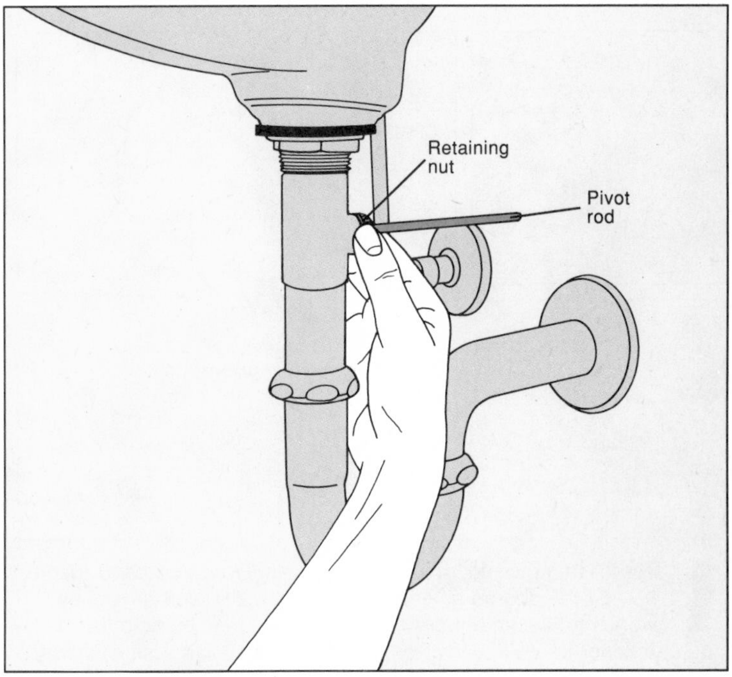

2 Installing the stopper and pivot rod. In this example, the pivot rod fits into an eye in the bottom of the stopper. With one hand, position the stopper in the drain opening with the eye facing the pivot rod *(above, left)*. With the other hand, place the rod into the drain body, then screw on the retaining nut *(above, right)* with pliers or an adjustable wrench. If you are installing a slotted stopper, rotate it until the slot catches the pivot rod. Some stoppers simply rest on the pivot rod.

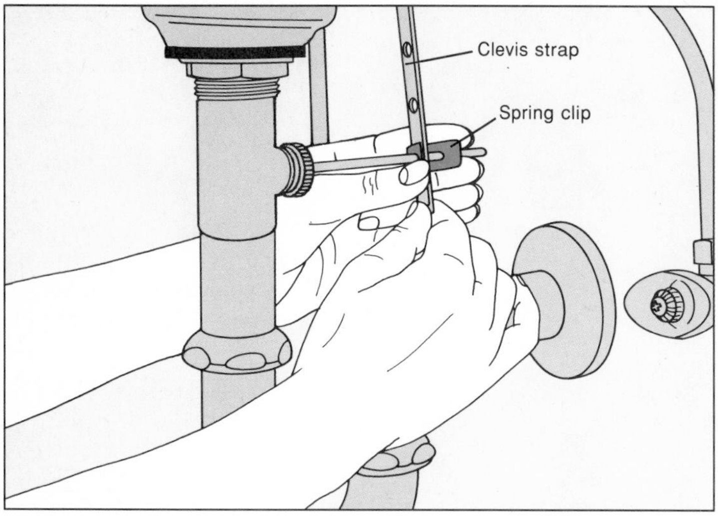

3 Connecting the lift rod. Pinch the spring clip to slide the free end of the pivot rod into a hole in the clevis strap *(above)*, then release the clip to hold it in place. From above, slip the lift rod into the hole between the faucet handles. From below, feed the lift rod into the clevis strap and tighten the clevis screw. Test the lift assembly and adjust if the stopper is not properly seated *(next step)*.

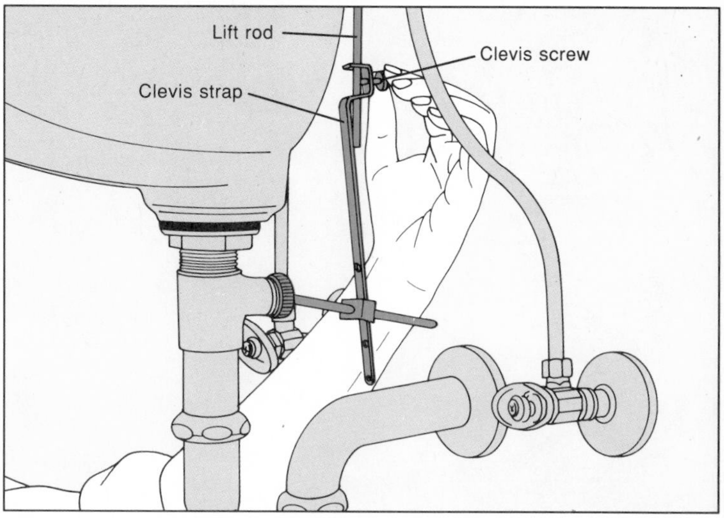

4 Adjusting the assembly. Loosen the clevis screw with pliers, then unscrew it by hand, as shown. To keep water from seeping out of the sink when the stopper is in place, push the clevis strap up the lift rod to shorten the assembly. Pulling the strap down will lengthen the assembly, and allow faster draining. Tighten the clevis screw and test, readjusting its length if necessary. If the stopper assembly is difficult to operate, pinch the spring clip to slide the pivot rod out of the clevis strap and move it to a higher or lower hole.

REPLACING THE DRAIN BODY

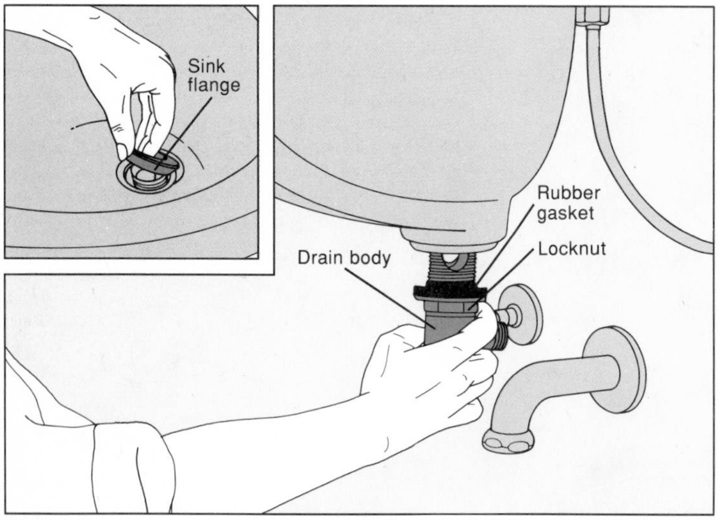

1 Removing the old drain body. Remove the trap bend *(page 48)* and the pop-up lift mechanism *(page 51)*. Use a monkey wrench to loosen the locknut holding the drain body to the sink. You can keep the drain body from rotating by wedging a screwdriver down through the drain opening and into a slot in the drain body. Push the drain body up into the basin and unscrew or lift off the sink flange *(inset)*, then pull the drain body down and out. If all else fails, use a hacksaw to cut through the drain body between the locknut and drain. Scrape away the putty from the drain opening and clean the basin.

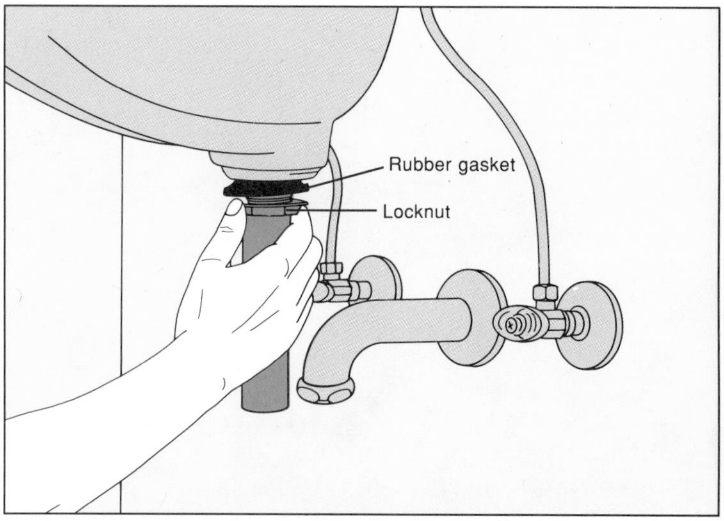

2 Installing the new drain body. Separate the pivot rod from the new drain body. Before installing, apply a strip of putty to the underlip of the sink flange or set a washer between the flange and the sink. Then drop the drain body through the opening in the sink. On models with a separate flange and drain body, set in the flange from above, push up the drain body from below and screw them together. Then push the rubber gasket tightly against the underside of the sink *(above)*. Line up the drain body so that the pivot hole faces the wall, hold the drain body steady and tighten the locknut with a wrench. If there is a separate tailpiece, wrap the threads with pipe tape and screw it to the drain body. Reconnect the trap bend *(page 50)* and install the pop-up lift mechanism *(page 51)*.

STOPPING LEAKS UNDER THE SINK

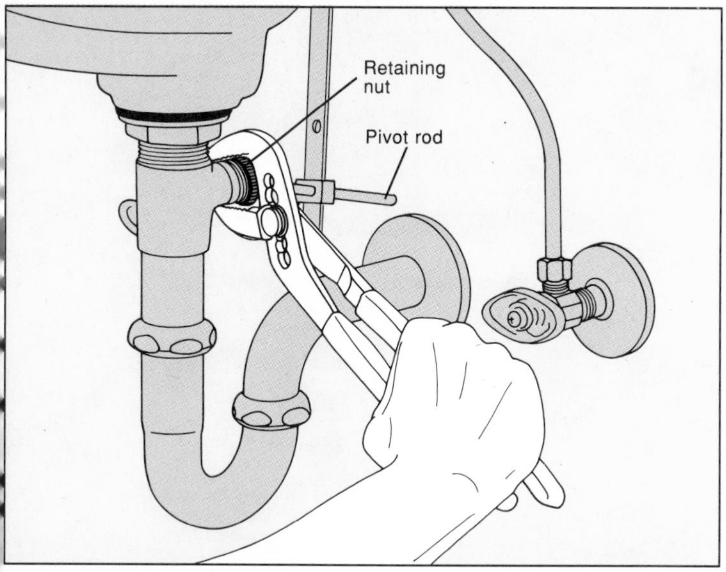

1 Stopping leaks at the drain body. If water is leaking from the point where the pivot rod enters the drain body, tighten the retaining nut with an adjustable wrench or pliers *(above)*. If leaks continue, unscrew the retaining nut, slide the pivot rod out of the drain body, and replace the washer or gasket under the retaining nut. If leaks persist and the lift mechanism is badly worn, replace its parts *(page 51)* or the drain body and pop-up assembly *(page 52)*.

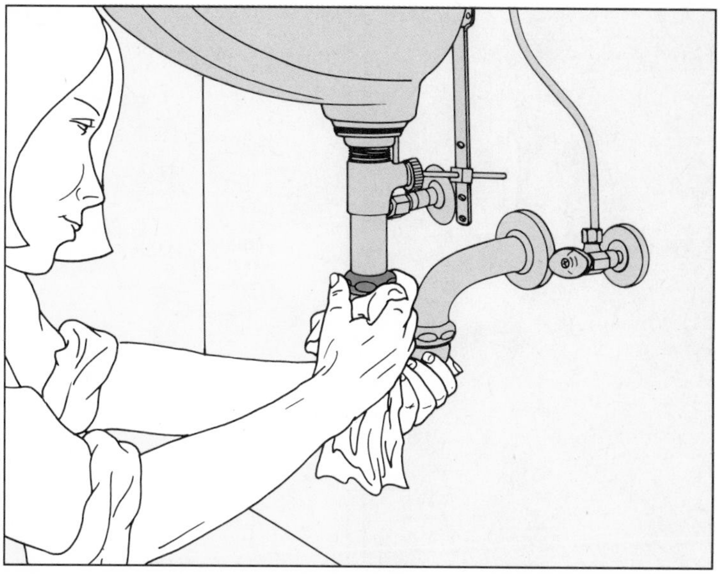

2 Stopping leaks at the trap. Tighten all slip nuts and locknuts a quarter-turn. Wipe the entire trap assembly with a dry cloth *(above)*, then run a basin full of water through the drain to test it. If leaks appear around the slip nuts, remove the trap bend *(page 48)*, then the trap arm *(page 49)*, and replace the washers. If the bend or arm is damaged or corroded, replace it. If the tailpiece or sink flange also leaks, go to step 3 before installing the trap. If a new trap arm is too long to fit into the drainpipe, trim off the excess length with a hacksaw to fit. (The trap arm should extend about 1 1/2 inches into the adapter or stub-out.)

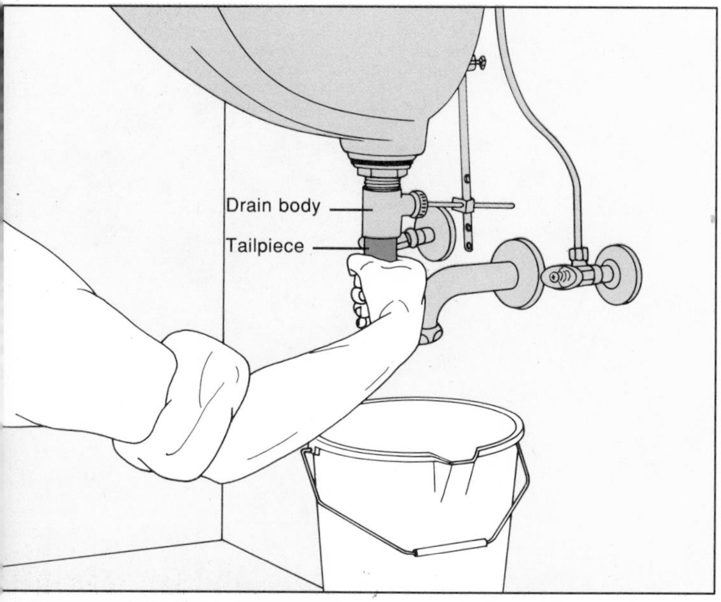

3 Stopping leaks at the tailpiece. Since the tailpiece is usually made of brass and easily crushed, unscrew it by hand, as shown. For a better grip, slip a wide rubber band onto the tailpiece. If the tailpiece is stubborn, remove it with a pipe wrench and buy a replacement of the same length. An old tailpiece may crumble; use long-nose pliers to twist out any remaining pieces from the threads of the drain body. Before installing a new tailpiece, wrap pipe tape around the threads, and replace the trap's slip nut and washer.

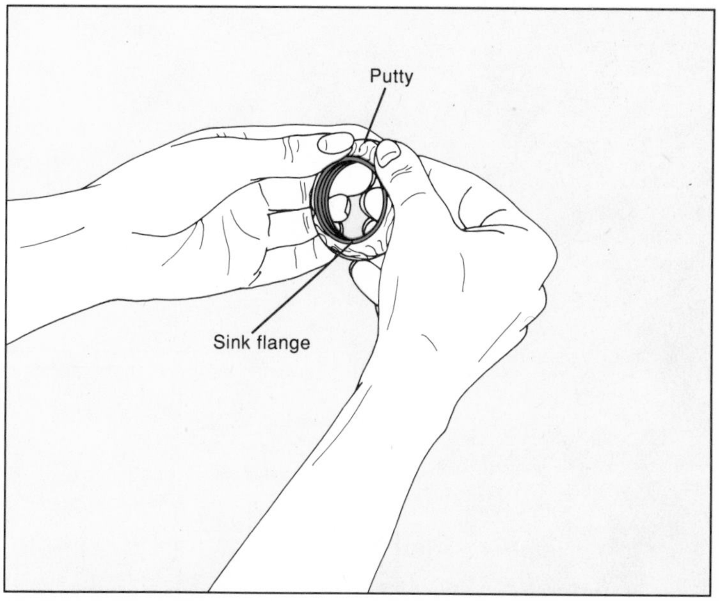

4 Stopping leaks at the sink flange. A worn washer or cracked putty under the sink flange may allow water to seep below the sink. Remove the trap bend, if you have not already done so. Then remove the drain body and free the flange from the sink *(page 52)*. Apply a thin strip of plumber's putty under the lip of the flange, as shown. Reinstall the drain body and pop-up assembly *(page 52)*, and wipe away any excess putty around the flange. If leaks persist, tighten all connections a quarter-turn.

REPLACING A DROP-IN SINK

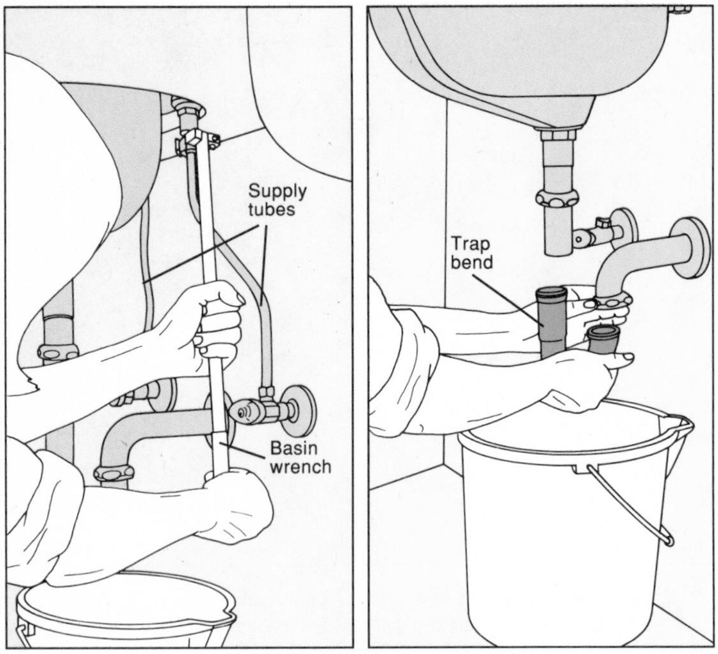

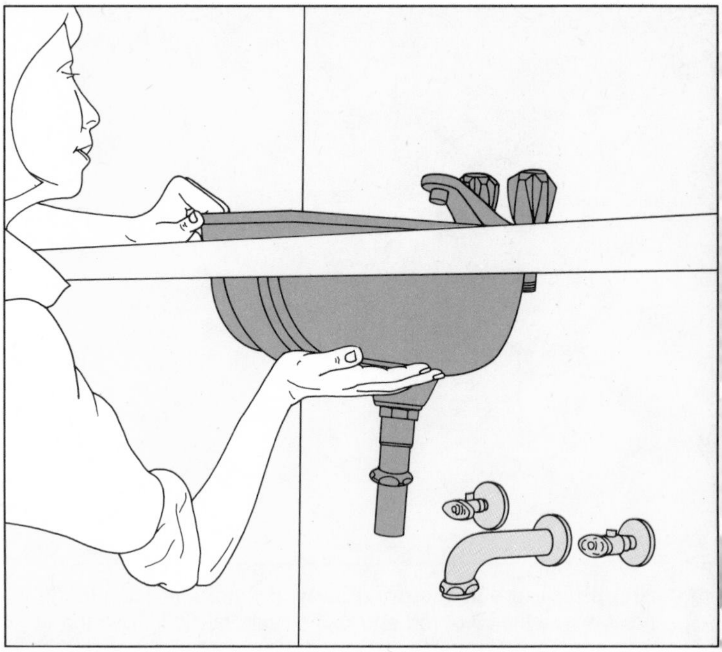

1 Disconnecting the supply tubes and trap. Turn off the water supply at the shutoff valves below the sink or at the main shutoff valve. Open the faucets (at the bathtub as well if you have closed the main shutoff) to drain the lines. Loosen the supply tube couplings at the faucet tailpiece with a basin wrench *(above, left)*. Push the supply tubes aside gently, or remove them, to avoid kinking. Remove the trap *(above, right)*. If the sink is frame-rimmed, support it from underneath before loosening it from the countertop.

2 Removing the old sink. Free the sink from the counter by scraping away putty or adhesive, or by unscrewing metal clamps or lug bolts from below. Then lift the sink up and out *(above)*, and set it on a rug or pad of newspapers and remove the faucet set. (To install a new faucet set, see page 31). Remove the pop-up lift mechanism *(page 51)*, then unscrew and remove the sink flange, drain body and tailpiece *(page 52)*.

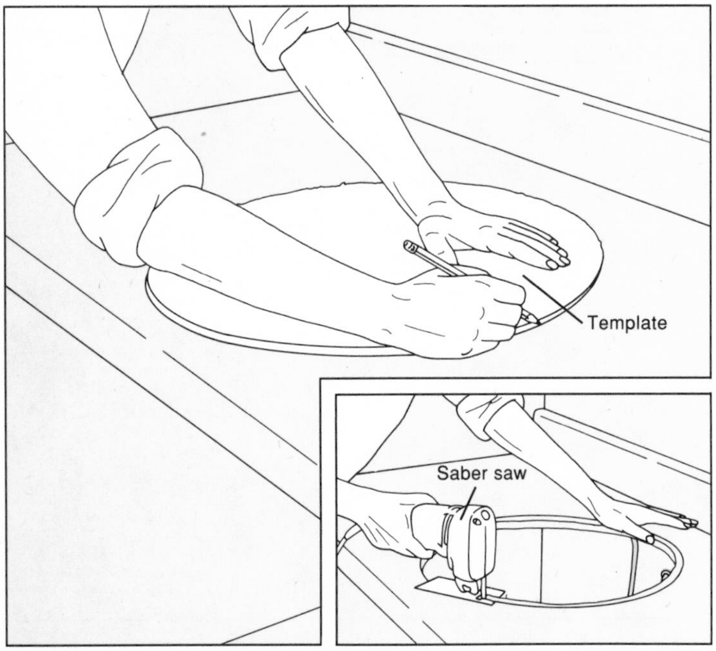

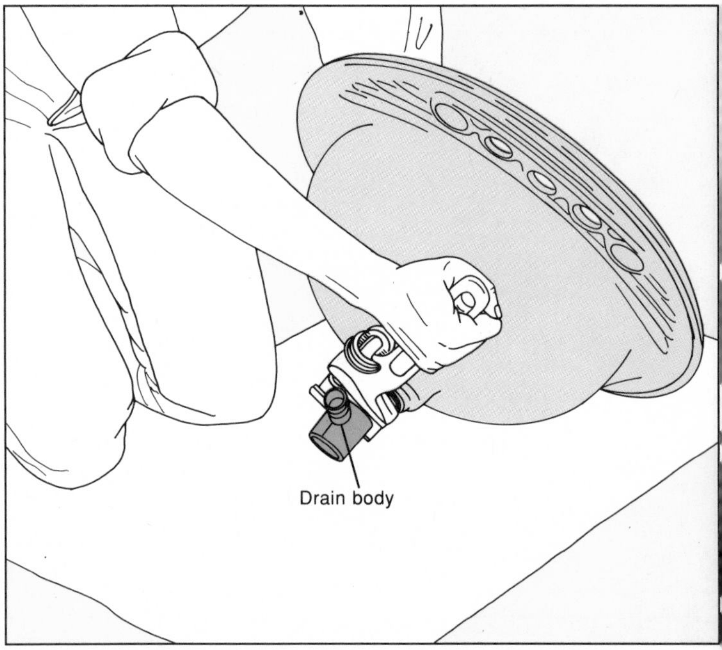

3 Preparing the countertop. Buy a replacement sink the same size as or larger than the old one. (To install a smaller one, you must also install a new countertop.) If the old faucets are to be reused, choose a sink with the correct distance between faucet holes. Trace the template that comes with the new sink onto the countertop *(above)* to be sure of a correct fit, then enlarge the hole using a saber saw *(inset)*. Set the new sink into the hole to try for size and adjust if necessary.

4 Installing the drain. Install the supply and drain hardware before setting the new sink in the counter. First attach the drain body and sink flange *(page 52)*, making sure that the pivot hole will face the back of the sink *(above)*.

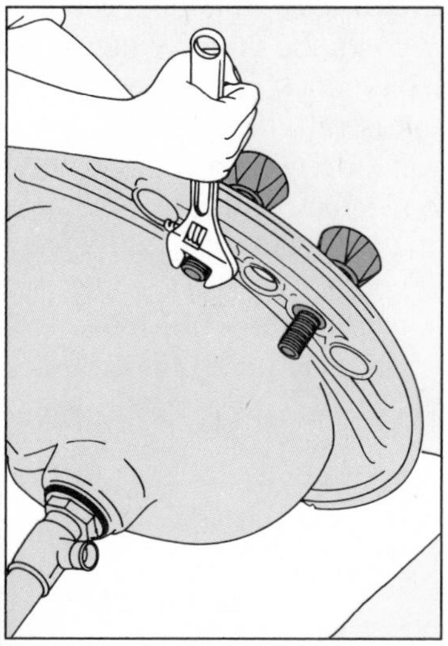

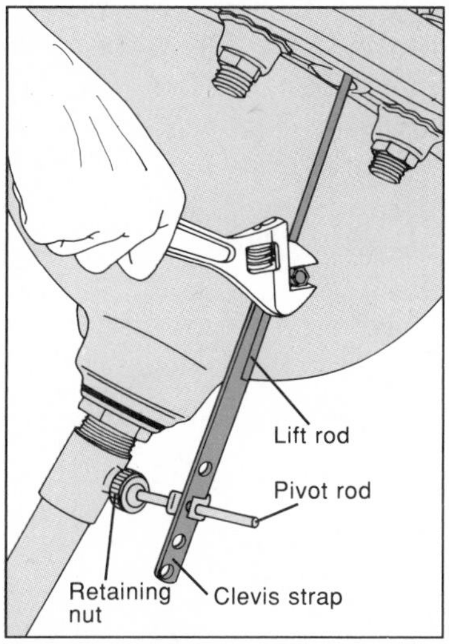

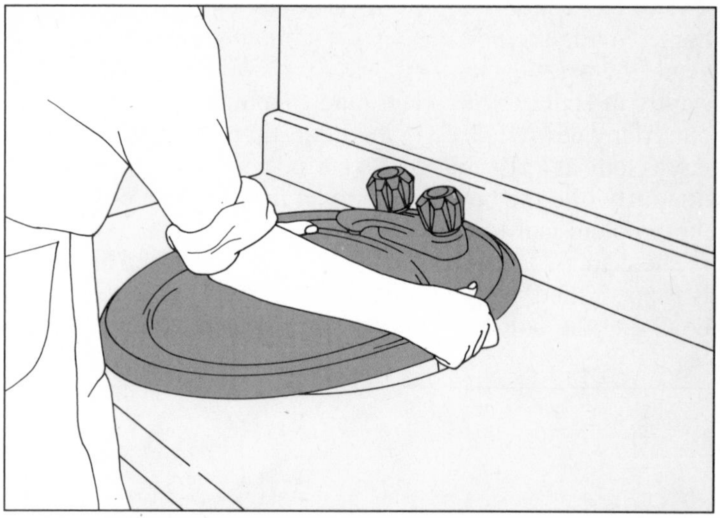

5 Attaching the faucet set and pop-up lift mechanism. Position the deck gasket over the faucet holes in the sink or apply putty around the holes. Insert the faucet tailpieces into the holes. Slip the washers and locknuts onto the tailpieces and tighten the nuts with a wrench *(above, left)*. Set the pivot rod into the drain body and tighten the retaining nut. Pinch the spring clip, slide the free end of the pivot rod into a hole on the clevis strap, then release the clip. Slip the lift rod into the hole between the faucet handles from above, then feed the rod into the clevis strap and tighten the screw *(above, right)*.

6 Mounting the sink. Apply putty or adhesive caulking under the rim of the sink or strip off the paper backing of a self-caulking sink. Lower the sink into place as shown, align it and press down. If lug bolts are provided, screw them tight from beneath while a helper holds the sink in place from above. Do not overtighten the bolts. Scrape away excess putty and reconnect the supply tubes *(page 31)* and trap *(page 50)*. Turn on the water supply, slowly at first, and check for leaks, tightening connections a quarter-turn if necessary. Remove the aerator, if any, and turn the water on full force to flush the lines before reinstalling it.

REPLACING A WALL-MOUNT SINK

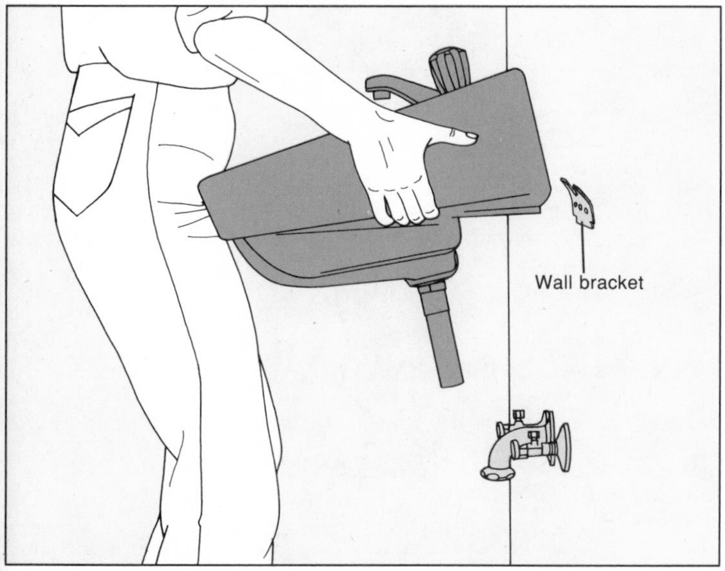

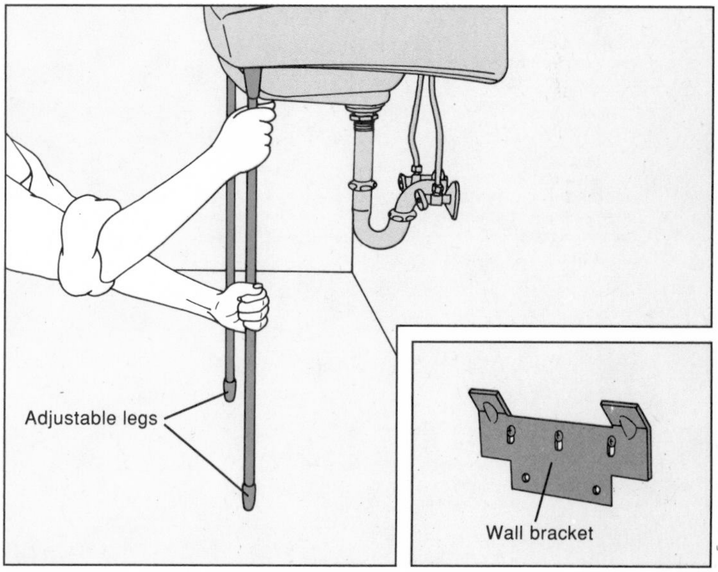

1 Removing the old sink. Turn off the water supply at the shutoff valves below the sink or at the main shutoff valve. Open the faucets to drain the lines. If you have turned off the main water supply, open the bathtub faucets as well. Remove the trap *(page 48)* and use a basin wrench to loosen the coupling nuts connecting the supply tubes to the faucet. Unscrew any bolts holding the sink in place, then lift the sink up and off the wall bracket, as shown. If you intend to reuse the faucets and drain assembly, set the sink on a rug or pad of newspapers and remove them *(page 54)*.

2 Installing the new sink. Buy a replacement sink that fits onto the existing wall bracket *(inset)*, or remove the old bracket and install the one that comes with the new sink. (Some wall-mount sinks may have two brackets.) Install the drain body *(page 52)*, then attach the faucet set and pop-up lift mechanism *(step 5, above)*. Set the sink onto the wall bracket. For better support, add adjustable legs. Place a level on the sink rim, adjust the legs until the sink is square, and tighten the legs, as shown. Apply caulk along the joint between the sink and wall. Reconnect the supply tubes *(page 31)* and trap *(page 50)*.

TOILETS

The toilet is the most heavily used plumbing fixture in the home. While the porcelain bowl and tank will last for many years, the working parts inside the tank will not. They are constantly in contact with water, and eventually corrode or wear out. Many new tanks contain all-plastic parts, which may not last as long as brass, but do not corrode. When water runs continuously into the bowl or the toilet fails to flush completely, the problem can usually be traced to the tank.

There are two basic mechanisms in a toilet. When the handle is tripped, the flush valve releases water from the tank to the bowl and the ball cock opens to fill the tank. Each time the toilet is flushed, the rushing water creates a siphoning action in the bowl that forces waste water down the drain. Water flows in to refill the tank when the ball cock is opened by the lowering float ball. Once the tank is full, the rising float arm closes the valve in the ball cock to stop water from overflowing. Adjusting or cleaning these mechanisms will solve most minor problems, but replacing the ball cock or flush valve is often the only lasting solution. Though most flush toilets use conventional tank components, certain new designs work more quickly, quietly and efficiently than their older counterparts. Consider replacing the ball cock with a plastic water-

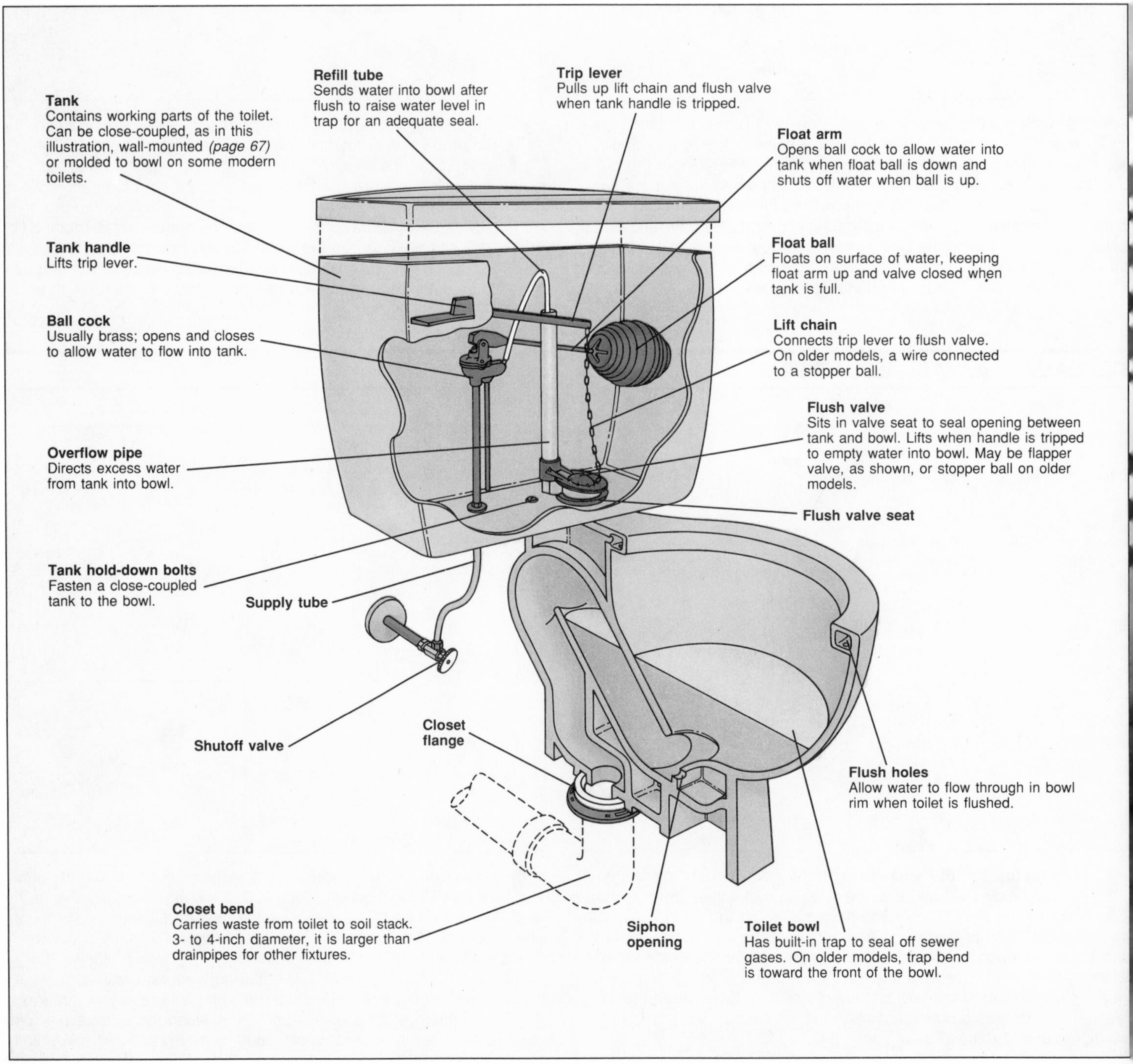
Tank
Contains working parts of the toilet. Can be close-coupled, as in this illustration, wall-mounted *(page 67)* or molded to bowl on some modern toilets.
Refill tube
Sends water into bowl after flush to raise water level in trap for an adequate seal.
Trip lever
Pulls up lift chain and flush valve when tank handle is tripped.
Float arm
Opens ball cock to allow water into tank when float ball is down and shuts off water when ball is up.
Tank handle
Lifts trip lever.
Float ball
Floats on surface of water, keeping float arm up and valve closed when tank is full.
Ball cock
Usually brass; opens and closes to allow water to flow into tank.
Lift chain
Connects trip lever to flush valve. On older models, a wire connected to a stopper ball.
Flush valve
Sits in valve seat to seal opening between tank and bowl. Lifts when handle is tripped to empty water into bowl. May be flapper valve, as shown, or stopper ball on older models.
Overflow pipe
Directs excess water from tank into bowl.
Flush valve seat
Tank hold-down bolts
Fasten a close-coupled tank to the bowl.
Supply tube
Shutoff valve
Closet flange
Flush holes
Allow water to flow through in bowl rim when toilet is flushed.
Closet bend
Carries waste from toilet to soil stack. 3- to 4-inch diameter, it is larger than drainpipes for other fixtures.
Siphon opening
Toilet bowl
Has built-in trap to seal off sewer gases. On older models, trap bend is toward the front of the bowl.

intake assembly that eliminates the need for a float ball and float arm. An old-fashioned stopper ball can be replaced with a simple device called a flapper valve.

When a small clog causes a toilet to drain slowly, use a plunger or manual auger to clear the blockage before an emergency develops. (Never use a power auger in a toilet drain or closet bend.) A serious blockage may require that you remove and reseat the toilet to get at it from underneath—perhaps a full day's job. Never use chemical drain openers to clear a clogged toilet. Chemicals are usually ineffective in a toilet and can be dangerous since the heat they produce can crack the bowl or eat away drainpipes. To prevent clogs, put only toilet paper down the toilet and keep dental floss, cotton swabs and paper towels out.

Most toilets manufactured today are insulated to prevent leaks caused by condensation. You can stop an older tank from sweating by lining it with flexible foam *(page 66)*. Instead of placing a brick in the tank to conserve water, bend the float arm down an inch or so. When working on a toilet, take care to place the tank cover in a safe place, and do not overtighten bolts. Porcelain cracks very easily, which usually means replacing the entire toilet.

TANK VARIATIONS

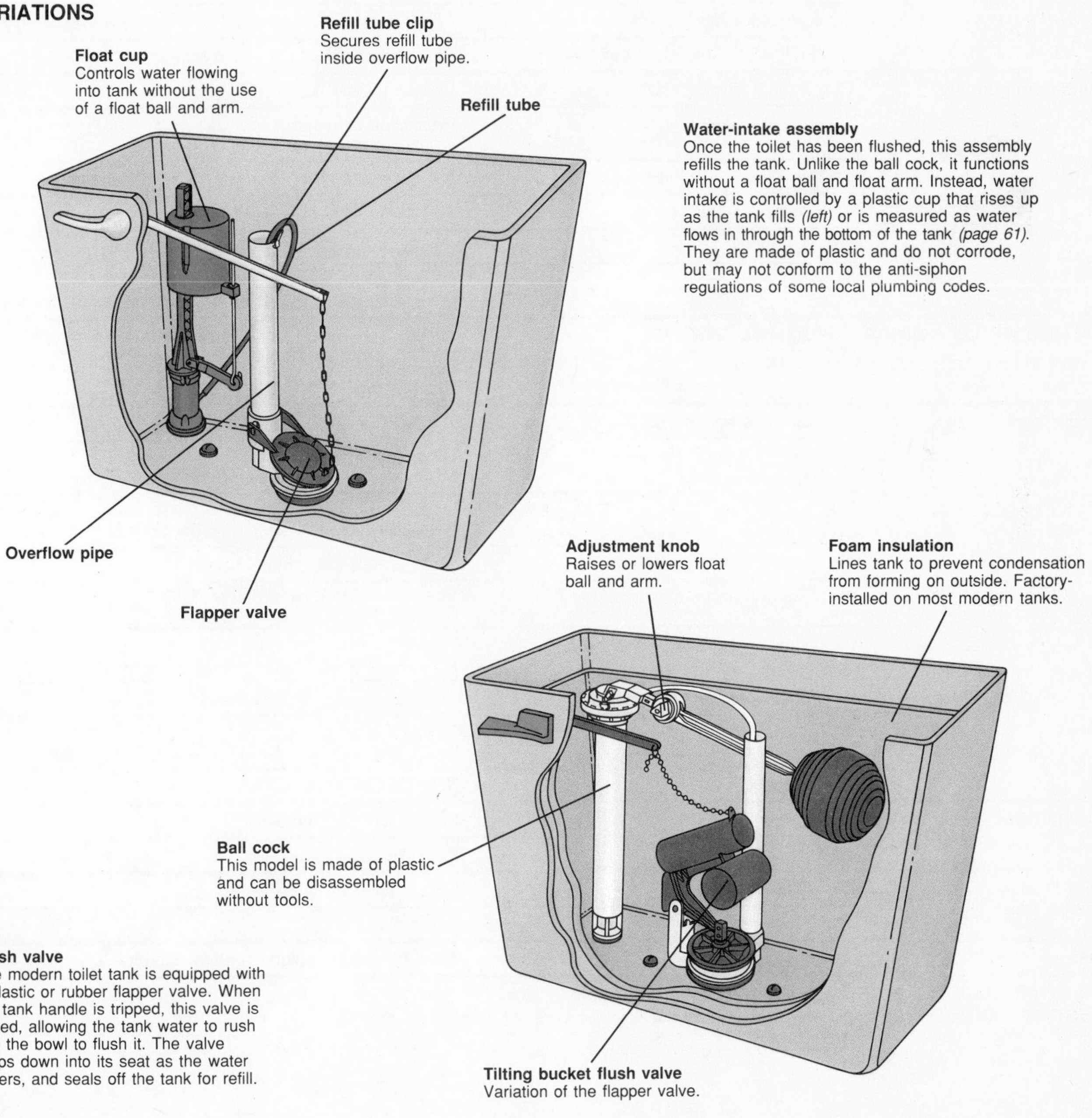
Refill tube clip
Secures refill tube inside overflow pipe.
Float cup
Controls water flowing into tank without the use of a float ball and arm.
Refill tube
Overflow pipe
Flapper valve
Adjustment knob
Raises or lowers float ball and arm.
Foam insulation
Lines tank to prevent condensation from forming on outside. Factory-installed on most modern tanks.
Ball cock
This model is made of plastic and can be disassembled without tools.
Tilting bucket flush valve
Variation of the flapper valve.

Water-intake assembly
Once the toilet has been flushed, this assembly refills the tank. Unlike the ball cock, it functions without a float ball and float arm. Instead, water intake is controlled by a plastic cup that rises up as the tank fills *(left)* or is measured as water flows in through the bottom of the tank *(page 61)*. They are made of plastic and do not corrode, but may not conform to the anti-siphon regulations of some local plumbing codes.

Flush valve
The modern toilet tank is equipped with a plastic or rubber flapper valve. When the tank handle is tripped, this valve is raised, allowing the tank water to rush into the bowl to flush it. The valve drops down into its seat as the water lowers, and seals off the tank for refill.

TROUBLESHOOTING GUIDE

SYMPTOM	POSSIBLE CAUSE	PROCEDURE
Toilet bowl overflows	Blockage in bowl or drain	Use plunger or auger *(p. 63)* □○ Remove toilet to unblock it *(p. 64)* ⬓●
Toilet does not flush	Water supply turned off	Turn on water supply
	Handle loose, broken or disconnected	Tighten or replace handle *(p. 62)* □○
	Lift chain broken or disengaged	Service lift chain *(p. 62)* □○
	Lift chain too loose	Shorten lift chain *(p. 62)* □○
	Ball cock faulty	Service ball cock *(p. 59)* ⬓◒ Replace ball cock *(p. 61)* ⬓◒
Toilet bowl drains sluggishly	Blockage in bowl or drain	Use plunger or auger *(p. 63)* □○ Remove toilet to unblock it *(p. 64)* ⬓●
	Float ball too low or rubbing against tank wall	Reposition float ball *(p. 59)* □○
	Water level too low	Adjust float arm *(p. 59)* □○ or water-intake assembly *(p. 61)* ⬓◒
	Flush holes blocked under bowl rim	Clear holes with unbent coat hanger and clean bowl □○
Water runs continuously	Tank handle stuck	Service handle *(p. 62)* □○
	Water level in tank high	Reposition or replace float ball *(p. 59)* □○ Adjust water-intake assembly *(p. 61)* ⬓◒
	Lift chain short or tangled	Adjust or replace lift chain *(p. 62)* □○
	Lift wire bent or corroded	Service lift wire *(p. 62)* □○
	Ball cock faulty	Service ball cock *(p. 59)* ⬓◒
	Flush valve leaking	Service or replace valve *(p. 62)* ⬓◒ Clean valve seat or replace flush-valve assembly *(p. 62)* ⬓◒
Vibration when toilet tank fills	Incorrect water level	Adjust float arm *(p. 59)* □○ or water-intake assembly *(p. 61)* ⬓◒
	Ball cock faulty	Service ball cock *(p. 59)* ⬓◒ Replace ball cock *(p. 61)* ⬓◒
Water under tank	Tank loose or washers worn	Tighten nuts or replace washers *(p. 66)* □○
	Water leaks through handle	Lower float ball *(p. 59)* □○ Shorten overflow pipe *(p. 62)* ⬓◒
	Flush valve leaks	Replace gasket *(p. 66)* ⬓◒ Replace flush-valve assembly *(p. 62)* ⬓◒
	Water spraying from refill tube	Replace refill tube □○
	Water spraying from ball cock	Service ball cock *(p. 59)* ⬓◒ Replace ball cock *(p. 61)* ⬓◒
	Condensation on tank	Insulate tank *(p. 66)* □◒ Install factory-insulated tank *(p. 67)* ⬓●
	Wall-mounted tank connections leak	Replace packing or flush elbow *(p. 67)* ⬓◒
	Supply tube leaks	Tighten coupling nut at shutoff valve *(p. 66)* □○ Replace supply tube *(p. 31)* ⬓◒ or shutoff valve *(p. 111)* ⬓◒
	Crack in tank	Replace tank *(p. 67)* ⬓●
Floor around bowl wet	Wax gasket faulty	Reseat toilet *(p. 64)* ⬓●
	Crack in bowl	Replace toilet *(p. 64)* ⬓●
Dampness or discoloration on ceiling below toilet	Wax gasket faulty	Reseat toilet *(p. 64)* ⬓●
	Cracked drainpipe	Call a plumber
Seat loose	Seat bolts corroded or seat cracked	Tighten bolts or replace seat *(p. 65)* □○

DEGREE OF DIFFICULTY: □ **Easy** ⬓ **Moderate** ■ **Complex**
ESTIMATED TIME: ○ **Less than 1 hour** ◒ **1 to 3 hours** ● **Over 3 hours**

SERVICING THE FLOAT ASSEMBLY

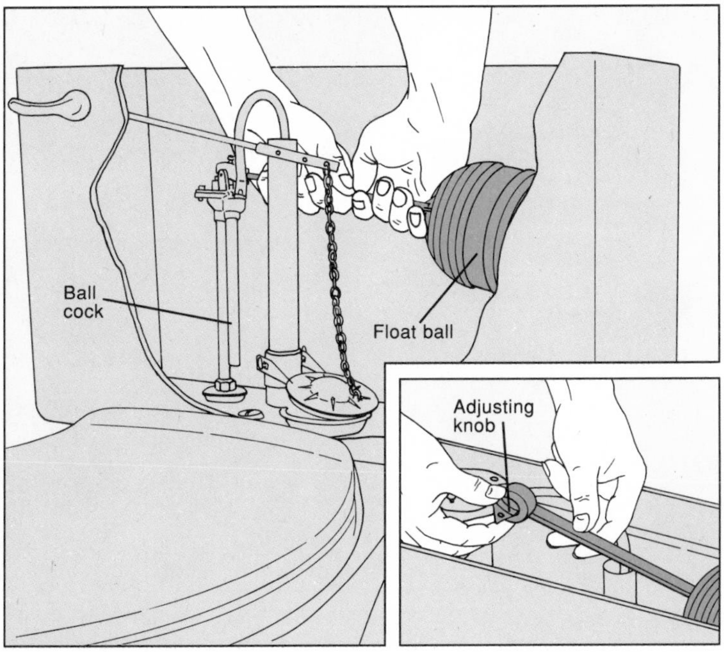

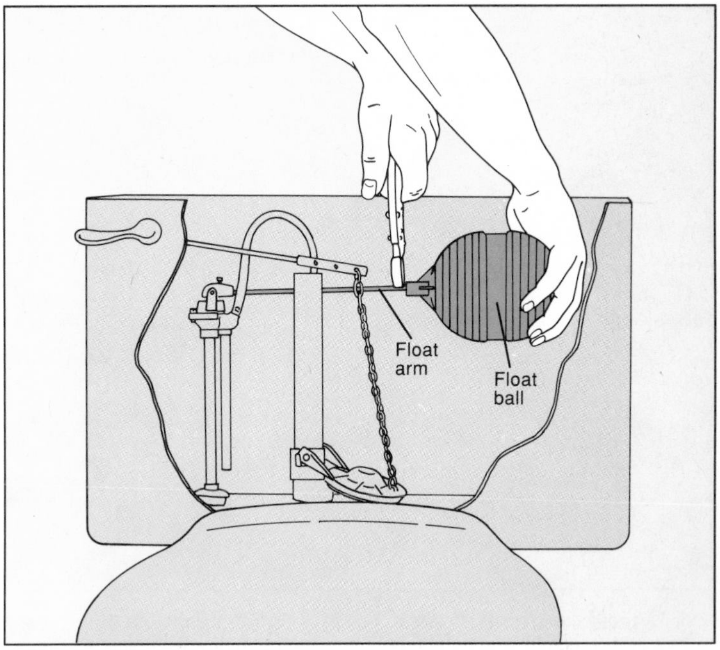

1 Adjusting the float arm. Water running over the top of the overflow pipe, or leaking through the handle, means that the water level is too high. Gently lift the float arm and bend it down slightly, as shown, to keep the water level 1/2 to 1 inch below the top of the overflow pipe. If water continues to run, go to step 2. An incomplete flush means that the water level is too low. Bend the float arm to raise the float ball, making sure the ball does not rub the tank. To raise or lower a plastic float arm, turn the knob at the ball cock *(inset).*

2 Replacing the float ball. A cracked float ball contains water that prevents it from rising high enough to close the ball cock. To remove the ball, grasp the float arm with locking pliers *(above)* and twist the float ball counterclockwise. If it will not come off, use the pliers to unthread the float arm from the ball cock. Replace the ball, coating the threads of the float arm with petroleum jelly, and screw the arm back onto the ball cock. Adjust the float level as in step 1. If water still runs, repair or replace the ball cock *(below).*

SERVICING THE BALL COCK

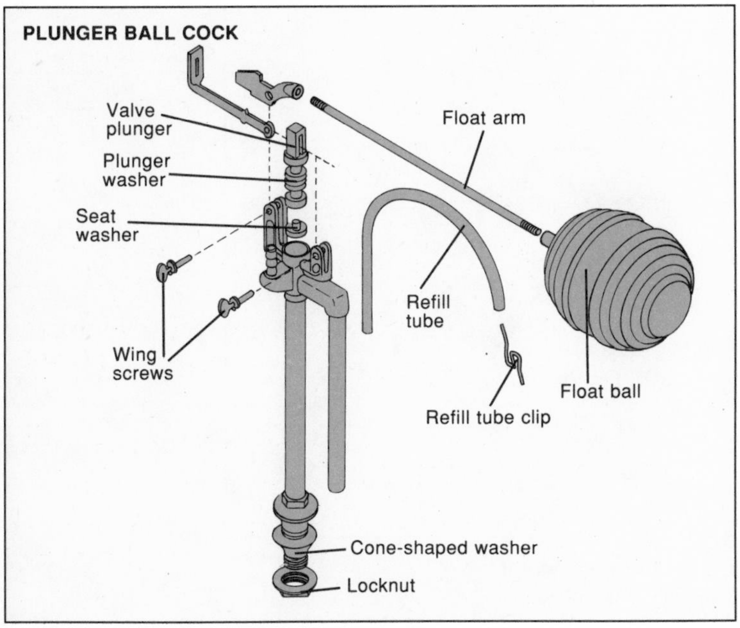

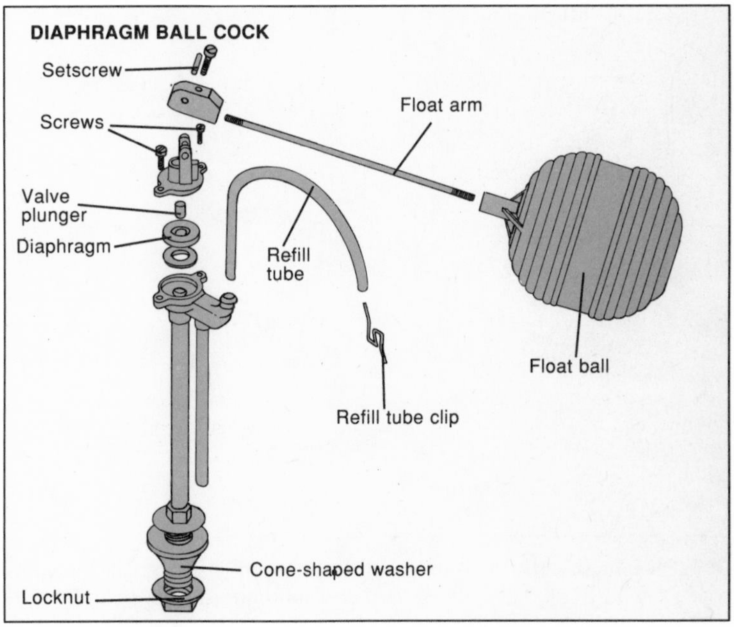

Ball-cock assemblies. Ball cocks allow water to flow into the tank when the toilet is flushed and the float ball and arm lower, and close when the float ball and arm rise. In the plunger-type ball cock *(above, left)*, the float arm applies pressure on a valve plunger and washer to seal off incoming water. In a diaphragm type *(above, right)* the plunger presses on a rubber diaphragm. In water-intake assemblies *(page 61)* a plastic cup or fill valve controls water flow.

SERVICING THE BALL COCK (continued)

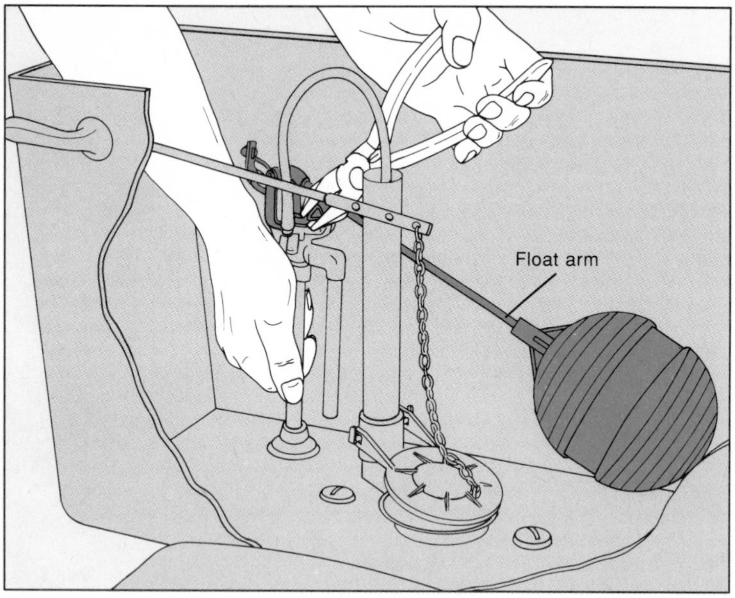

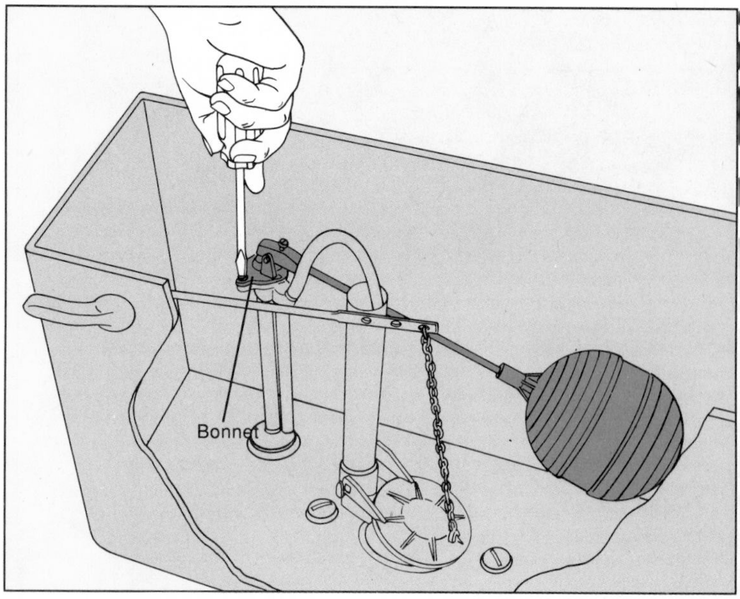

1 Access to the ball-cock assembly. When possible, buy replacement parts for your make and model of toilet before disassembling it. Turn off the water supply and flush the toilet to empty the tank. Remove the tank cover and carefully lay it aside. To disassemble a plunger-type ball cock *(above, left)*, remove the wing screws by hand, or with pliers if stubborn. Slide out the float arm, then pull up on the valve plunger to remove it. To access a diaphragm-type ball cock, remove the screws on the bonnet *(above, right)*, then lift up the float arm with the bonnet attached. This will expose the diaphragm and valve plunger. If the assembly is too corroded to service, replace it *(page 61)*.

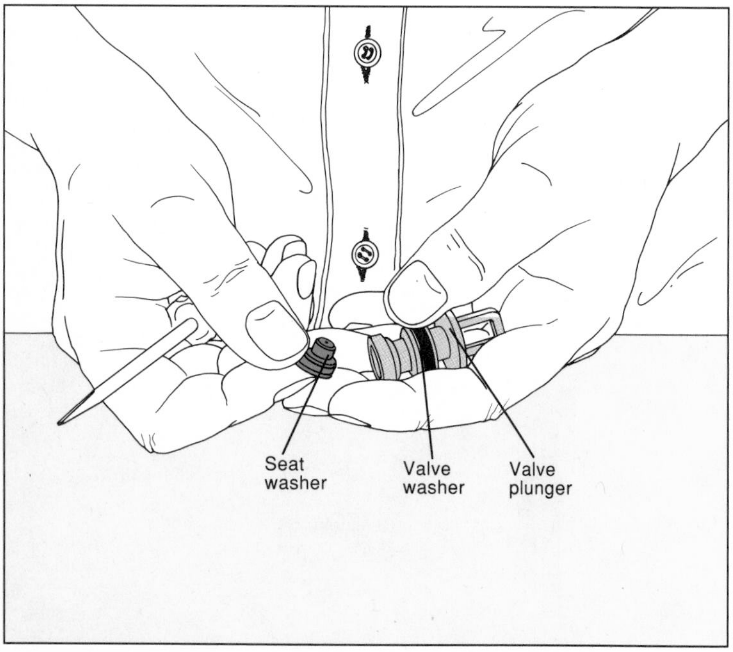

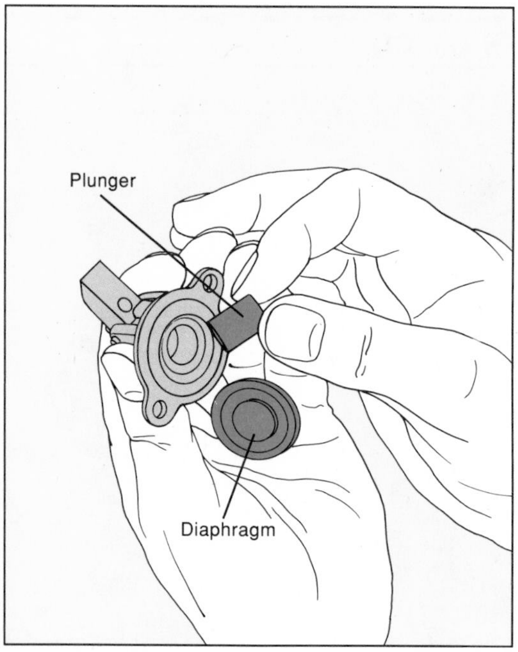

2 Replacing the washers and diaphragm. Use a small screwdriver to pry the washers off the valve plunger *(above, left)* or diaphragm *(above, right)* from the ball-cock assembly. Scrape away sediment inside the ball cock with a small knife or toothbrush and vinegar. Then replace the washers and plunger, or diaphragm. Reassemble the ball cock, turn on the water and flush to test the repair. If leaks persist, replace the ball cock *(page 61)*.

REPLACING THE BALL COCK

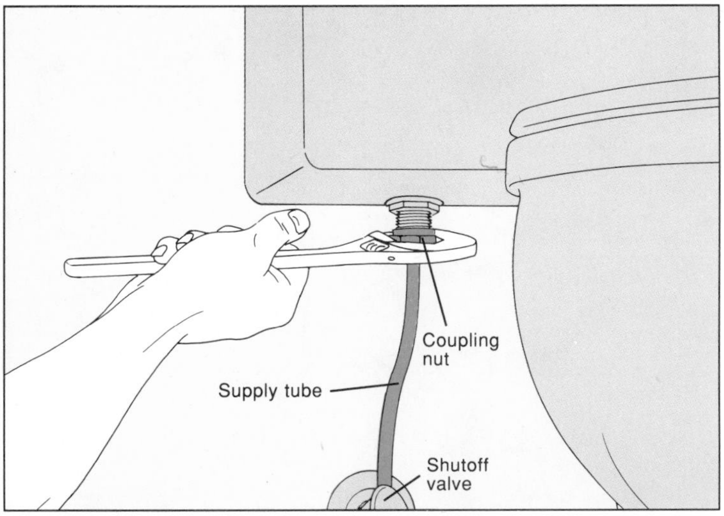

1 Preparing the tank. You can replace a worn or broken ball cock with an identical assembly *(next step)* or with a plastic water-intake assembly *(step 3)*. Remove the tank cover and carefully lay it aside. Place a container on the floor beneath the tank to catch water runoff. Shut off the water supply and flush the toilet, holding the handle down to drain as much water from the tank as possible. Remove the float arm from the ball cock *(page 60)* and pull the refill tube from the overflow pipe. Sponge up any remaining water in the bottom of the tank. With an adjustable wrench, disconnect one end of the supply tube from the tank *(above)* and the other end from the shutoff valve. Gently push the supply tube aside or remove it to avoid kinking.

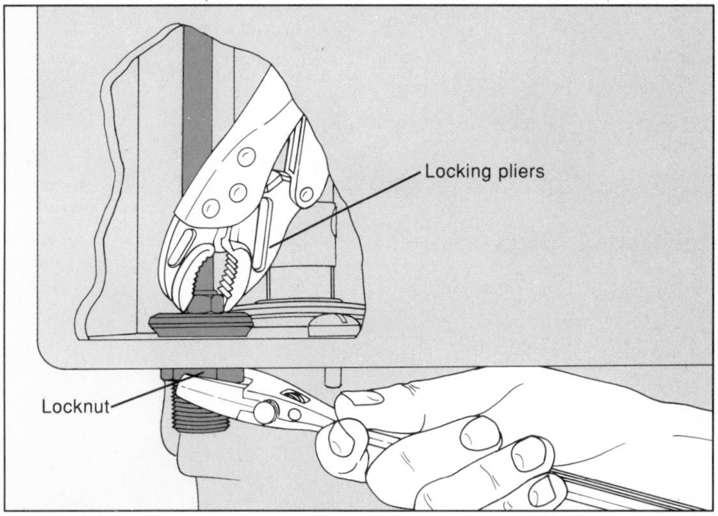

2 Replacing the ball cock. To prevent the ball cock from turning inside the tank, attach locking pliers at its base, then wedge them against the tank wall, as shown. Loosen the locknut under the ball cock with an adjustable wrench. If necessary, apply penetrating oil, wait fifteen minutes, then try again. If all else fails, cut through the shaft of the ball cock between the locknut and the tank, protecting the tank with plastic tape. Pull the ball cock up and out. Place a strip of plumber's putty around the cone-shaped washer of the new ball cock, then set the assembly firmly into the tank opening. Hold the ball cock with one hand while you tighten the locknut, first by hand, then one-half turn with an adjustable wrench. Place the refill tube into the overflow pipe and reinstall the float arm and ball. Reconnect the supply tube and slowly turn on the water. Flush and check the water level, adjusting the float arm if necessary *(page 59)*.

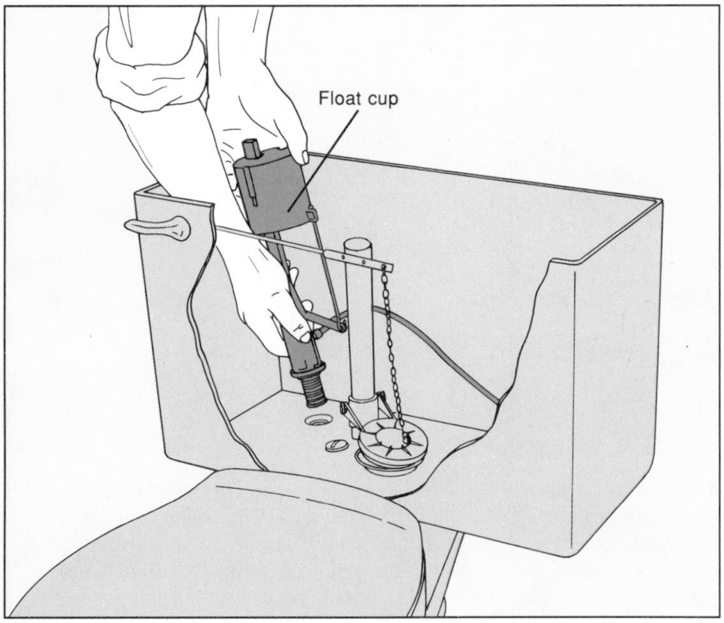

3 Installing a water-intake assembly. If you prefer to replace a conventional ball cock with a water-intake assembly, install it in the same way. Use caution when tightening the retaining nut on the underside of the tank; the plastic threads on the assembly shaft may strip and crack under excessive force. Reconnect the supply tube using the washers provided with the new assembly and turn on the water supply, slowly at first, to check for leaks.

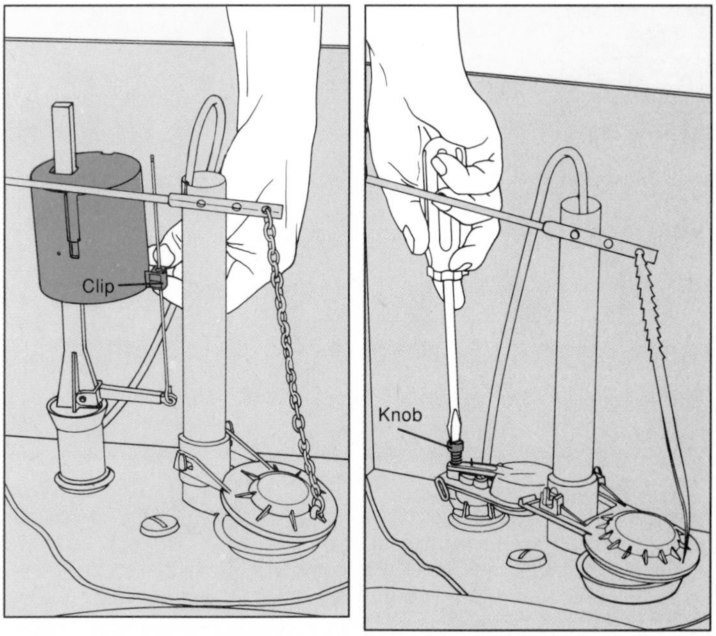

4 Adjusting the water level. Flush the toilet and check the water level in the tank. It should be 1/2 to 1 inch below the top of the overflow pipe, and just below the handle. To adjust the water level for a float cup *(above, left)*, pinch the clip and slide the cup 1/2 inch at a time: up to raise the water level, or down to lower it. Adjust the water level for a metered fill valve *(above, right)* by turning the adjustment knob by hand or with a screwdriver, one half-turn at a time: clockwise to raise the water level, counterclockwise to lower it.

SERVICING THE FLUSH ASSEMBLY

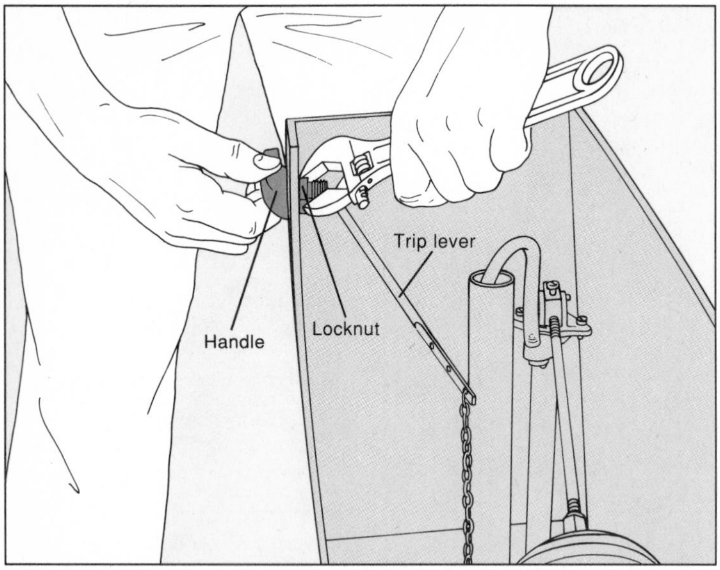

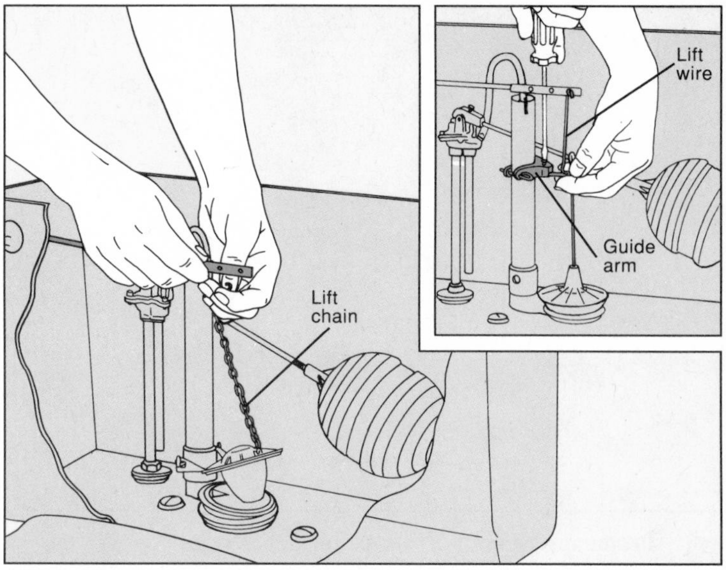

1 Adjusting the handle. Remove the tank cover and carefully lay it aside. For a loose tank handle, tighten the locknut inside the tank counterclockwise with an adjustable wrench *(above)*. If sediment and corrosion are blocking the handle, apply penetrating oil to the threads, wait 15 minutes, then unscrew the locknut. When the nut is impossible to remove, cut through the handle shaft with a hacksaw and replace the handle and trip lever. Unhook the chain from the trip lever and slide the trip lever, with the handle attached, through the hole in the tank. Scrub the handle threads with a toothbrush and vinegar. Reinstall the entire assembly, tightening the locknut counterclockwise.

2 Adjusting the lift chain. When the handle must be held down to flush the toilet, the chain may be too long. To shorten it, try hooking the chain through a different hole in the trip lever *(above)*, or use long-nose pliers to open and remove chain links. A slow flush may indicate a short or broken chain. To lengthen it, you must replace the chain. Do not try to repair or lengthen the chain by adding a safety pin or piece of wire; a second metal will promote corrosion. To adjust the length of an older flush assembly, loosen the screw on the guide arm *(inset)*, then slide the guide arm up to shorten, or down to lengthen it.

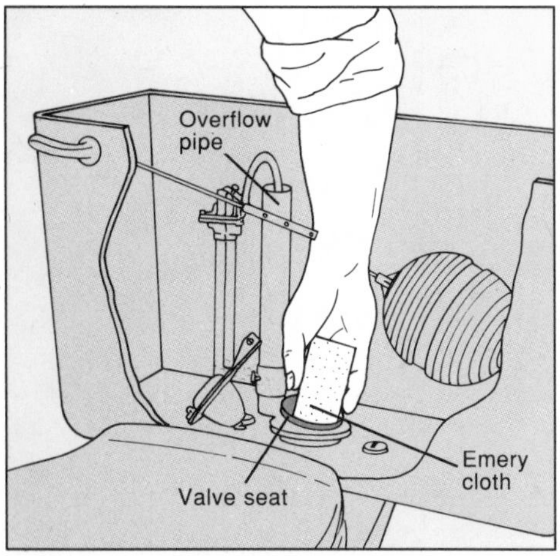

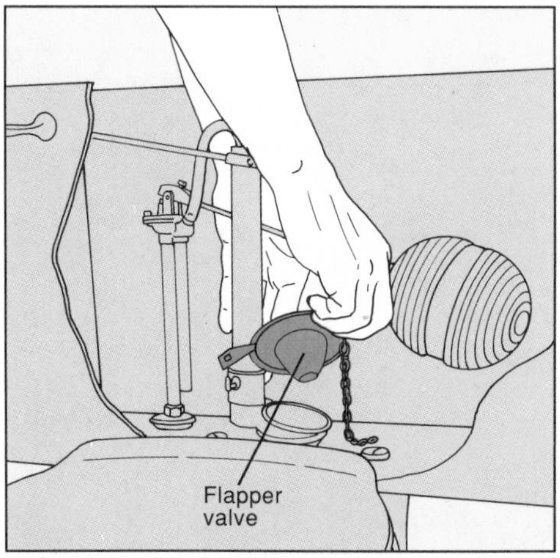

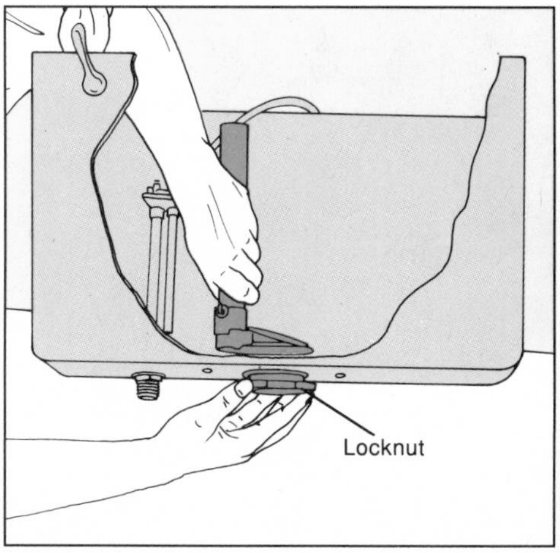

3 Cleaning the valve seat. Mineral deposits and sediment on the valve seat can prevent the flapper valve from sealing tightly and permit water to run continuously into the bowl. Turn off the water supply, or use a coat hanger to hold up the float ball *(page 10)*. Flush the toilet to drain the tank, then remove the flapper valve by unhooking it, or by sliding it up the overflow pipe. Gently scour inside the seat and its rim with emery cloth *(above)*. Then turn on the water supply and flush to check for leaks.

4 Replacing the flapper valve. A flapper valve that is soft or distorted allows water to leak into the bowl between flushes. Unhook the lift chain and remove the flapper valve *(above)*. Buy a replacement that fits the valve seat in your tank and install it. If you cannot find a valve that fits your seat, or if leaks persist after replacing the flapper valve, replace the entire assembly *(step 5)*.

5 Replacing the flush-valve assembly. For this repair you must first remove the tank *(page 67)*. Cut the new overflow pipe with a hacksaw to 1/2 inch below the tank handle. Apply a thin strip of plumber's putty to the cone-shaped washer on the flush-valve assembly. Then fit the assembly snugly into the tank opening and scrape away excess putty. Thread on the retaining nut, as shown, tightening it with a spud wrench or monkey wrench. Then push a new conical gasket up over the locknut. Ease the tank onto the bowl and reconnect them.

UNCLOGGING THE TOILET

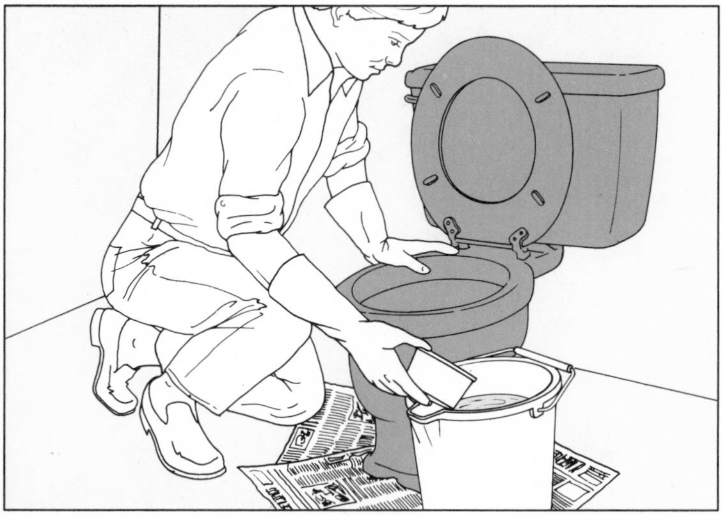

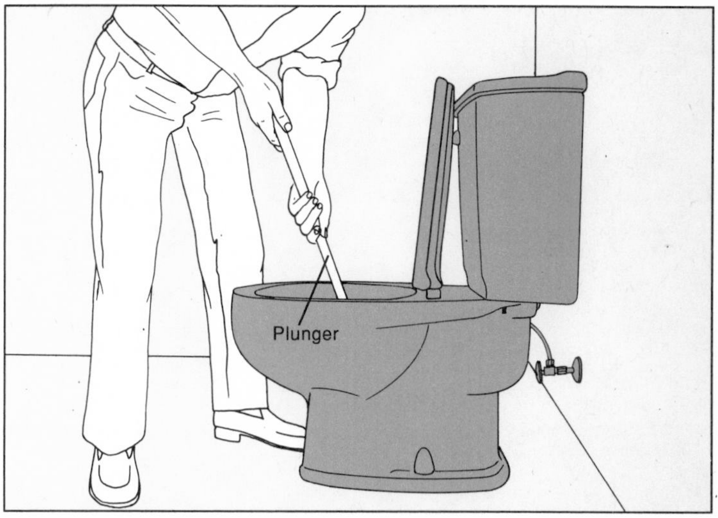

1 Preparing to clear the blockage. Do not flush the toilet. Spread newspaper around the base of the bowl. If the bowl is overflowing, put on rubber gloves and use a plastic container to bail out half the water *(above)*. If the bowl is empty, add water to half-full, then go to step 2 to clear the blockage with a plunger.

2 Using a plunger. A flange-type plunger fits into the toilet drain and exerts more pressure than the regular type. Place the rubber cup squarely over the drain opening—the larger one if there are two. Keep the cup below the water level, pump up and down rapidly 8 to 10 times, then pull the plunger up sharply. If the water rushes away, you may have released the blockage. Use the plunger once again, to be sure the water is draining freely. Pour in a pail of water to test and repeat it necessary.

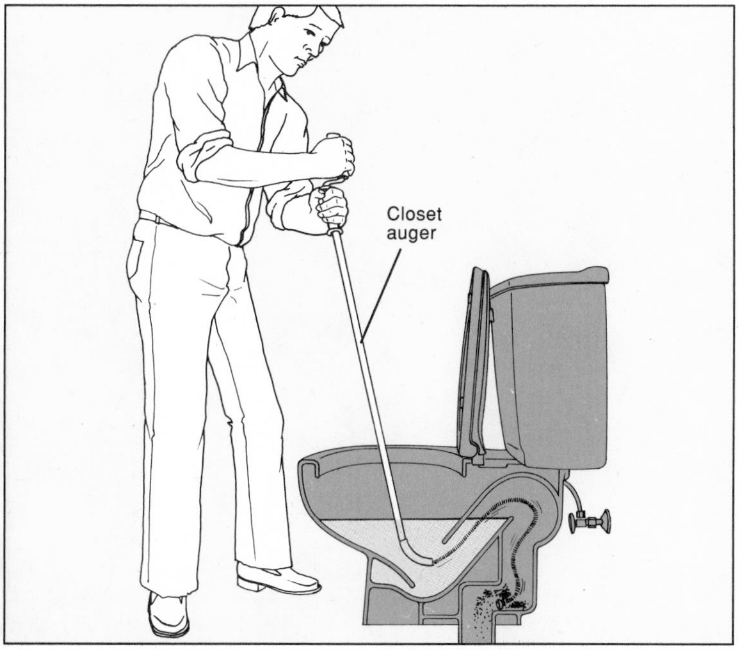

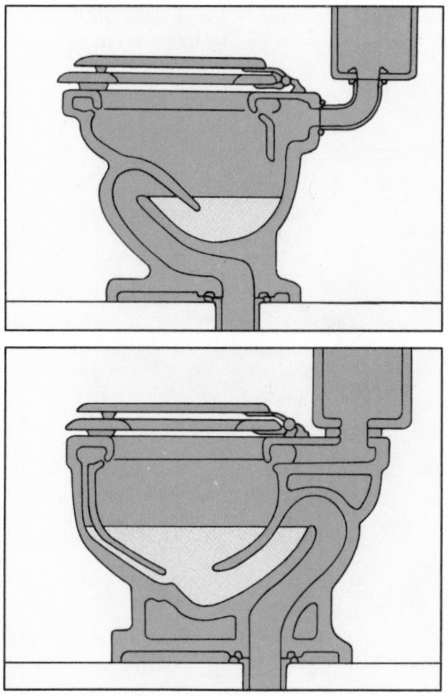

3 Using an auger. If possible, use a closet auger to clear a clogged toilet. Its long sleeve is curved to help start the coil into the toilet bend without scratching the porcelain. With the help of the illustrations above, determine the direction in which to guide the auger in your toilet. Feed the curved tip into the drain opening *(above)*. Crank clockwise until the auger tightens up, then continue cranking in the other direction. When the auger tightens again, reverse the direction until the auger is as far in the drain as it will go. Then pull the handle up and out to remove the auger. If it jams, push gently, then pull again. You may have to turn the handle as you pull up. Then use a plunger to ensure that the drain runs freely. Repeat with the auger if necessary. As a last resort, you may have to remove the toilet to try to reach the blockage from underneath *(page 64)*.

REMOVING AND RESEATING A TOILET

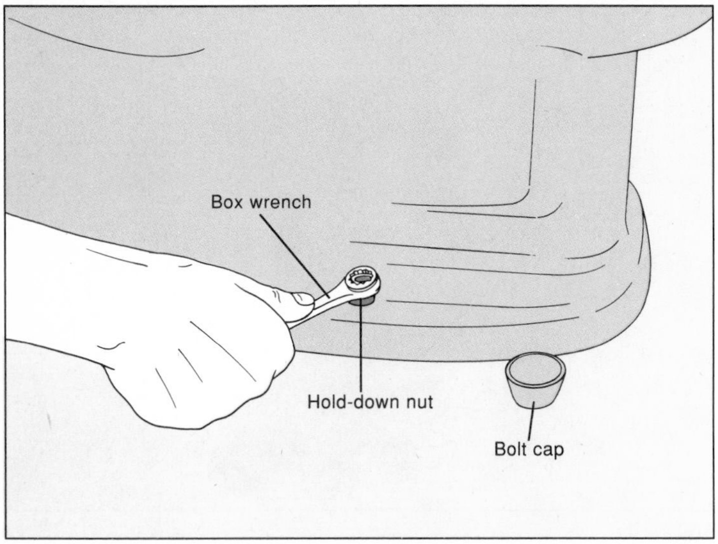

1 Removing the hold-down bolts. Although replacing a toilet is not difficult, it may take several hours to complete this job. Turn off the water supply. If you do not have a shutoff valve at the toilet, consider installing one before removing the toilet *(page 111)*. Remove the tank cover and carefully lay it aside. Flush the tank, holding down the handle to drain as much water as possible, then sponge out the tank and bowl. With an adjustable wrench, loosen the supply tube coupling nuts and gently push aside or remove the supply tube to avoid kinking it. Pry off the bolt caps with a putty knife, or break and replace them with plastic caps. Remove the nuts under the caps with a socket wrench or box wrench *(above)*. If impossible to remove, cut through the bolt with a hacksaw.

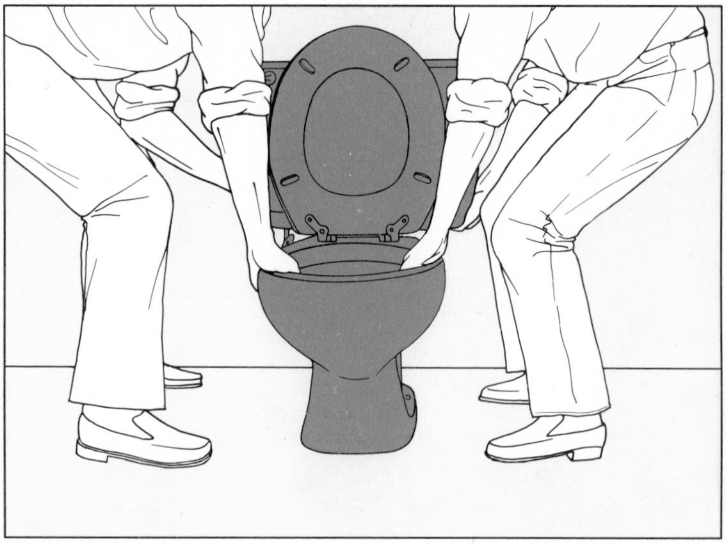

2 Lifting the toilet. If the toilet has a wall-mounted tank, first remove the flush elbow *(page 67)*. Otherwise, you and a helper can lift the toilet without removing its tank. (To remove the tank, see page 67). Standing over the bowl, twist and rock it to break its seal. Then, with knees bent and back straight, lift the toilet straight up and off the hold-down bolts *(above)*. Let the water from the trap run into a pan. Then gently lay the toilet on its side on a pad of newspapers or an old blanket. Cover the drain hole to prevent loose parts from falling in and to contain sewer gases.

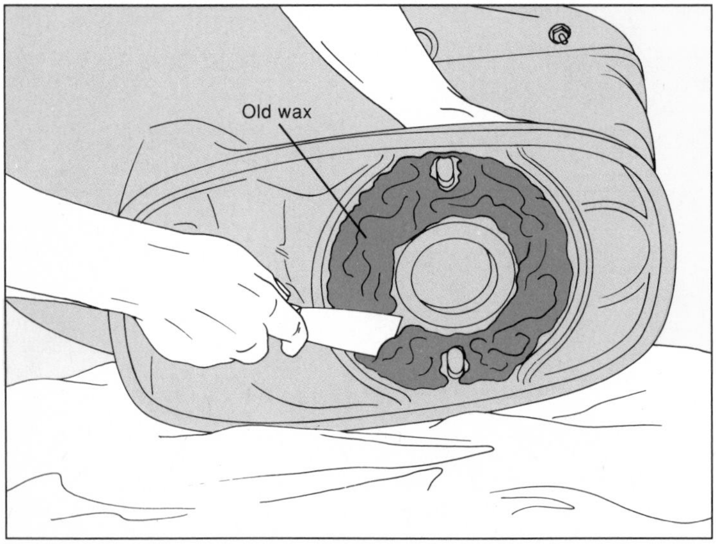

3 Removing the old seal. Use a putty knife to scrape off the old wax seal from around the toilet opening *(above)*, and from the closet flange in the floor. Do not reuse this seal, but keep some of it to pack around the hold-down bolts before reseating the toilet. Wearing rubber gloves, reach up inside the toilet bend from underneath and try to remove any blockage. You can also try to reach a blockage with an auger or unbent clothes hanger. If there appears to be nothing in the toilet, try augering into the closet bend.

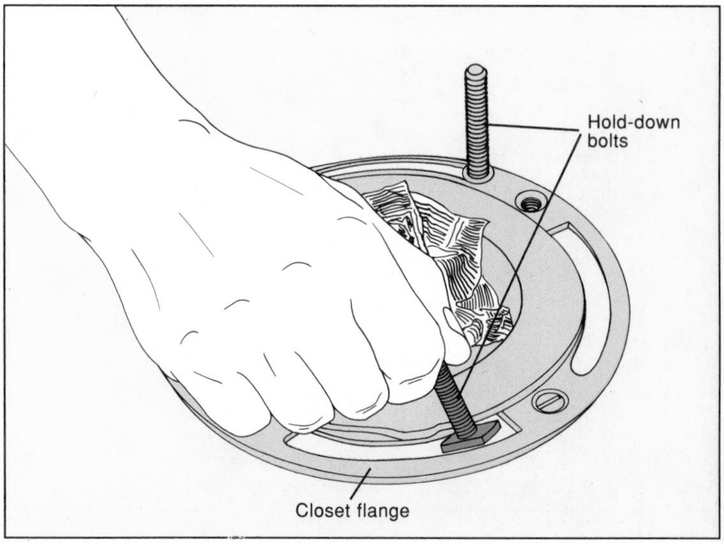

4 Replacing the hold-down bolts. Mark the exact distance of both hold-down bolts from the wall. Remove the bolts by sliding them out from the floor flange *(above)*. Older toilets may have two additional bolts in front that screw into the floor. Remove these bolts with channel-joint pliers or saw them off with a hacksaw. Coat new hold-down bolts with petroleum jelly to make future removals easier, then slide them into position, making sure they are parallel and in the same position as the old ones. Pack some old wax or putty around the base of each bolt to hold it in place while seating the toilet.

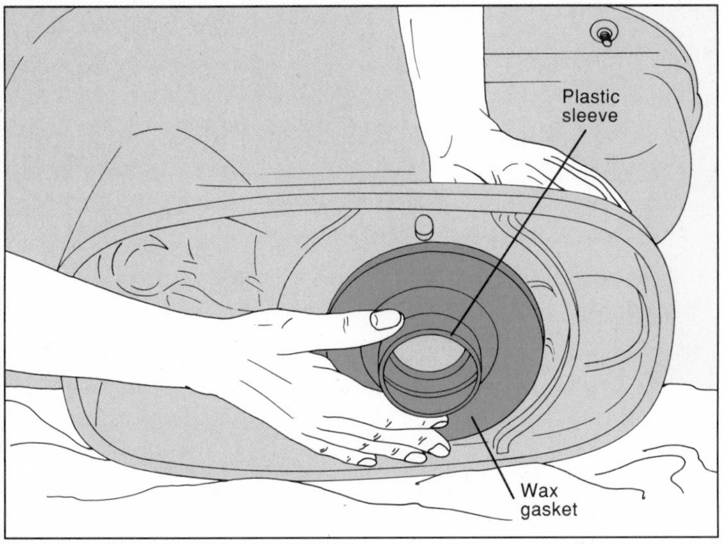

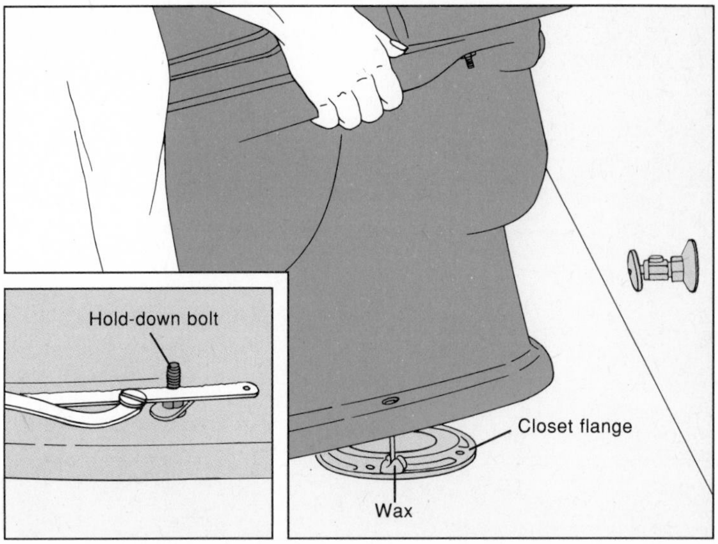

5 **Replacing the wax gasket.** Place the new wax gasket, beveled side up, over the drain opening. For a 4-inch drainpipe, you can install a gasket with a plastic sleeve that extends down into the drain and seals any cracks in the closet bend. Install it with the plastic sleeve facing away from the bowl *(above)*. Then knead some plumber's putty into a thin rope about 1/4 inch in diameter, and place it around the toilet rim, or apply silicone sealant after you have reseated the toilet on its flange.

6 **Reseating the toilet.** Lower the bowl gently over the hold-down bolts, as shown. Press down firmly, then sit on the bowl, twisting and rocking to spread the wax evenly. Position new washers and nuts over the hold-down bolts and tighten one-quarter turn with an adjustable wrench. Scrape away excess putty, then pour several pails of water into the bowl to test for leaks. Pull up the toilet and reseat with a new gasket if necessary. Trim new hold-down bolts with a hacksaw *(inset)* and replace the bolt caps. Reconnect the supply tube, slowly turn on the water, and flush to check for leaks.

REPLACING THE TOILET SEAT

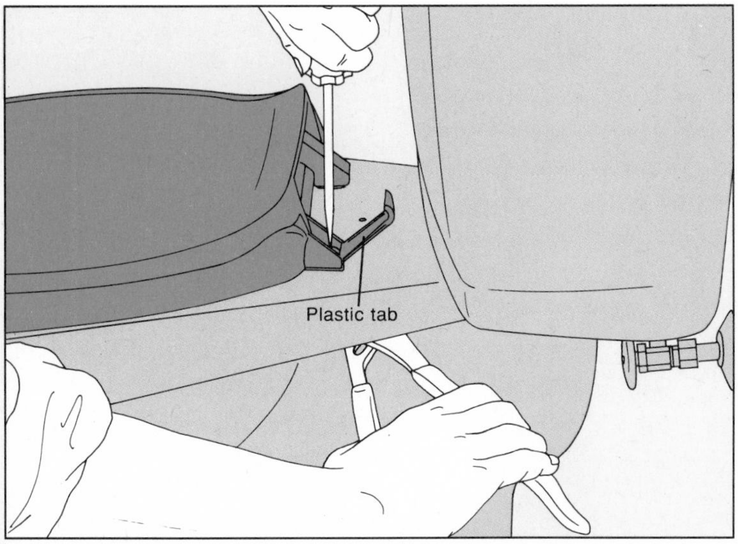

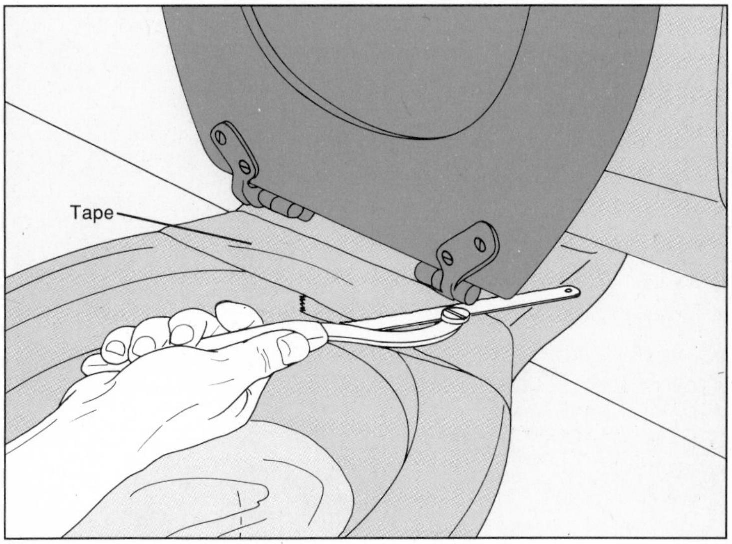

Removing the seat bolts. Measure both the width and length of the seat before buying a replacement. To remove the seat bolts, which may be hidden under plastic tabs, hold one bolt steady with a screwdriver, then unscrew the nut with a socket wrench or pliers *(above, left)*. To loosen corroded bolts, apply penetrating oil, wait overnight, then try again. If necessary, protect the bowl with tape or a thin piece of cardboard and cut through the bolts with a hacksaw *(above, right)*. Replace the seat and bolts, and hand-tighten the nuts. Check that the seat aligns with the bowl, and tighten the nuts one-quarter turn with a wrench.

INSULATING THE TANK

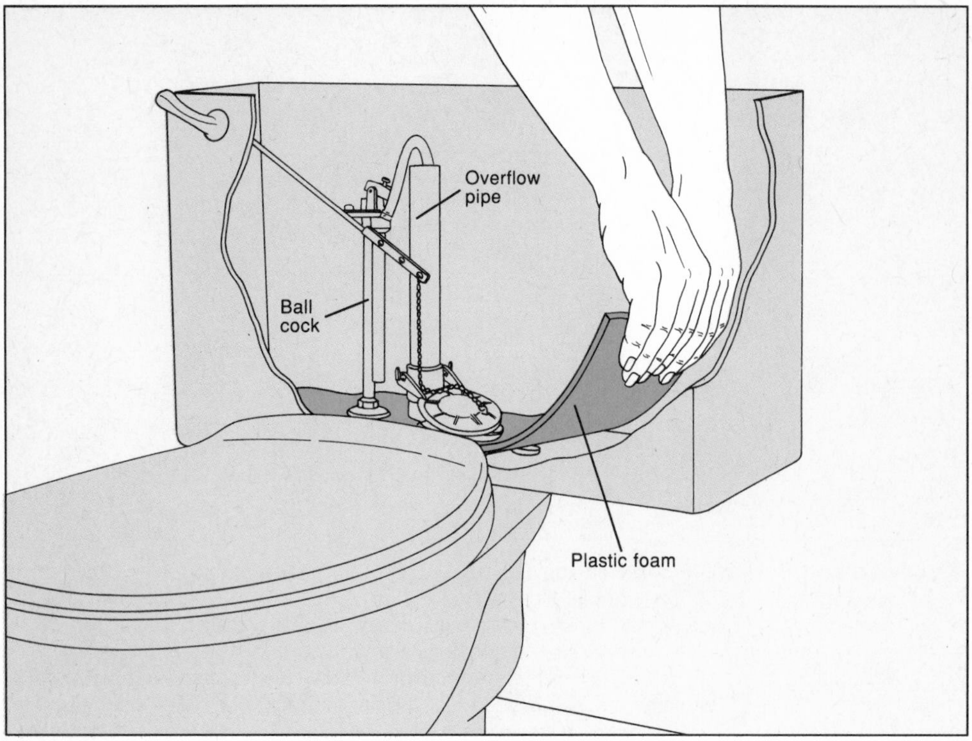

1 Lining the tank with foam. Lining a sweating tank will often stop leaks caused by condensation, especially in hot weather. Remove the tank cover and carefully lay it aside. Mark the water level inside the tank; the foam should cover the tank to 1/2 inch above this line. Shut off the water supply and flush, holding the handle to drain as much water as possible. For easier access, remove the float ball and float arm *(page 59)*. Sponge up any remaining water and use a hair dryer to dry the tank walls completely. Cut and install the foam according to instructions in the kit, applying a bead of adhesive over all edges and seams. Replace the float ball and arm. Leave the tank cover off and allow the adhesive to set overnight before turning on the water supply and filling the tank.

STOPPING LEAKS FROM THE TANK

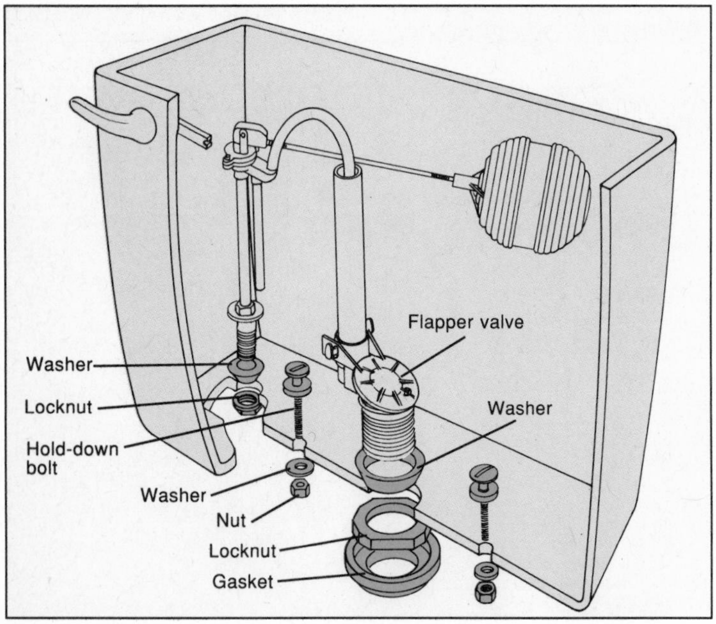

1 Finding the source of a leak. The diagram above shows where leaks occur if seals are not watertight. First look through the Troubleshooting Guide on page 58 to check for obvious causes of leaks and ensure that no tank parts are at fault. If you are unable to find the leak, pour a few drops of food coloring into the tank, and wait—up to a day—to see where the coloring leaks out.

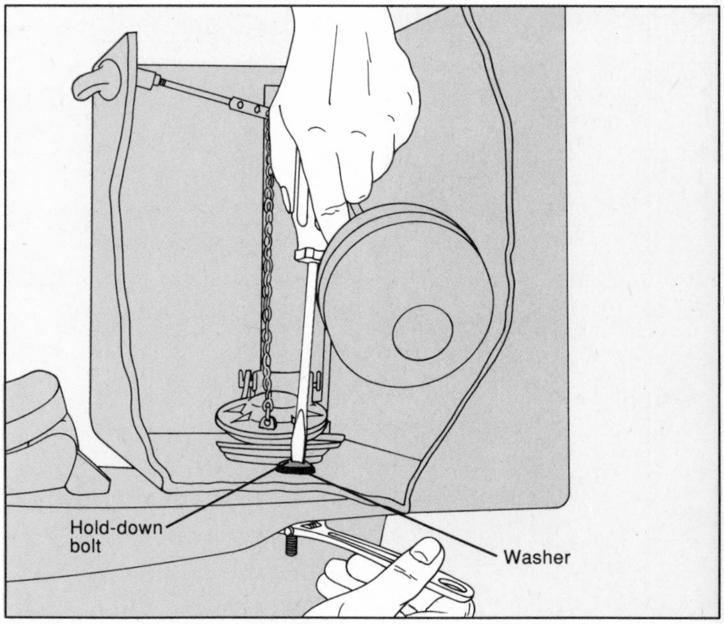

2 Stopping leaks under the tank. To tighten a tank hold-down bolt, secure it inside the tank with a screwdriver, as shown, and tighten the nut with an adjustable wrench. If leaks persist, loosen and remove the hold-down bolts and replace the washers. For a hairline crack in the tank, replace the tank. Tighten a loose locknut at the ball cock counterclockwise a quarter-turn with an adjustable or open-end wrench, holding the ball cock inside the tank steady with the other hand.

REPLACING THE TANK

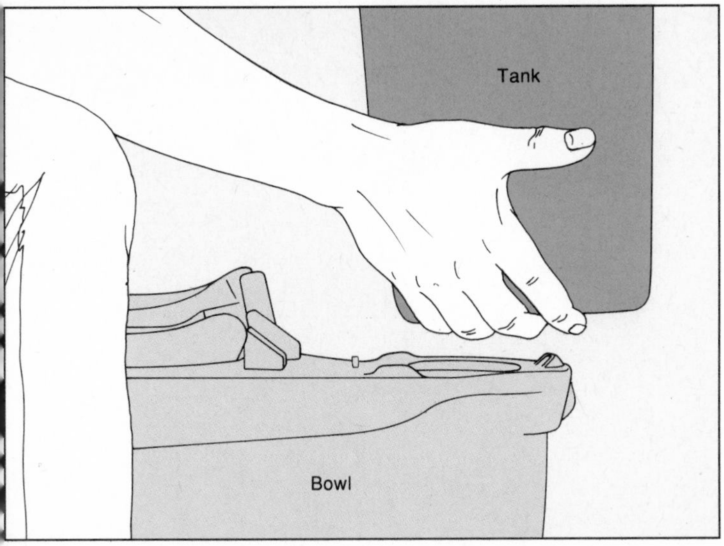

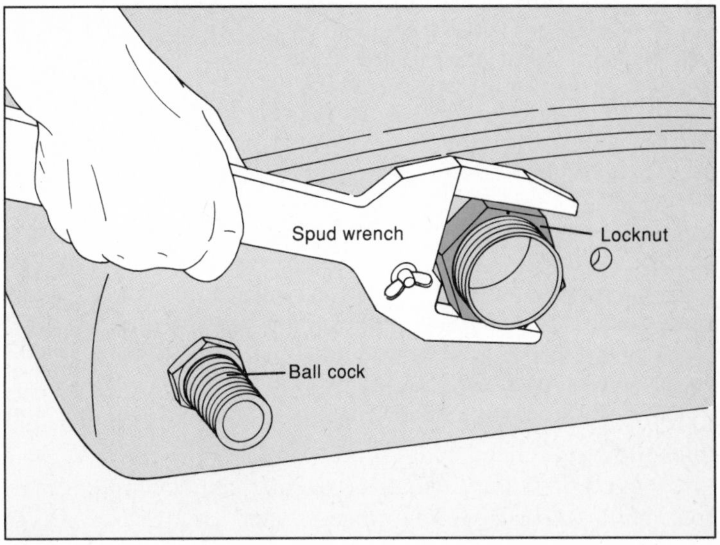

1 Disconnecting the tank. Shut off the water supply. Remove the tank cover and carefully lay it aside. Flush, drain and sponge out any remaining water in the tank. With an adjustable wrench or channel-joint pliers, loosen the coupling nuts at the supply tube. Gently push the tube aside or remove it to avoid kinking. To remove the hold-down bolts, hold the bolt on the inside of the tank with a flat-tipped screwdriver and loosen the nut with an adjustable wrench *(page 66)*. Lift the tank straight up and off the bowl, twisting carefully if there is resistance from the tank gasket. Gently lay the tank on a rug or pad of newspapers. Pry off or scrape the old gasket from the tank and bowl.

2 Replacing the tank. If you are reusing the old tank, first remove the exposed locknut with a spud wrench, turning it counterclockwise, as shown. If the nut is badly corroded, apply penetrating oil and wait at least 15 minutes before trying again. Hold on to the overflow pipe and valve seat, and pull the flush-valve assembly free. Install the new assembly as described on page 62. Place the tank back on the bowl, keeping it parallel to the wall, install new hold-down bolts and washers. Reconnect the supply tube and turn the water on slowly. Flush to test for leaks, and tighten the connections one-quarter turn if necessary.

SERVICING A WALL-MOUNTED TANK

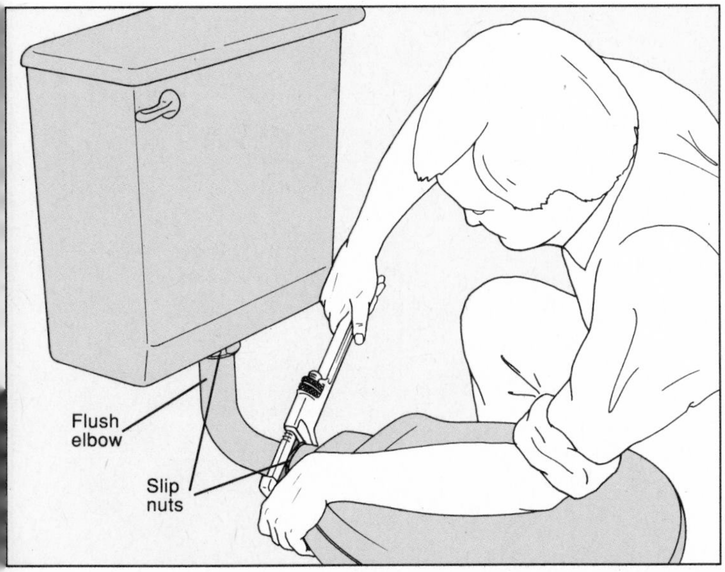

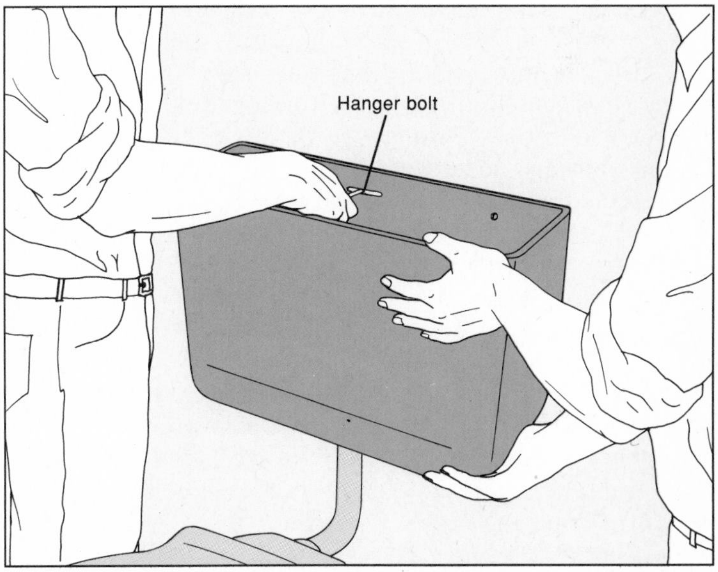

1 Stopping leaks at the flush elbow. Corrosion may cause leaks where the toilet connects to the flush elbow. Turn off the water supply, flush and drain the tank. With a spud wrench or monkey wrench, loosen the slip nut at the bowl *(above)*, then at the tank, holding the flush-valve assembly steady inside the tank to keep it from turning. Remove the old packing and clean off the threads of the locknuts with a wire brush or steel wool. Wrap new packing string or plumber's putty around the flush elbow and against the locknut and thread the slip nuts back on. If leaks continue, replace the elbow and slip nuts.

2 Removing a wall-mounted tank. Have a helper support the tank *(above)*, or sit on the bowl and support the tank with your knees, then remove the hanger bolts securing the tank to the wall. Rock the tank gently to loosen it, then pull it off the bowl. Lay the tank on a rug and remove the slip nuts and flush elbow. If necessary, cut through the nuts with a hacksaw, taking care not to damage the locknut threads. Replace the flush elbow, trimming it first with a hacksaw to the same size as the old one. Replace the packing and slip nuts and tighten all connections with a spud wrench.

BATHTUBS AND SHOWERS

Since one-third of a typical household's water pours through the shower head and tub spout—and down the bathtub drain—it's not surprising that these fixtures can call for a variety of repairs. On the supply side, tub spouts may drip, faucet handles leak and diverters refuse to send water to the shower head. On the drainage side, the tub may empty too slowly or water seep away annoyingly during a bath.

Leaks around the faucet handles, diverter or shower head call for a while-you're-at-it approach to repair. While the fixture is removed and disassembled, clean it thoroughly with vinegar and a wire brush, and replace any parts that appear worn, pitted or heavily corroded with exact duplicates. In the case of older or less expensive fixtures, it may be less trouble to replace the entire faucet, spout or shower head rather than hunt for replacement parts.

Clogged drains are usually caused by an accumulation of soap and hair, or by foreign objects such as stray hairpins or plastic toys. But don't reach for the chemical drain opener as soon as the drain backs up. A plunger or auger is much safer for the trap and drain than caustic chemicals.

Simple maintenance can prevent most drain problems. Use a drop-in basket strainer to catch hair and stray objects. Clean the strainer or stopper regularly (and put the debris in the garbage, not down the drain). And don't cure a slow drain by removing the stopper to let the water out; this will invite more serious problems farther down the system.

One of the challenges of tub and shower repair is gaining access to the parts that need work. The pipes that supply hot and cold water to the faucet may be walled in behind tile or wallboard; if there is no access panel, you will have to create one as part of the repair *(page 79)*. Likewise, a clogged or leaking trap may lie beneath a bathroom floor; an otherwise simple repair may require knocking a hole in the ceiling below. Patience is the key to working in these typically cramped locations. Fortunately, their awkward placement behind walls and under floors protects these pipes from wear and tear and accidental damage.

When tackling repairs to the supply pipes and fixtures, be sure to first turn off the water. To locate the shutoff valves, look for an access panel on the other side of the wall where the faucets are located. The shutoff valves should be behind this panel or in the basement directly below the tub. If not, turn off the main shutoff valve. Most tub and shower repairs require only a screwdriver, adjustable wrench, pipe wrench and locking pliers. Protect chromed parts from the bite of a wrench by wrapping them with friction or plastic tape. And be sure to have a pail and rags close at hand to sop up any water that spills from the system.

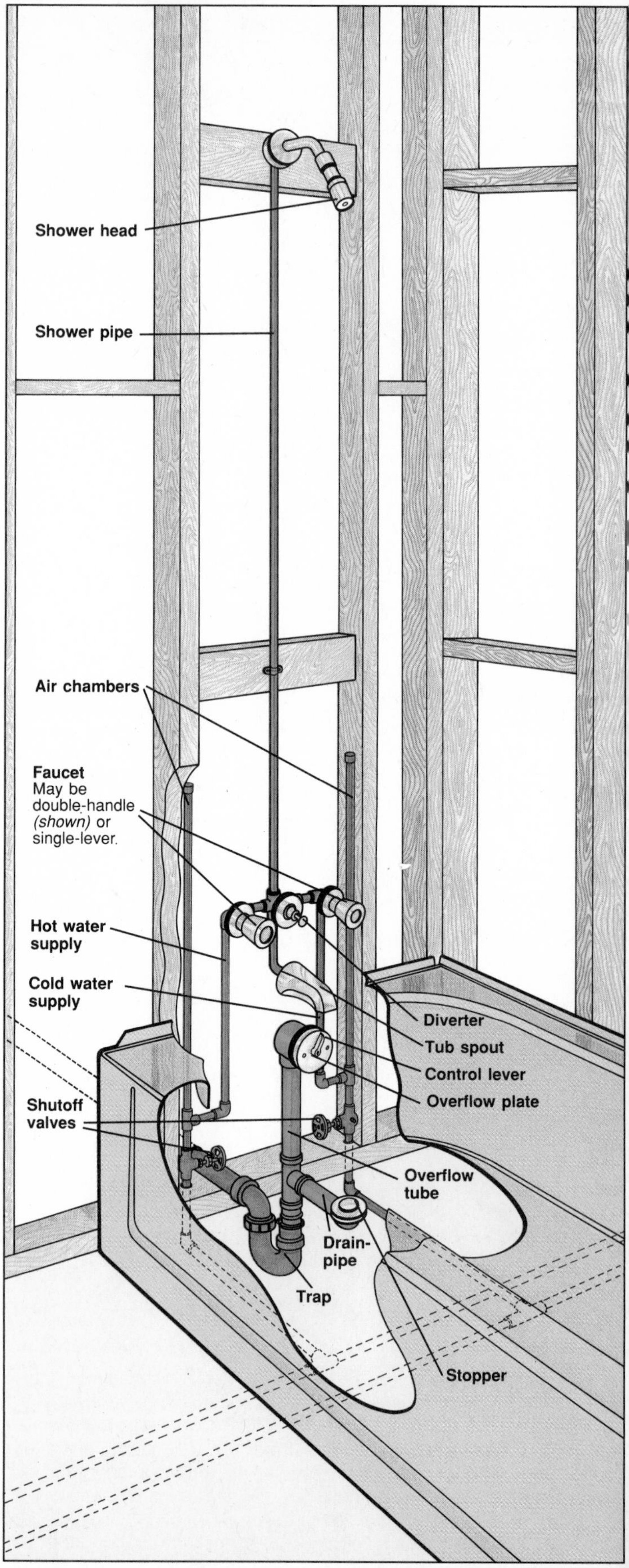

TROUBLESHOOTING GUIDE

SYMPTOM	POSSIBLE CAUSE	PROCEDURE
Water seeps from bathtub (pop-up drains)	Accumulation of hair or soap around stopper	Remove and clean stopper *(p. 70)* □○
	Accumulation on lift assembly	Remove and clean lift assembly *(p. 70)* □○
	Lift assembly too long	Adjust lift assembly *(p. 70)* □○
	Lift assembly worn or damaged	Replace lift assembly *(p. 70)* □○
Water seeps from bathtub (trip-lever drains)	Accumulation on plunger	Remove and clean lift assembly *(p. 70)* □○
	Lift assembly too short or plunger worn	Adjust or replace lift assembly *(p. 70)* □○
Control lever won't stay in position	Lever mechanism corroded	Adjust or replace lift assembly *(p. 70)* □○
Water drains too slowly from bathtub (pop-up drains)	Accumulation around stopper	Clean stopper *(p. 70)* □○
	Lift assembly clogged, corroded or too short	Clean and adjust lift assembly *(p. 70)* □○
Water drains too slowly from bathtub (trip-lever drains)	Lift assembly clogged, corroded or too long	Clean and adjust lift assembly *(p. 70)* □○
	Lift assembly worn or damaged	Replace lift assembly *(p. 70)* □○
Drain slow or blocked	Drainpipes clogged	Use a chemical drain cleaner with caution *(p. 13)* Open drainpipe with a plunger, auger or hose *(p. 72)* □○
	Trap blocked	Auger through cleanout plug and trap *(p. 73)* ⬓◒
Water leaks around shower or tub	Leaky joint between tub/shower or wall/floor	Seal joints with caulking *(p. 131)* □○
Faucet handle leaks (older double-handle faucets)	Packing nut loose	Tighten nut *(p. 22)* ⬓◒
	Packing or packing washer worn	Replace worn parts *(p. 22)* ⬓◒
Faucet handle leaks (double-handle faucets)	O-rings or stem worn	Replace O-rings or stem *(p. 74)* ⬓◒
Faucet handle leaks (single-lever faucets)	Adjusting ring loose	Tighten adjusting ring *(p. 76)* □○
Water leaks from tub spout or shower head (double-handle faucets)	Stem washer worn	Replace washer *(p. 74)* □○
	Seat worn	Replace or dress seat *(p. 74)* ⬓◒▲ Replace faucet set *(p. 78)* ■●
Water dripping from tub spout or shower head (single-lever faucets)	Ball-type: worn ball assembly or seats and springs	Replace worn parts *(p. 76)* ⬓◒
	Cartridge-type: worn cartridge	Replace cartridge *(p. 77)* ⬓◒
	Faucet set worn	Replace faucet set *(p. 80)* ■●
Water incompletely diverted from spout to shower (tub-spout diverters)	Diverter worn	Replace tub spout *(p. 81)* □○
Water leaks around diverter handle (push-pull diverters)	O-rings worn	Replace O-rings or diverter *(p. 82)* ⬓◒
	Diverter worn	Replace diverter *(p. 82)* ⬓◒
Water incompletely diverted from spout to shower (push-pull diverters)	O-rings worn	Replace O-rings or diverter *(p. 82)* ⬓◒
	Diverter coated with sediment	Clean diverter *(p. 82)* ⬓◒
	Diverter worn	Replace diverter *(p. 82)* ⬓◒
Weak or uneven pressure from shower head	Mineral buildup on shower head	Clean or replace shower head *(p. 83)* ⬓◒
Shower head leaks	Connections between shower head and arm loose	Tighten shower head to arm or seal joint with pipe tape *(p. 83)* □○

DEGREE OF DIFFICULTY: □ **Easy** ⬓ **Moderate** ■ **Complex**
ESTIMATED TIME: ○ **Less than 1 hour** ◒ **1 to 3 hours** ● **Over 3 hours**

▲ **Special tool required**

BATHTUB DRAINS

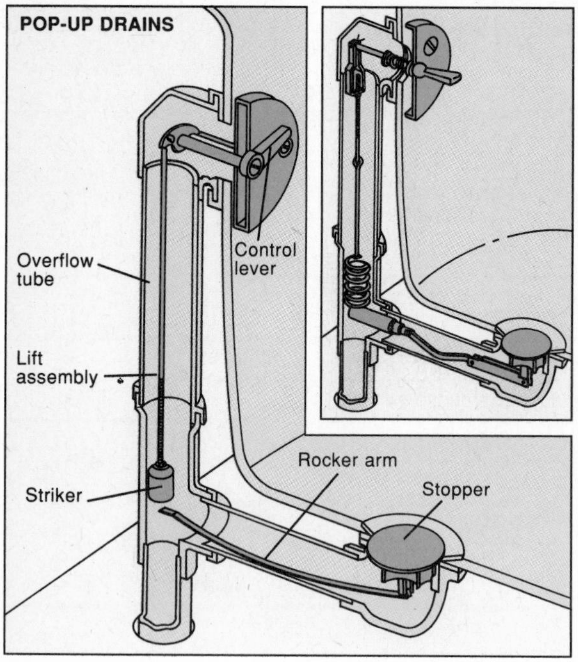

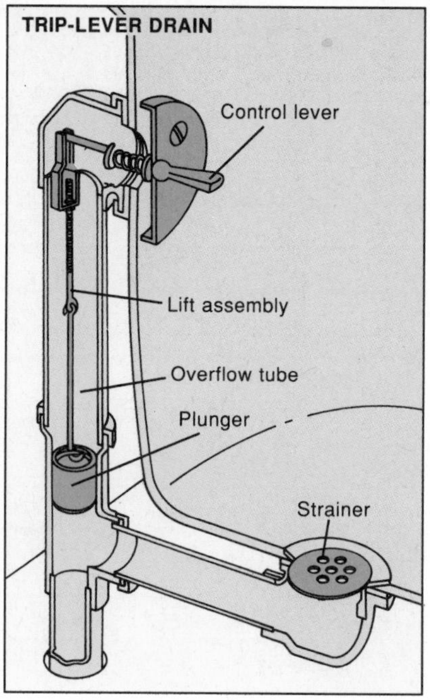

Pop-up drains. The distinguishing feature of a pop-up drain *(far left)* is a curved rocker arm attached to the stopper. The arm extends to the intersection of the drain, where a lift assembly in the overflow tube rests. The control lever raises the assembly so the striker at the bottom lifts off the rocker. As that end of the rocker springs up, the stopper falls to close the drain. When the lift assembly is lowered, the striker presses down on the rocker, pushing up the stopper to open the drain.

Trip-lever drain. Tubs with trip-lever drains *(near left)* usually have a strainer in the drain, but no stopper. Inside the overflow tube a hollow brass plunger is suspended from the lift assembly. The control lever lowers the plunger down the tube, blocking the intersection of the drain. When the plunger is raised, the water runs freely down the drain.

SERVICING THE DRAIN ASSEMBLY

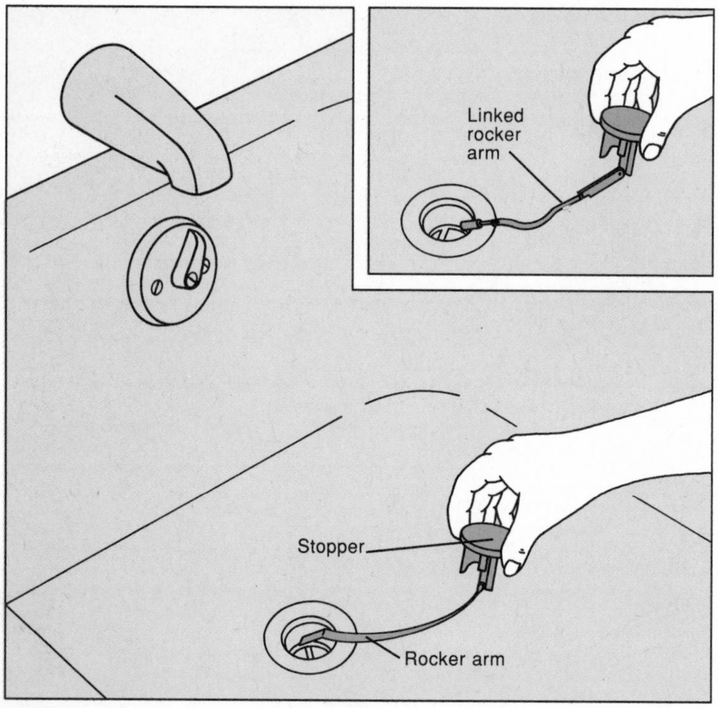

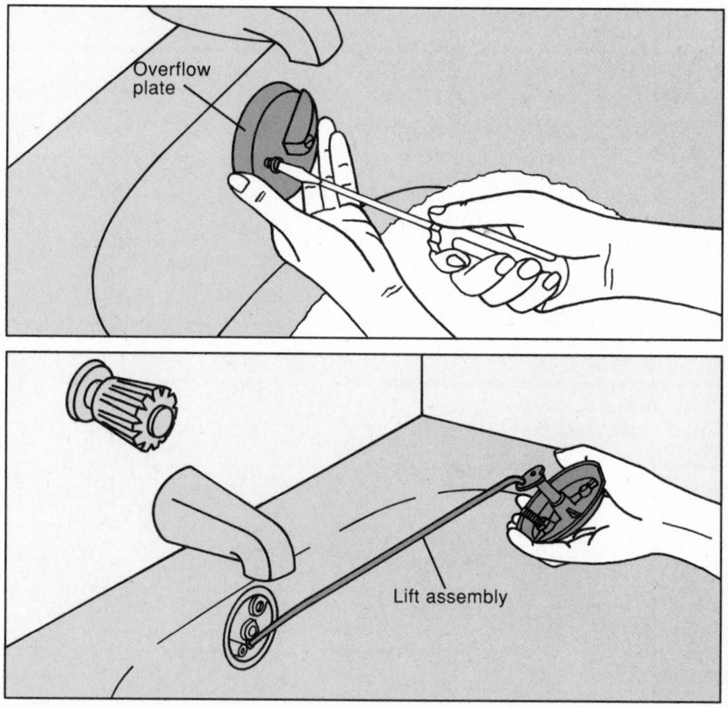

1 Cleaning the pop-up stopper. If water seeps down the bathtub drain when it is closed, or the tub drains too slowly, the pop-up stopper may be clogged. To remove the stopper, turn the control lever to open the drain, pull up the stopper, and work the rocker arm clear of the drain opening *(above)*. Scour the stopper assembly with fine steel wool. Feed it back into the drain with the rocker arm curving downward in the drainpipe. Wiggle a linked rocker arm *(inset)* back and forth until it sits back in place. If either problem persists, go to step 2.

2 Removing the pop-up or trip-lever assembly. Cover the drain with a bath mat to protect the tub and prevent loss of parts. Remove the screws on the overflow plate and pull it away from the tub, as shown. To remove a stubborn lift assembly, spray penetrating oil down the overflow tube and wait 15 minutes before trying again. Pull the lift assembly up through the overflow opening. Clean the assembly *(step 3)*, adjust it *(step 4)*, or replace a badly worn assembly *(step 5)*.

SERVICING THE DRAIN ASSEMBLY (continued)

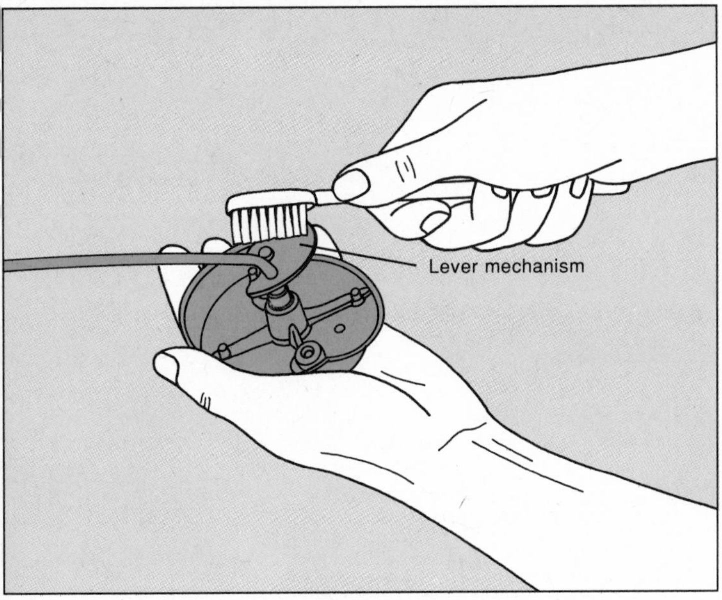

3 Cleaning the lift assembly. Wash off debris around the striker, spring or plunger. Remove corrosion from the assembly using vinegar and steel wool or an old toothbrush. Clean the lever mechanism on the back of the overflow plate *(above)* to ensure that the control lever stays in position. Lubricate the entire mechanism lightly with petroleum jelly or silicone lubricant.

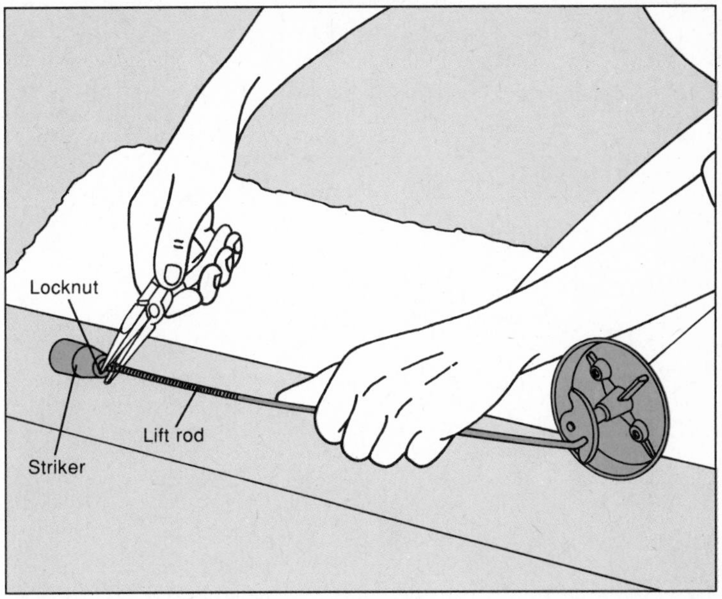

4 Adjusting the lift assembly. If the stopper does not fit snugly into the drain, or does not rise far enough to drain properly, the lift assembly may need adjustment. Loosen the locknut that holds the striker in place. Rotate the nut down the threaded rod to lengthen the lift assembly, or up to shorten it, then retighten *(above)*. Reassemble and test. If problems persist, replace the drain assembly *(next step)*.

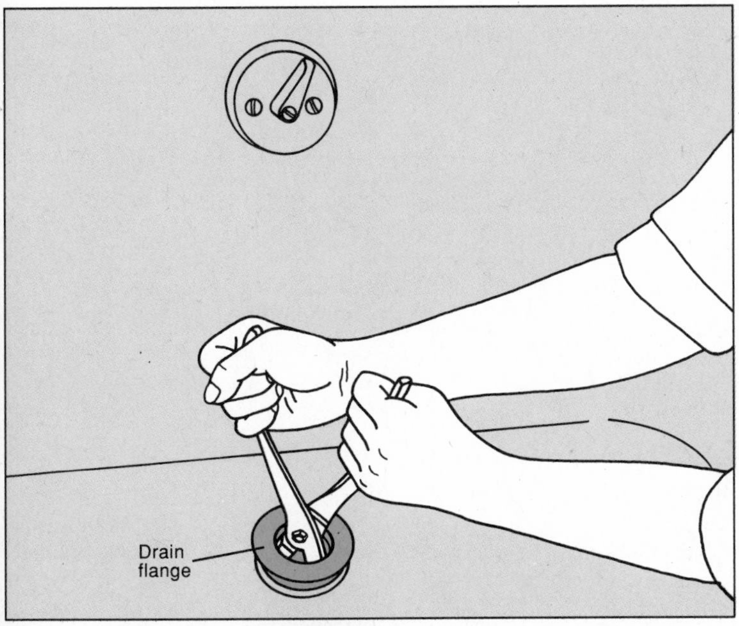

5 Replacing the assembly. When a drain assembly of the same make and model as the old one is not available, simply replace the assembly with a rubber plug and drop-in basket strainer. Pull out the pop-up stopper or pry up the trip-lever strainer and remove the overflow plate with the lift assembly attached. Remove the cotter pin to release the lift assembly or cut the linkage. Screw the overflow plate back in place. If the lift assembly cannot be disconnected from the back of the overflow plate, buy a replacement overflow plate with screws that line up with the holes in the overflow tube.

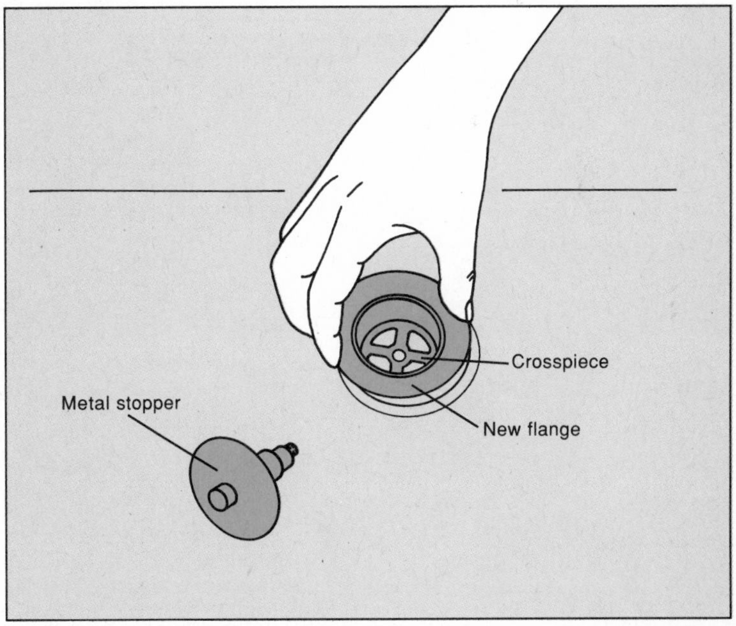

You can also use a metal stopper that opens and closes the drain at a touch. With the stopper or strainer removed, unscrew the drain flange counterclockwise, using pliers *(above, left)*. To increase leverage, insert the handle of a wrench between the handles of the pliers. Take the flange to a plumbing supply store; its threads must match those of the stopper's flange. Apply a strip of plumber's putty under the lip of the new flange. Screw the flange into the drain opening *(above, right)* and thread the metal stopper into the crosspiece.

CLEARING A CLOGGED DRAIN

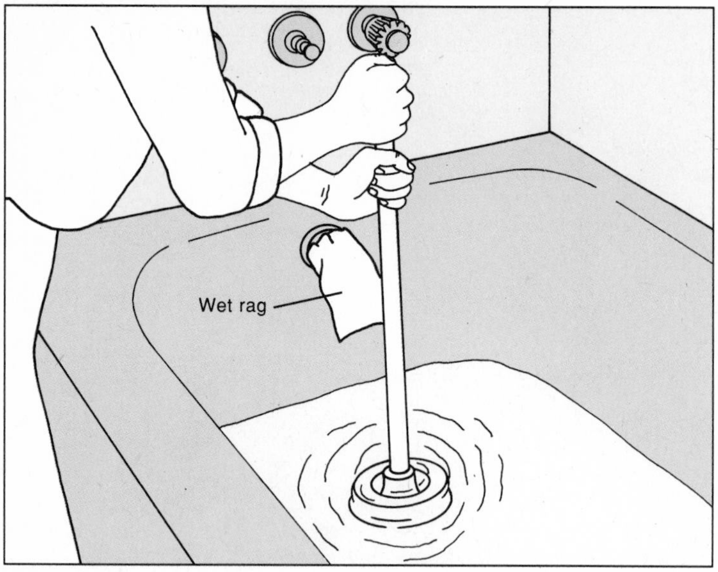

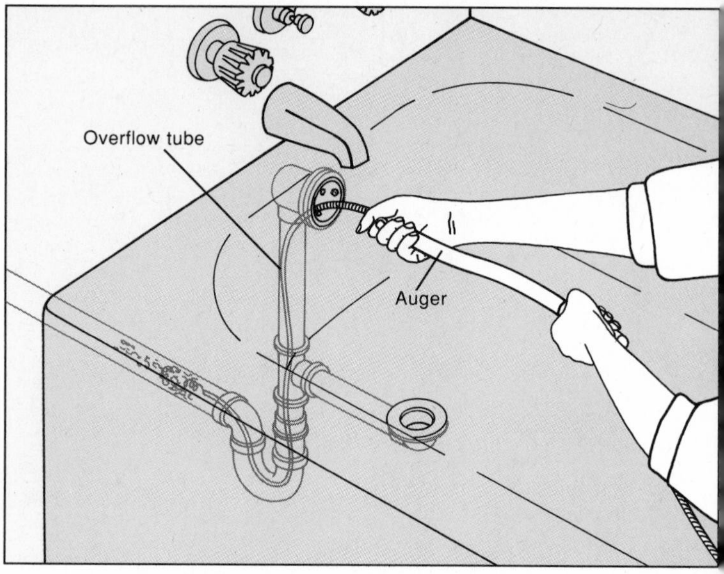

1 Using a plunger. If there is a tub stopper, remove it and the overflow plate *(page 70)*. Plug the overflow opening with tape, a large, wet rag or a rag wrapped in plastic. Coat the plunger rim with petroleum jelly and run enough water into the tub to cover the plunger cup. Insert the plunger at an angle so that no air remains trapped under it. Push down and pull up forcefully, keeping the plunger upright and the cup sealed over the drain opening *(above)*. Repeat the process many times; patience is the key to plunging. If the clog remains, use an auger *(step 2)* or, if local codes permit, a hose *(steps 3-4)*.

2 Using an auger. Have a pail ready to catch any debris snagged by the end of the auger. In a shower stall, pry up or unscrew the strainer and work through the drain opening. In a bathtub, remove the stopper and the lift assembly *(page 70)*, and feed the auger down the overflow tube *(above)*. Maneuver the auger around the corners in the drain, rotating it clockwise to break up the clog. Remove the auger slowly and run water to test the drain. If the clog remains, use a hose *(steps 3-4)* or work directly on the trap *(page 73)*.

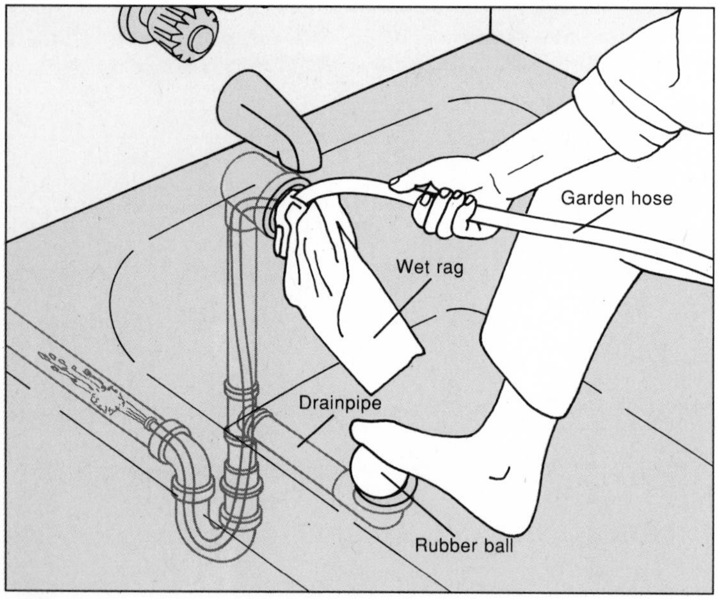

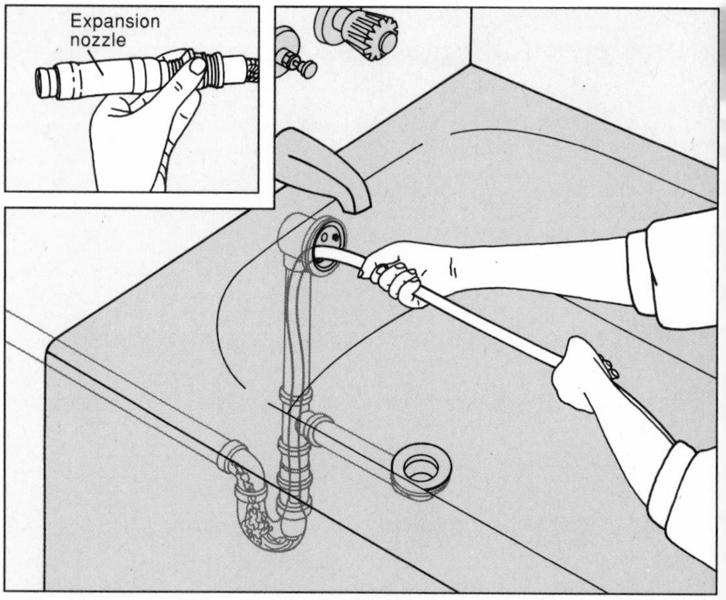

3 Using a hose. A hose attached to an outdoor faucet may reach the tub through a window. If not, attach the hose to an indoor faucet using a threaded adapter. Close all nearby drains, feed the hose down the overflow tube and pack rags tightly around the hose. (When the overflow tube is too narrow for the hose, feed the hose down the drain opening as far as possible.) Press down firmly on the plug or a rubber ball *(above)* to seal the tub drain. Hold the hose firmly while someone turns the water on full force and then off again several times to flush the blockage.

4 Using an expansion nozzle. This attachment, available at plumbing supply stores, expands to seal the overflow tube and force water through the drain in jets, increasing the effectiveness of the hose. Measure the opening to determine which nozzle will fit that diameter of pipe. Attach the nozzle to the hose *(inset)* and the hose to a faucet. Seal off all nearby drains. Insert the hose as in step 3, and turn the water slowly to full force *(above)*. Turn the water off, detach the hose from the faucet, and wait for the nozzle to deflate before removing it.

WORKING ON THE BATHTUB TRAP

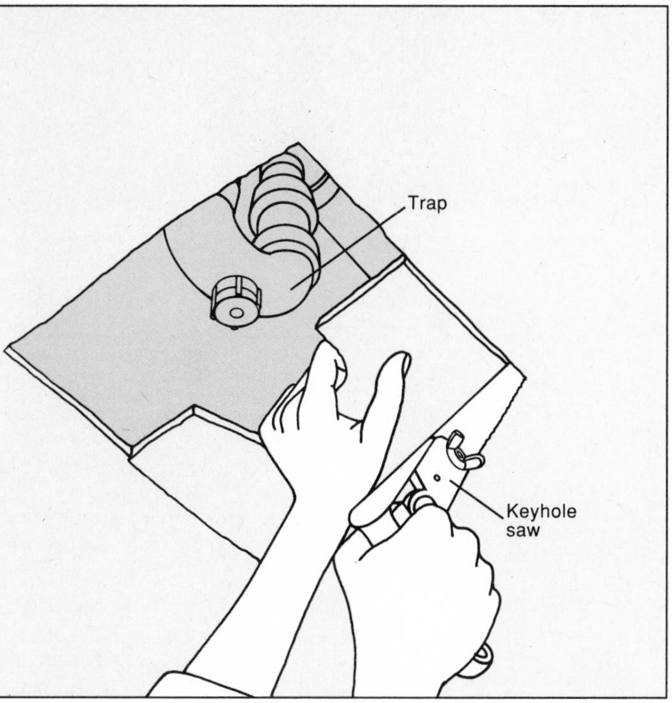

1 Getting at the trap. If the techniques described on page 72 do not unclog the drain, work directly on the trap. The shower trap is usually directly below the drain; the bathtub trap is usually ocated below the head of the bathtub. If there is an access panel on he wall behind the tub, remove it to see if you can reach the trap with short-handled or offset wrench and go to step 2. If there is insufficient pace, you must work through the ceiling below. Measure from the athtub wall to the drain and transfer that measurement to the ceiling elow. Make a small hole with a hammer and chisel to locate the trap, hen cut a neat, square hole with a drywall saw or keyhole saw *(above).*

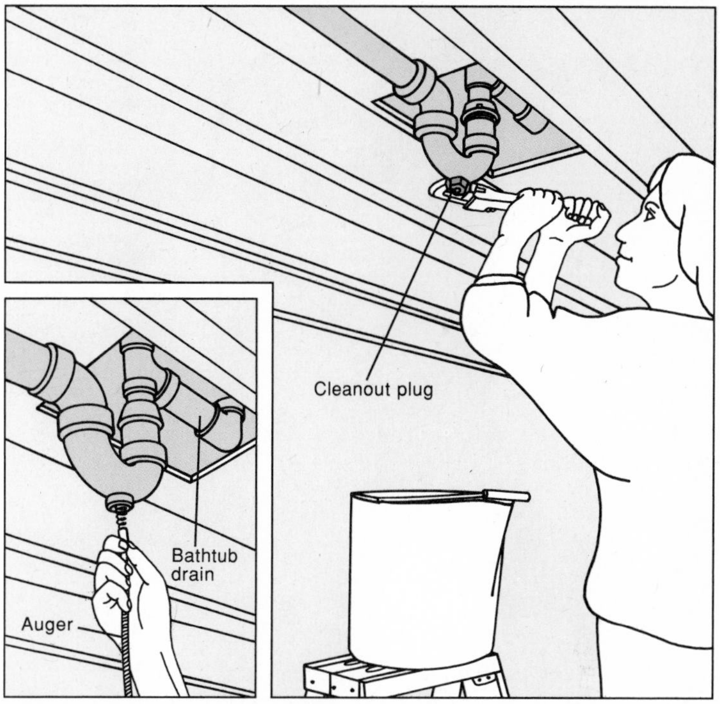

2 Working through the cleanout plug. Place a pail on the top of a ladder to catch water runoff. If there is no cleanout plug on the bottom of the trap, go to step 3. If there is a plug, use a pipe wrench to loosen it *(above).* Tap the edges of a stubborn cleanout plug with the wrench, or apply penetrating oil. Remove the plug by hand and catch the water and debris in the pail. Insert the auger through the cleanout opening *(inset).* Push the tip toward the bathtub drain first, rotating it to break up clogs. Then auger toward the main drain. If debris is removed, screw the cleanout plug back in place and run water to test the drain.

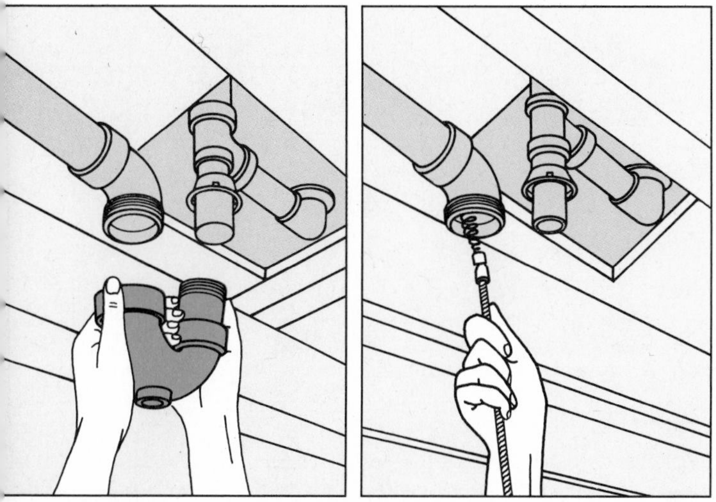

3 Working on the trap. Unscrew the two slip nuts that secure the trap to the drainpipe. If the nuts do not loosen, tap the edges with the wrench, or apply penetrating oil, and try again. Pull the rap down and off *(above, left).* Work the auger through it, then rinse it ut with water to clear any debris. First auger through the drainpipe oward the tub, then toward the main drain *(above, right).*

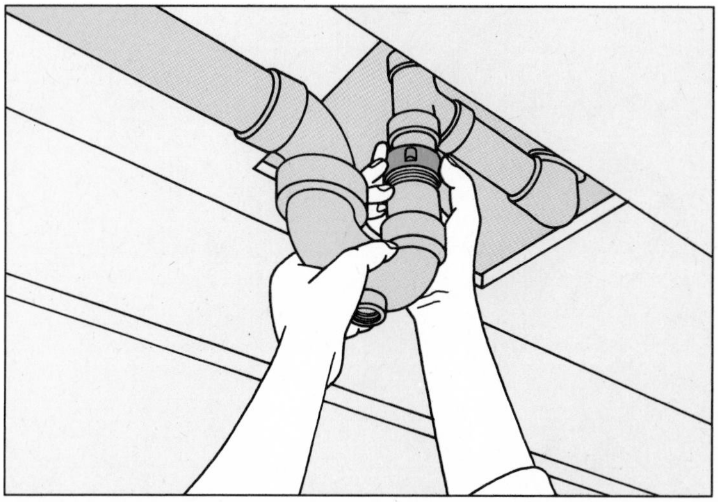

4 Reinstalling the trap. Fit new washers in the slip nuts, then lift the trap back in place *(above)* and tighten the slip nuts. Leave the access hole open for a few days. If leaks occur, tighten both nuts one-half turn with a wrench. If you have made a hole in the ceiling, install an access panel *(page 79)* or patch it over *(page 139).*

SERVICING DOUBLE-HANDLE FAUCETS

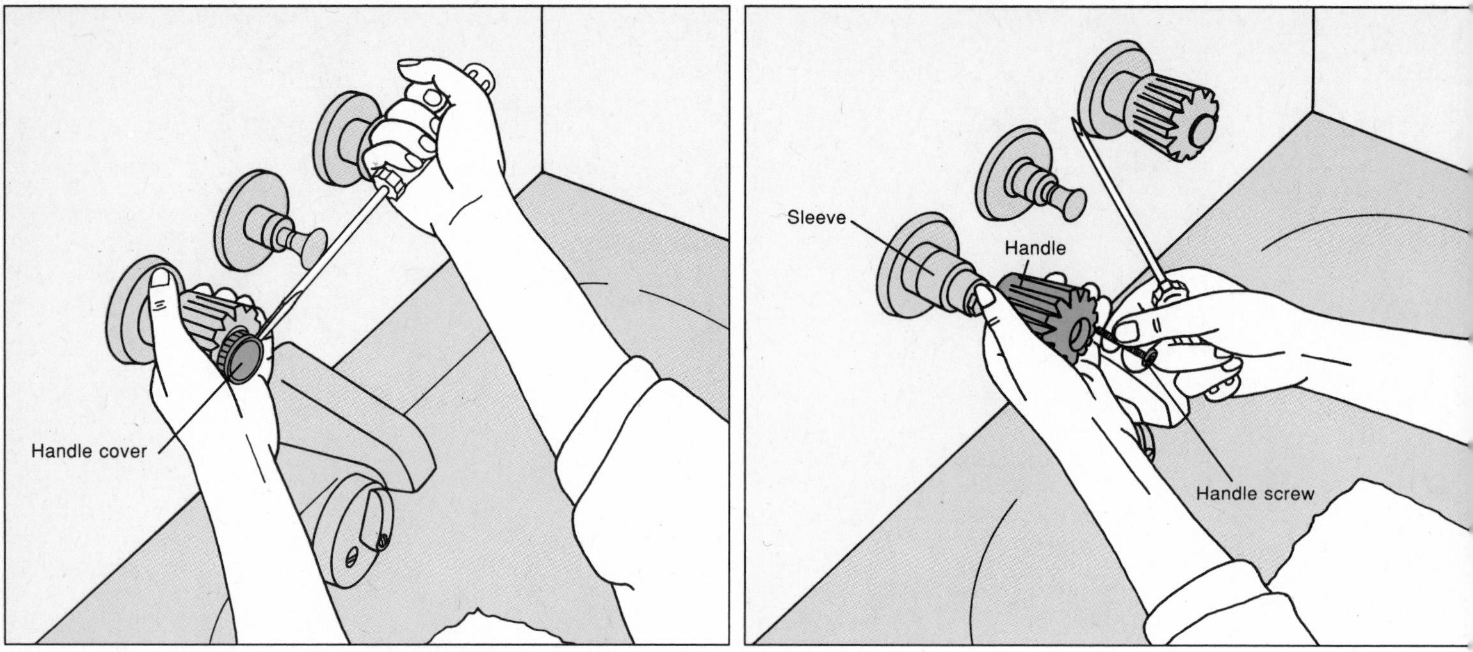

1 Removing the handle. Turn off the water supply and open the faucets. Close the drain and cover it with a bath mat to protect the tub and prevent loss of parts. Use a screwdriver or a knife to pry off the handle cover *(above, left).* Remove the screw, handle and sleeve *(above, right).* If the handle is stubborn, pour hot water over it and carefully pry up the base with a screwdriver. To facilitate reassembly, line up the parts on the mat (or along a strip of masking tape) in the same order as they are removed.

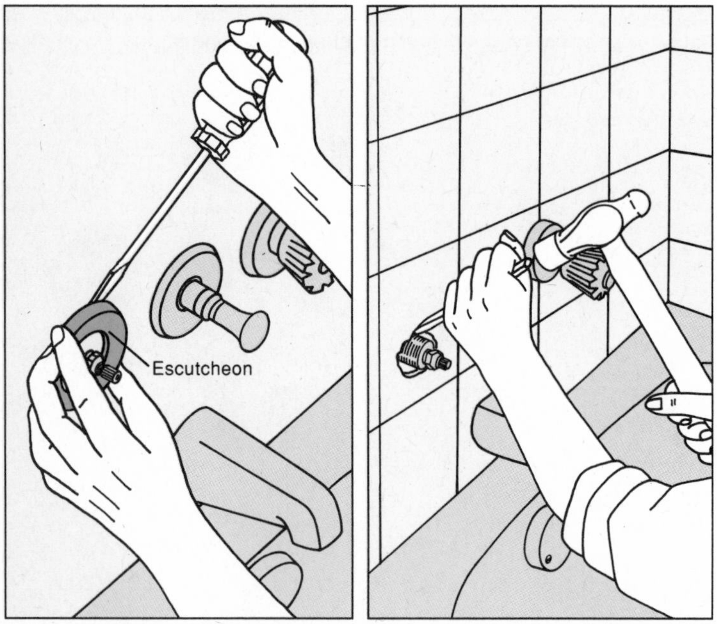

2 Getting at the stem. If there is a setscrew in the escutcheon, loosen it with a hex wrench. Use a screwdriver to pry the escutcheon from the wall *(above, left).* Tap a stubborn escutcheon with the screwdriver or spray it with penetrating oil and wait 15 minutes before trying again. If the bonnet nut on the stem is below the surface of the wall, trim back the tile, if any, and chip out any plaster or concrete *(above, right).*

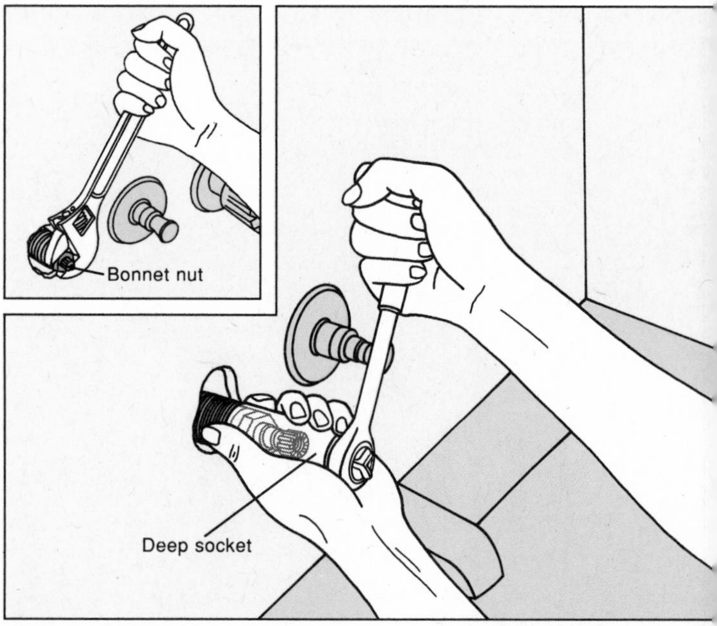

3 Removing the stem. Use a reversible ratchet with a deep socket to remove the bonnet nut and stem. Place the socket over the stem, gripping the nut and turning the ratchet counterclockwise *(above).* If the stem stays in the faucet, which it may in olde sets, unscrew and remove it. If necessary, spray on penetrating oil an wait 15 minutes before trying again. If you are using an adjustable wrench *(inset),* be careful not to damage the soft brass bonnet nut.

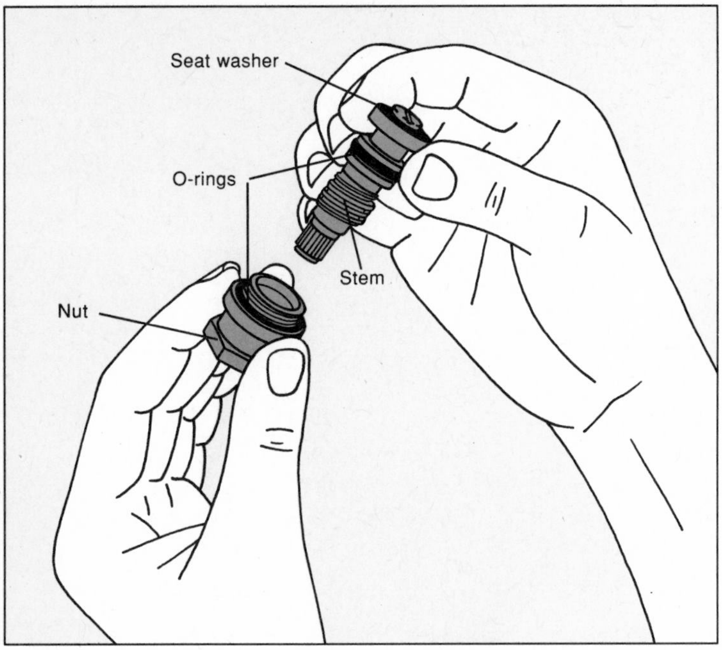

4 Curing leaks around the handle. Separate the stem from the nut, as shown. (For better leverage, reattach the faucet handle.) Pry off and replace the O-rings. If the stem appears worn, pitted or corroded, replace the entire assembly.

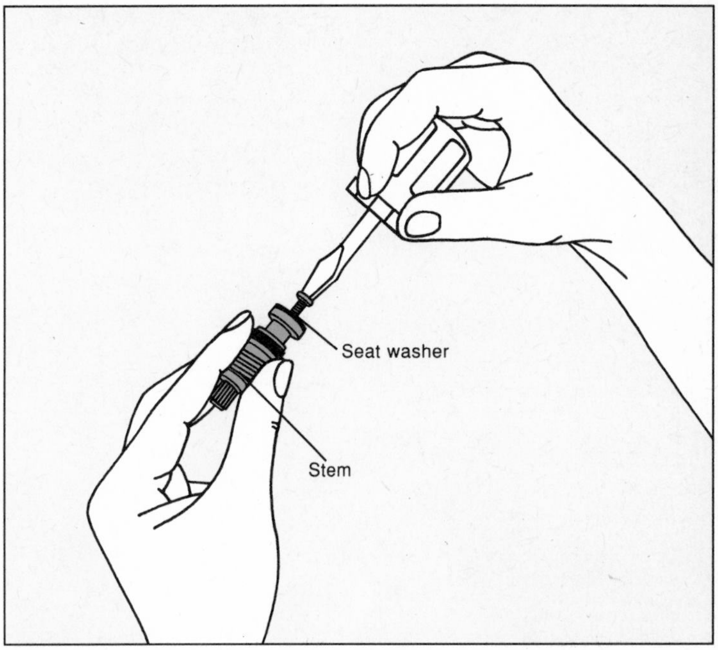

5 Curing leaks from the tub spout or shower head. Remove the screw that holds the seat washer to the stem *(above)*. Replace the old washer with a new one of the same size. While the stem is out of the faucet body, inspect the valve seat in the wall with a flashlight or feel it with your finger. If the seat is smooth and shiny, lubricate the stem lightly with petroleum jelly or silicone lubricant before setting it back in the faucet. If the seat appears damaged, replace it *(next step)*.

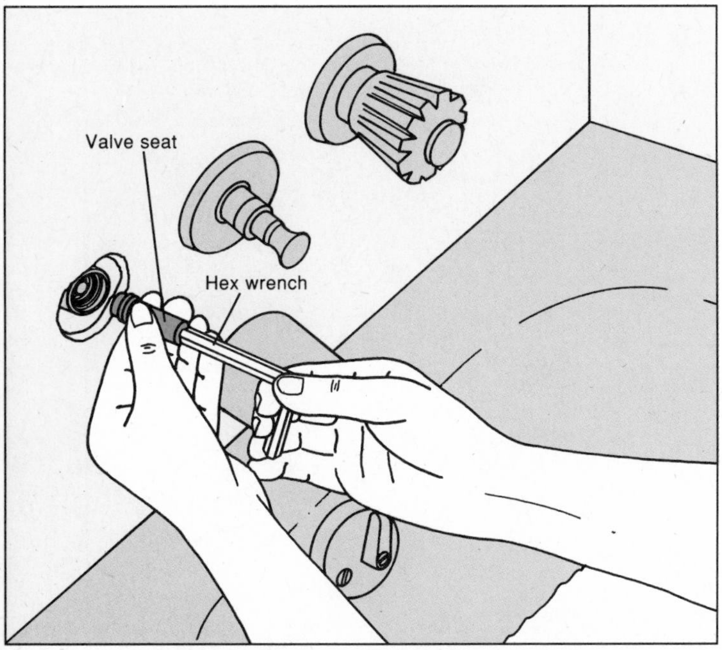

6 Replacing the seat. To remove a worn valve seat *(above)*, insert a valve-seat wrench or a large hex wrench into the faucet body, and turn it counterclockwise. Buy an exact replacement, lubricate the outside with pipe-joint compound, and screw it into the faucet body. If the seat cannot be removed (it may be built into the faucet), its surface must be ground smooth with a valve-seat dresser.

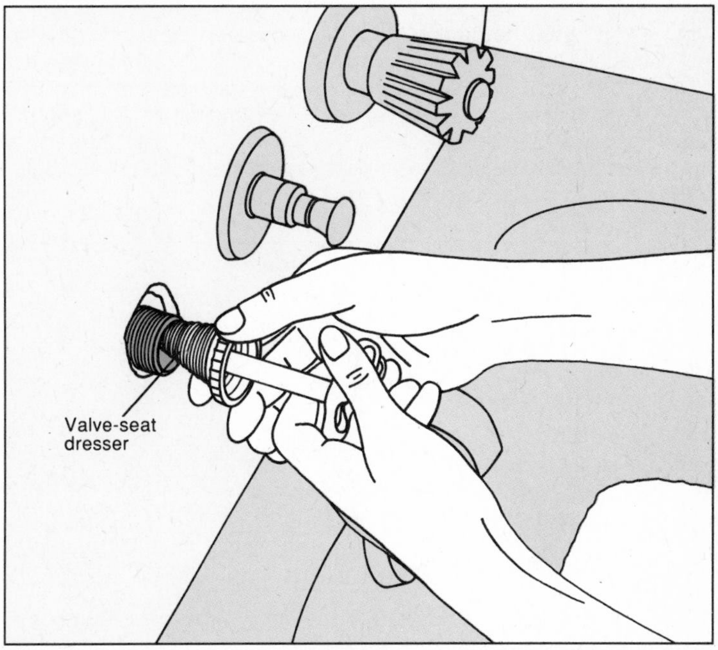

7 Dressing the seat. Insert the valve-seat dresser in the faucet until the cutting edge contacts the seat *(above)*. Turn the handle clockwise several times to grind the seat smooth. Remove metal filings with water and a damp cloth. Reassemble the faucet, putting back the parts in the reverse order in which they were removed. If water continues to leak from the spout or shower head for more than a few minutes, replace the faucet set *(page 78)*.

SERVICING SINGLE-LEVER FAUCETS (Ball-type)

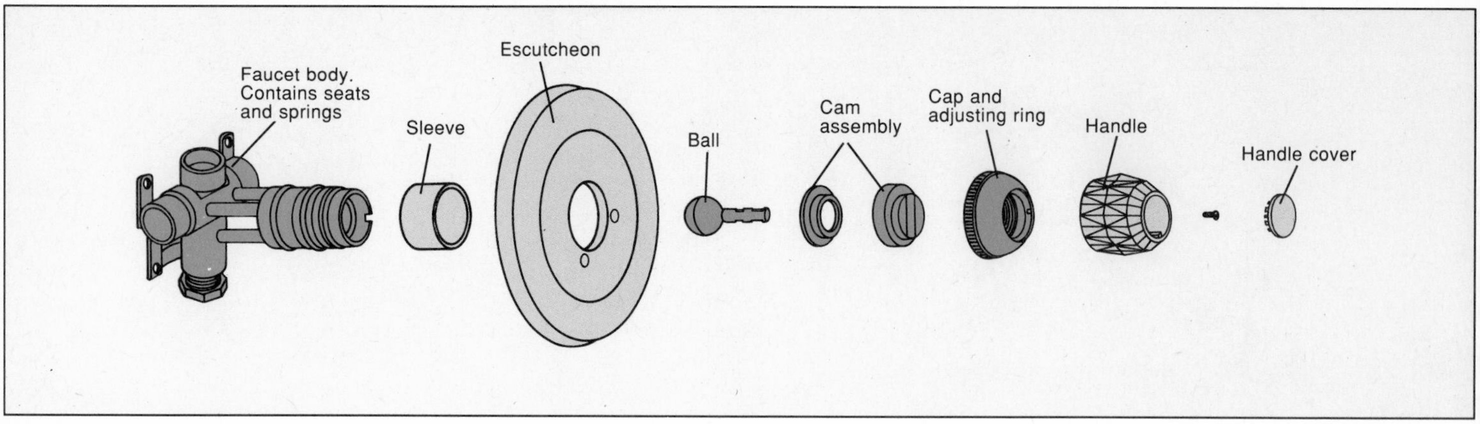

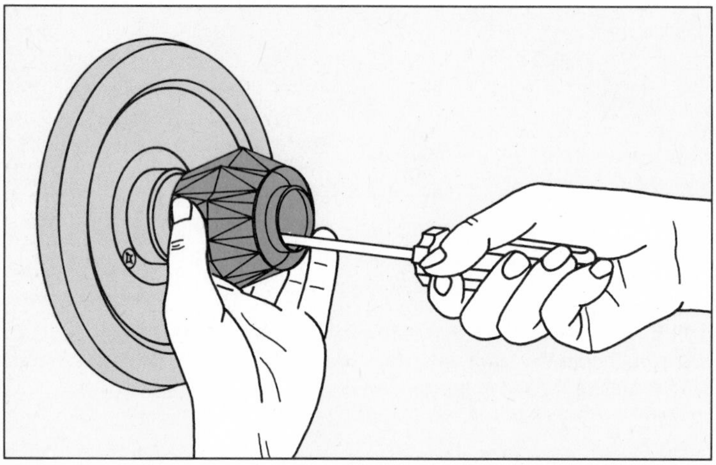

1 Removing the handle. Turn off the water supply and open the faucet. Close the drain and cover it with a bath mat to prevent loss of parts. Pry off the handle cover with a screwdriver or knife. Remove the handle screw *(above)*. and lift off the handle. To cure leaks from the spout or shower head, buy a kit of replacement parts for your faucet make and model, and replace the seats, springs and ball.

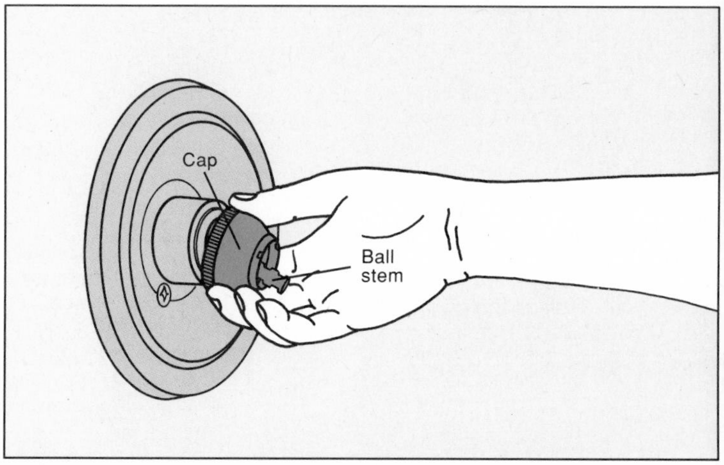

2 Removing the cap and ball. Unscrew and lift off the cap, *(above)*. Pull the ball stem to remove the cam-and-ball assembly that controls the mixture of hot and cold water.

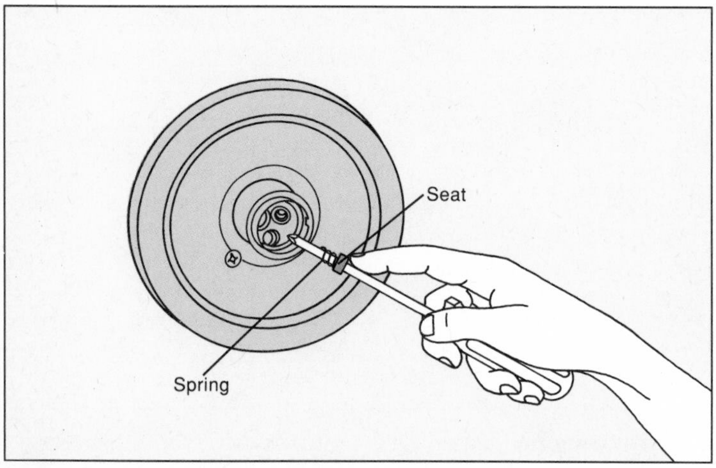

3 Replacing the seats and springs. Use a small screwdriver or a nail to lift the rubber seats and springs out of the two small sockets in the faucet body *(above)*. Set in new springs. If the springs are cone-shaped, insert the large end first. Place new seats over the springs. To reassemble the faucet, set the ball in the faucet body, adjust the cams over the ball, and screw on the cap.

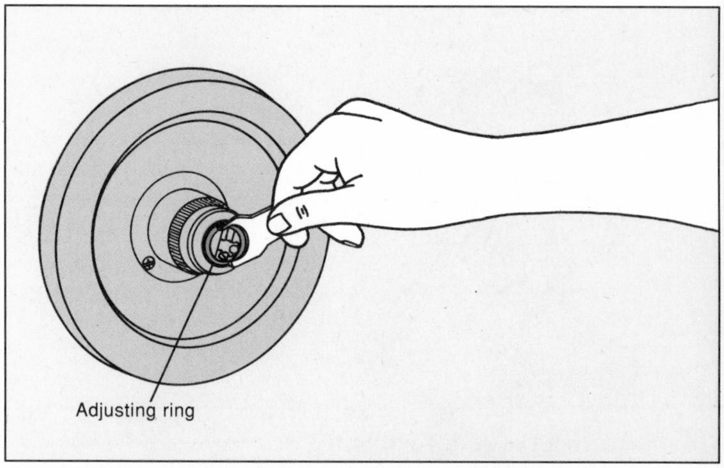

4 Tightening the adjusting ring. Using pliers or a special wrench often included in the ball-type replacement kit, turn the adjusting ring inside the cap clockwise *(above)*. Turn on the water supply. Use the ball stem to turn on the faucet. If water leaks from around the stem, tighten the ring. Replace the handle.

SERVICING SINGLE-LEVER FAUCETS (Cartridge-type)

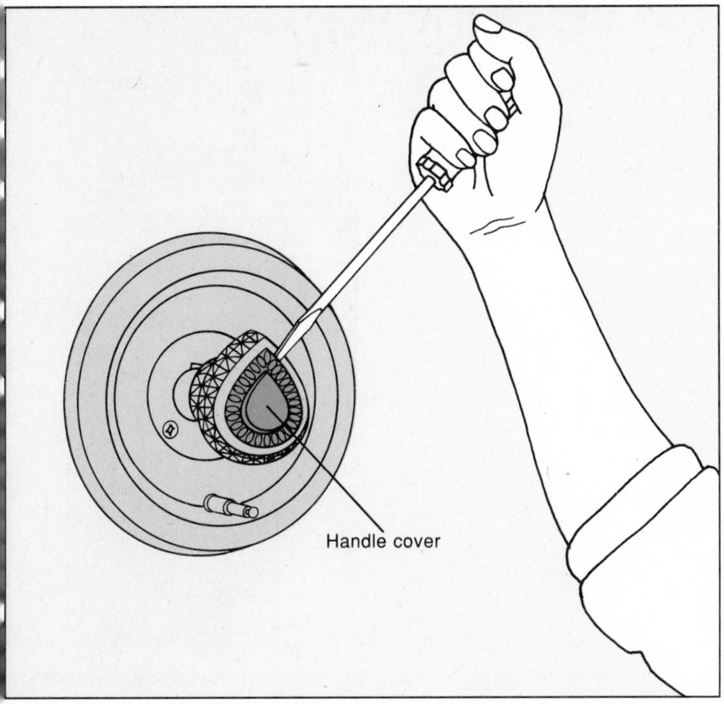

1 Removing the handle. Turn off the water supply and open the faucet to empty the supply lines. Close the drain and cover it with a bath mat to prevent loss of parts. Use a screwdriver or a knife to pry off the handle cover *(above)*, exposing the handle screw. Unscrew and remove the handle.

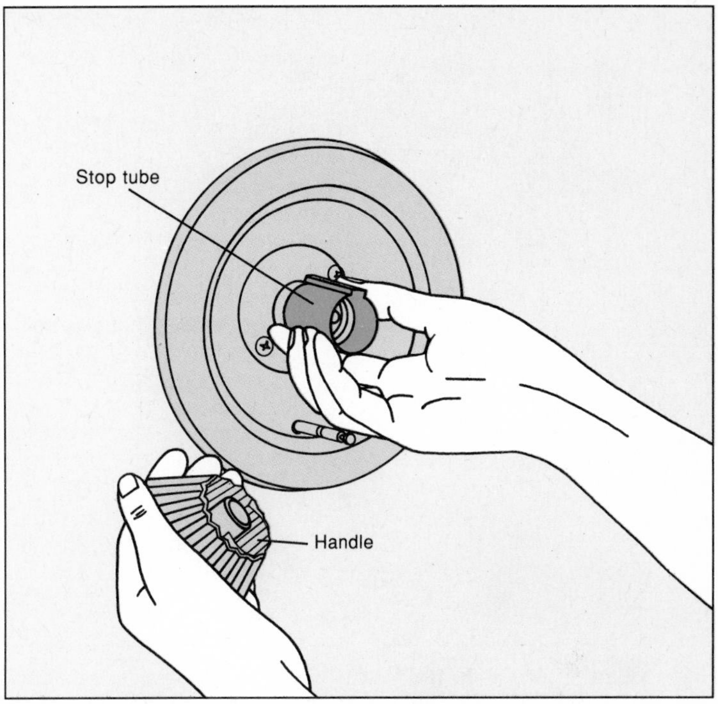

2 Getting at the cartridge stem. Lift off the handle and pull off the stop tube *(above)* to reveal the stem of the cartridge and its retainer clip, which secures the cartridge to the faucet body. (On some faucets, the escutcheon must first be removed.)

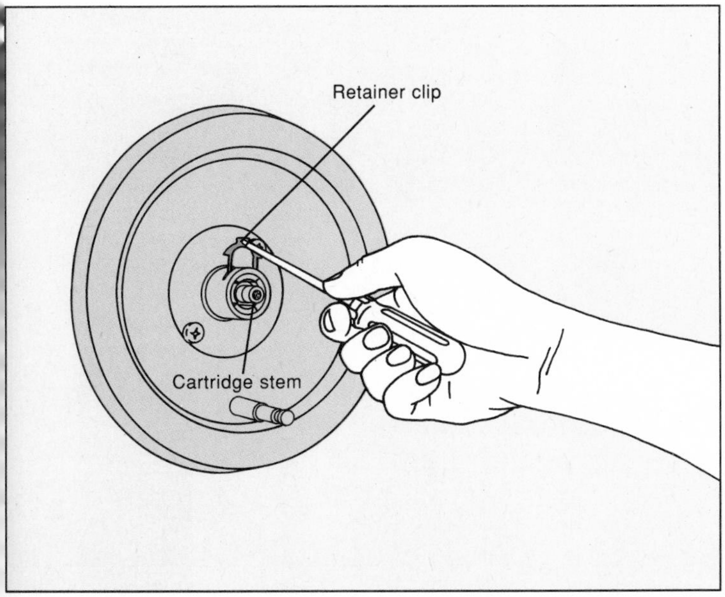

3 Freeing the cartridge. Using a small screwdriver or a nail, carefully pry up the cartridge retainer clip *(above)*. Grip the clip firmly, or it may spring away from the faucet.

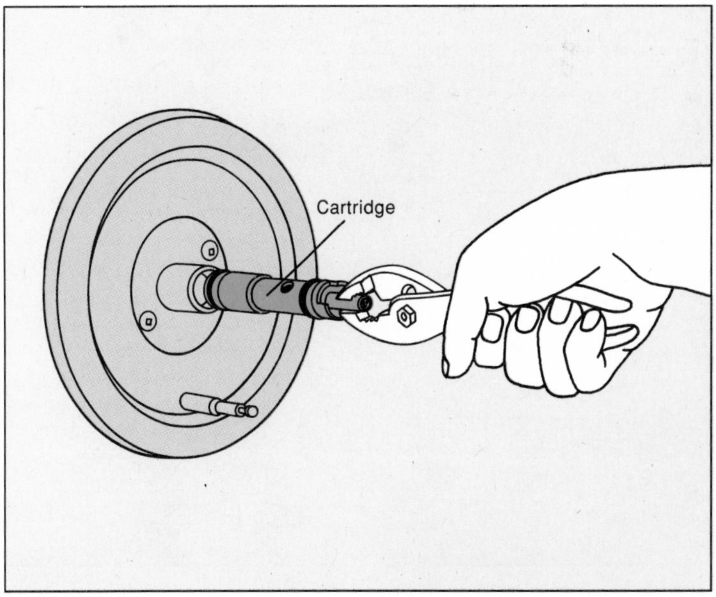

4 Replacing the cartridge. Grip the cartridge stem with pliers, then pull it free from the faucet body *(above)* by rotating it from side to side. Insert the new cartridge with the flat part of the stem facing up, or the hot and cold water will be reversed. Reinstall the retainer clip, stop tube and handle.

REPLACING A DOUBLE-HANDLE FAUCET

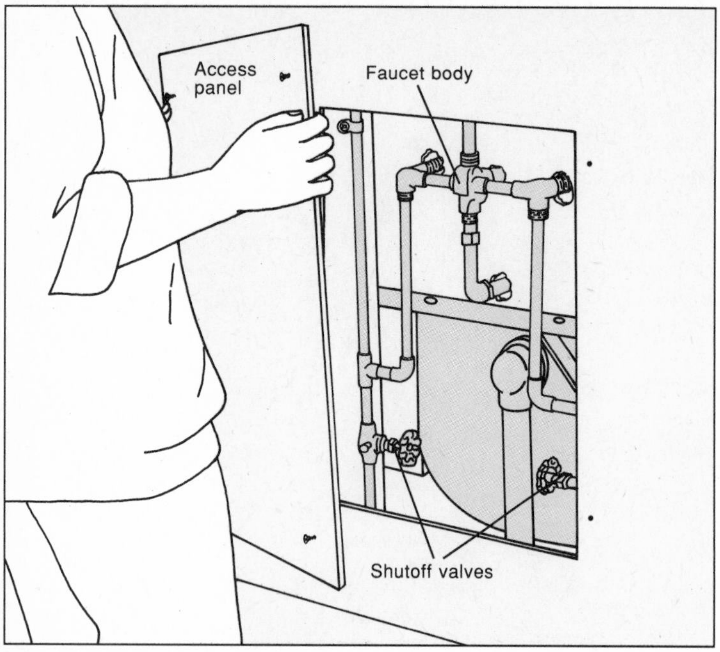

1 Gaining access to the faucet body. Replacing a faucet body is a full day's job that requires access to the plumbing as well as basic pipefitting skills *(page 93)*. Look for an access panel on the other side of the wall where the faucet is located, remove the screws and lift the panel away from the wall *(above)*. If there is no panel, you may be able to install one *(page 79)*. Otherwise, you will have to expose the faucet body from the tub side and repair the tile later. Turn off the water supply and open the faucets. Remove all handles, escutcheons, and the tub spout from the old faucet body.

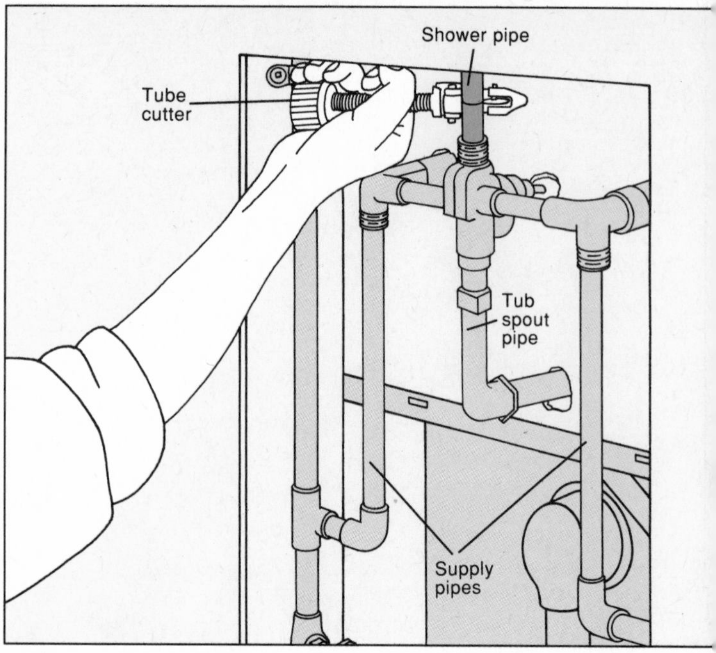

2 Cutting the pipes. When judging where to cut the copper pipe leading into the faucet body, choose points where the pipes can be easily reconnected using couplings and elbows. (Do not cut galvanized pipes; mark and unthread them instead.) Cut the shower pipe, as shown, and the hot and cold water supply pipes. Use a small tube cutter or a mini-hacksaw for cramped locations.

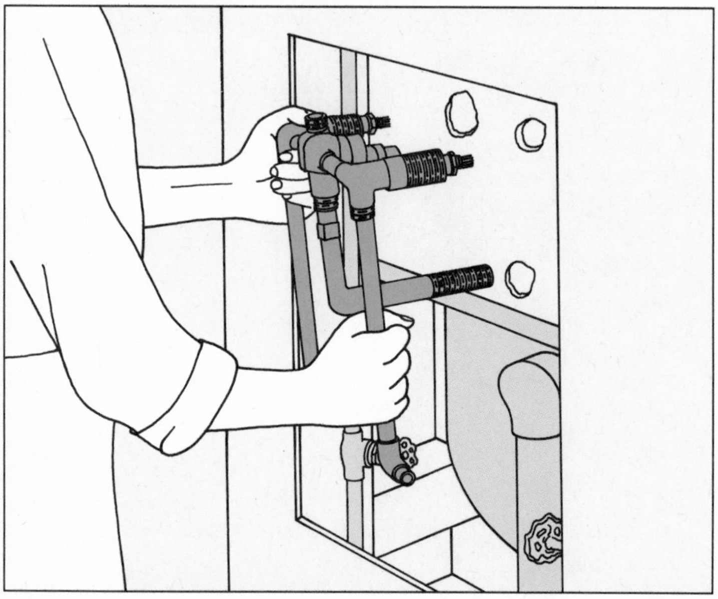

3 Removing the old faucet body. Ease the faucet body from the wall *(above)*. If it is stuck, chip away at grout and plaster from the front or remove silicone caulking with long-nose pliers. Take the faucet body, with the cut pipes still in it, to a plumbing supply store. Buy a replacement faucet body of the same make and model as the old one, or at least the same size. To calculate the amount of new pipe required, add the lengths exposed in the old faucet plus half an inch for each. Buy copper pipe and the necessary elbows and couplings.

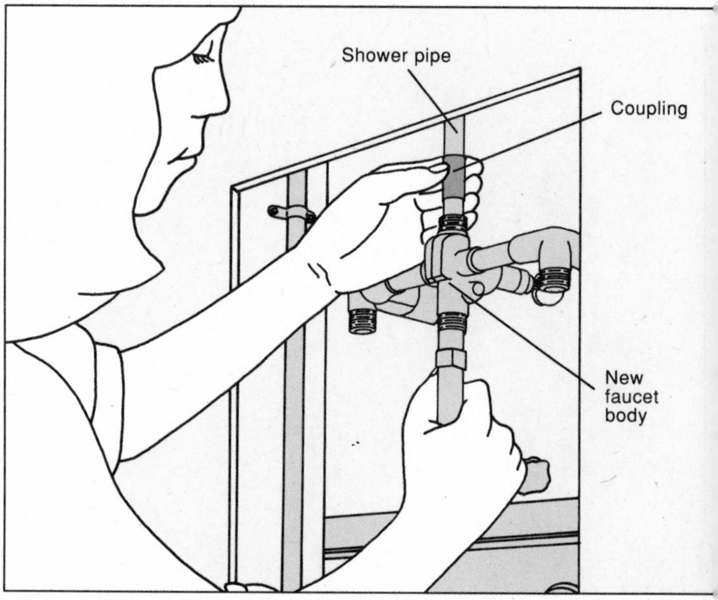

4 Connecting the shower pipe. Cut the lengths of pipe needed to connect the new faucet to the shower pipe and the supply pipes. Use a regular tube cutter, being careful not to cut any of the pipes too short. Set the pipes into the new faucet. Screw the spout nipple into the spout pipe. Place the faucet body inside the access panel, fitting the faucets and diverter, if there is one, through the holes in the wall. If any pipes are too long, cut off or file down as necessary. Use a coupling to attach the shower pipe to the faucet body *(above)*.

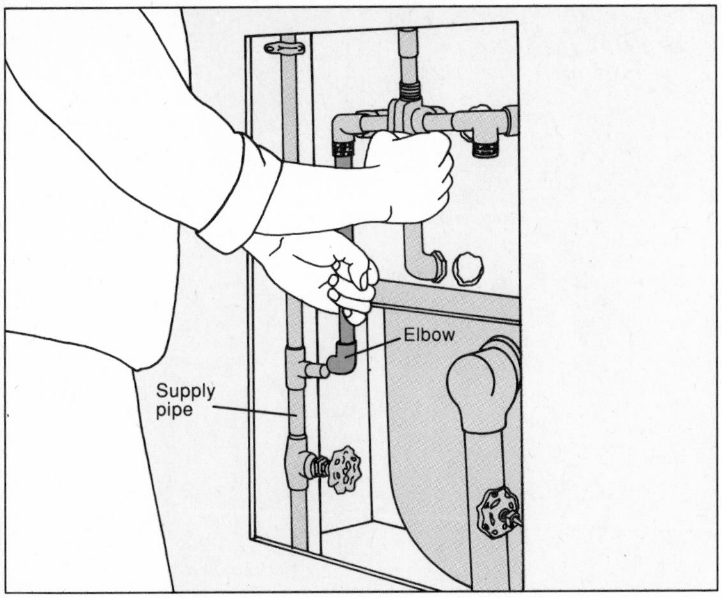

5 Connecting the supply pipes. Holding the faucet in place *(above)* connect one supply pipe to the faucet body, using a coupling or an elbow, depending on the configuration of the pipes. Then connect the other supply pipe.

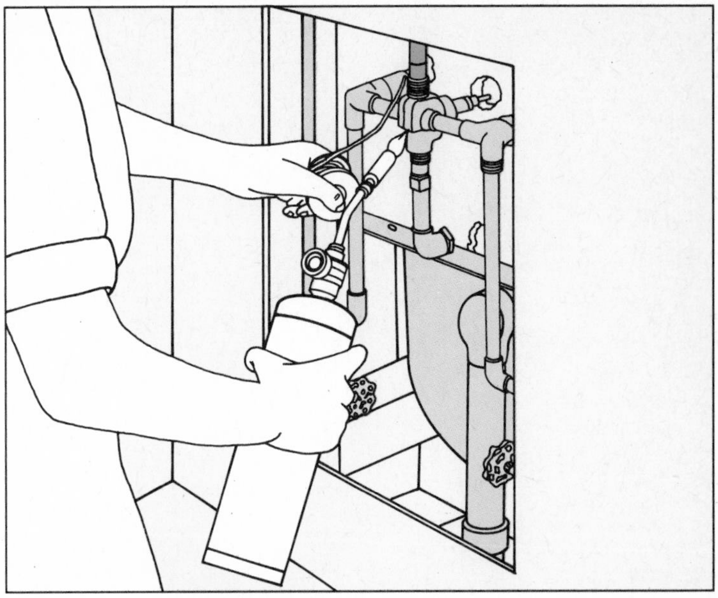

6 Soldering the pipes. Solder each of the joints *(page 140)*. From the front, slip on the escutcheons, attach the faucet handles and diverter knob, if any, and reconnect the tub spout. To prevent seepage behind the wall, seal the fixtures with plumber's putty or silicone. Leave the access panel open for a few days to check for leaks. If the pipes leak around the joints, resolder them.

ADDING AN ACCESS PANEL

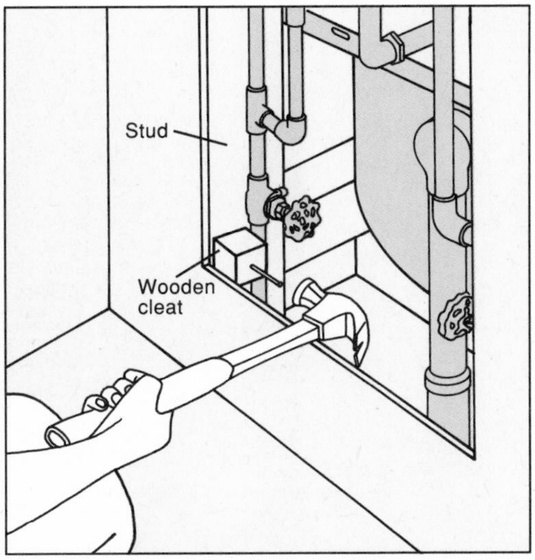

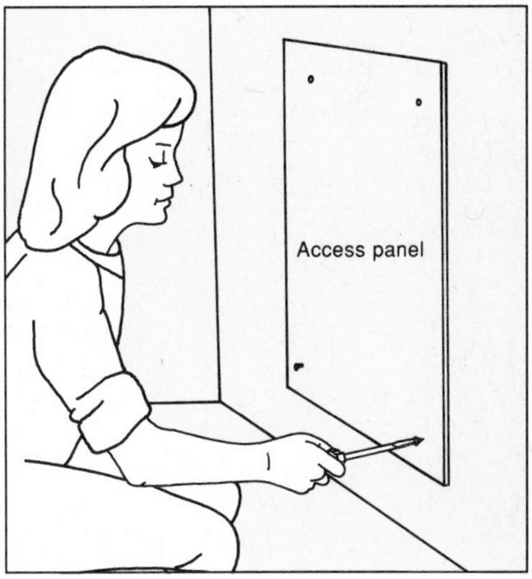

To determine the location of the faucet set in the wall, measure from the floor up to the faucet and from the corner of the tub wall out to the faucet. Transfer these measurements to the wall behind the bathtub. Use a hammer and chisel to make a hole at the appropriate spot in the wall. With a keyhole saw or drywall saw, cut from the hole out to the stud, then down *(above, left)*. Use a ruler to draw a rectangular box, stud to stud and about two feet high. Then continue cutting. Be careful not to cut through any pipes or wiring. Cut a plywood panel slightly larger than the hole in the wall. Nail wooden cleats to the studs *(above, center)* to form a lip for attaching the panel. Attach the access panel to the cleats with screws *(above, right)*.

REPLACING A SINGLE-LEVER FAUCET

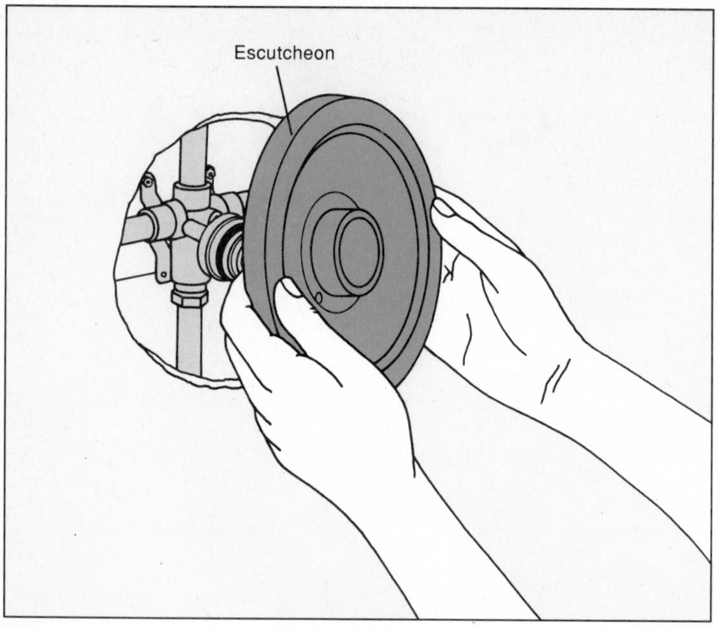

1 Gaining access to the faucet body. Turn off the water supply and open the faucet to empty the lines. Remove the handle *(page 77)*, then unscrew the escutcheon and lift it off *(above)*. If there is an access panel on the other side of the wall where the faucet is located, remove the panel and replace the faucet from the back. Follow the same procedure as for a double-handle faucet *(page 78)*. If the faucet set is inaccessible, as it sometimes is because of a crosspiece, or there is no access panel, it may be possible to replace the faucet set from the tub side.

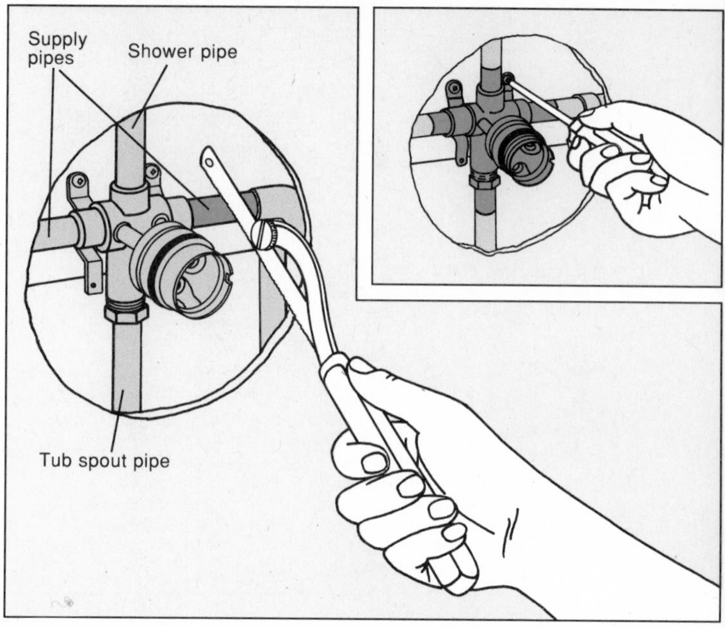

2 Removing the faucet set from the tub side. Cut the copper shower pipe, spout pipe, and supply pipes with a mini-hacksaw *(above)*. You may have to chip away tiles or wallboard for easier access. Remove the screws that secure the faucet body to the wooden crosspiece behind it *(inset)*. Lift the faucet set out of the wall. Buy a replacement faucet of the same make and model as the old one, or at least the same size. To calculate the total pipe length required, add the lengths of the four pipes in the old faucet plus at least half an inch for each length. Buy copper pipe and the necessary fittings.

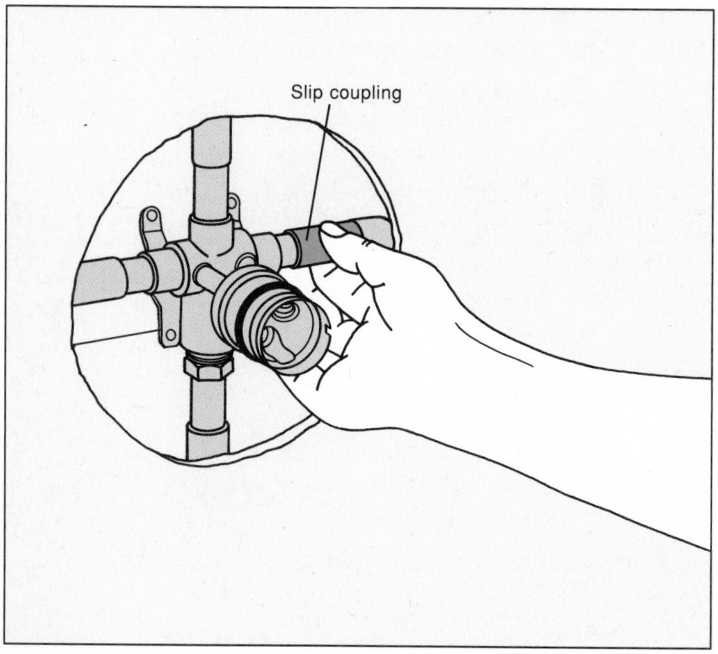

3 Reconnecting the pipes. To cut, connect and solder copper pipe (or to replace copper with plastic pipe), see page 93. Cut the new pipe into four lengths to connect the shower pipe, spout pipe and supply pipes, and ream them carefully. Position the faucet in the wall. Connect the pipes with slip couplings *(above)*.

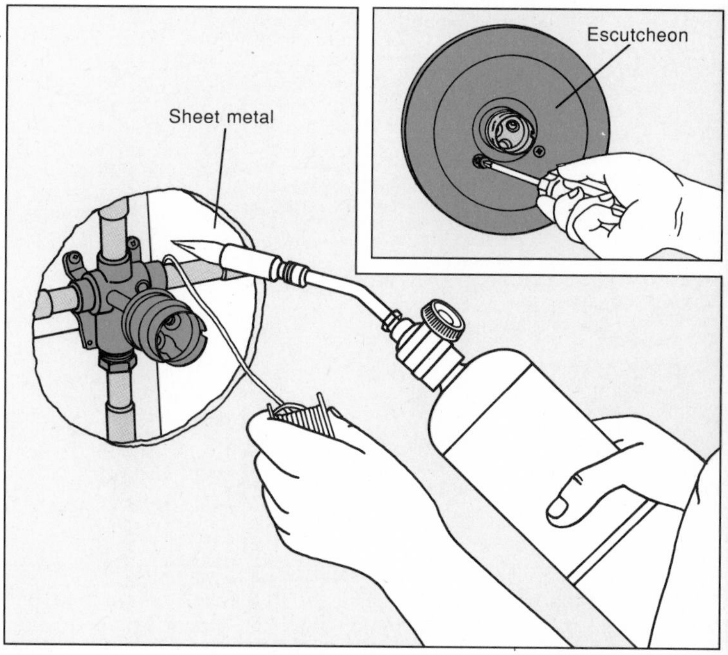

4 Completing the installation. Secure the faucet body by screwing it to the crosspiece. To avoid a fire hazard, slide a small strip of thin sheet metal or fireproof material behind the soldering points. Remove all plastic and rubber parts, then solder the joints *(above)*. Screw on the escutcheon *(inset)* and reconnect the handle. After using the faucet for several days, remove the handle and escutcheon to check for leaks and resolder if necessary.

REPLACING A SCREW-ON TUB SPOUT

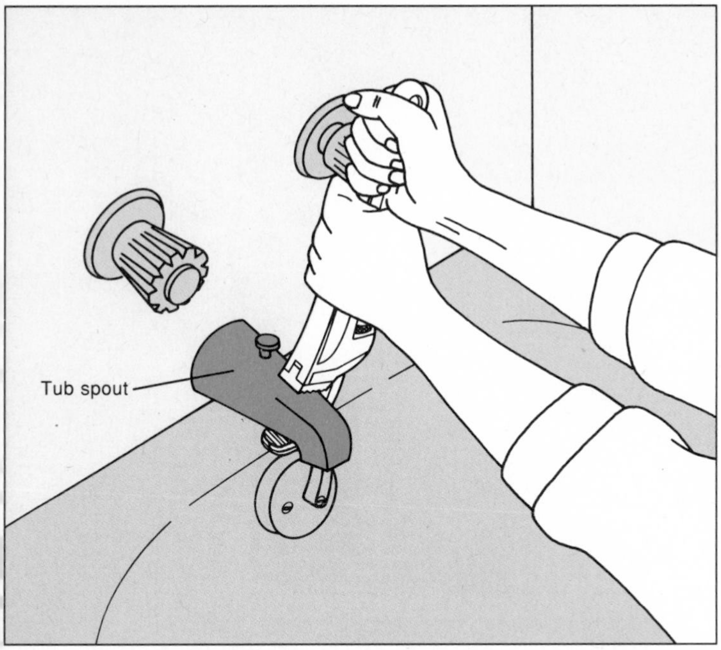

1 Loosening the spout. If the tub spout contains a diverter valve that is not completely diverting water from the spout to the shower head, replace the tub spout. Check under the spout for a setscrew; if there is one, it is a slip-fit spout *(page 82)*. Otherwise, grip the spout with a pipe wrench and turn counterclockwise to loosen it *(above)*. If the spout will not move and there is access from behind, apply penetrating oil, wait 15 minutes and try again. Do not use too much force, or you might damage the plumbing behind the wall.

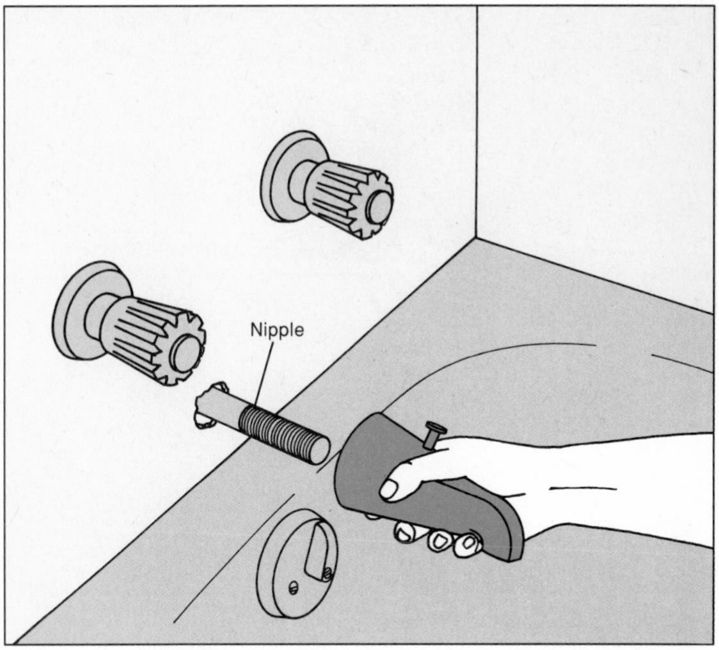

2 Removing the spout. Twist the loosened spout off the nipple by hand *(above)* or off the adapter at the end of a copper pipe. If you can find a replacement spout, go to step 4. If a compatible spout is unavailable, go to step 3.

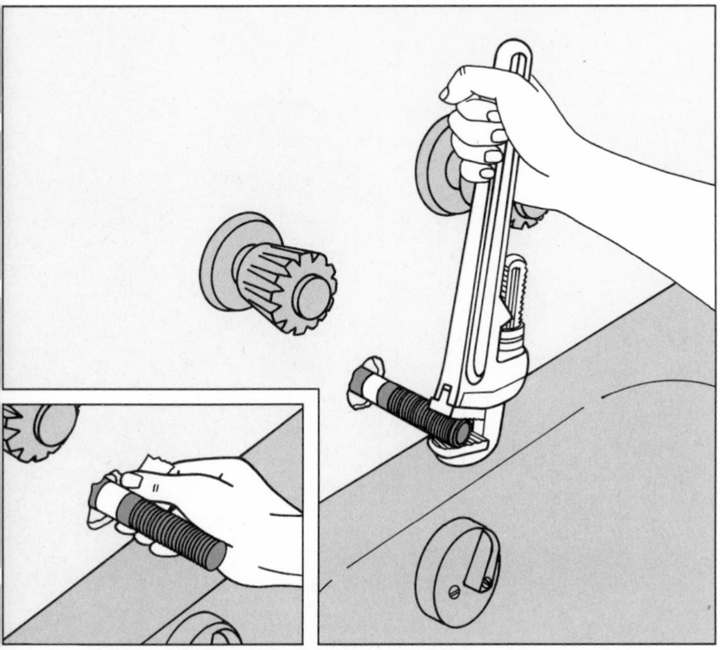

3 Removing the nipple. Wrap masking tape around the nipple to mark the point where it protrudes from the wall *(inset)*, then unscrew it with a pipe wrench *(above)*. Take the nipple and tub spout to a plumbing supply store. Fit the nipple into a new spout and measure the difference between the tape and the edge of the spout. Buy a threaded brass nipple that is the length of the old one minus this difference, or have one cut to that length. Apply pipe tape to the threads, then screw the nipple into the wall by hand.

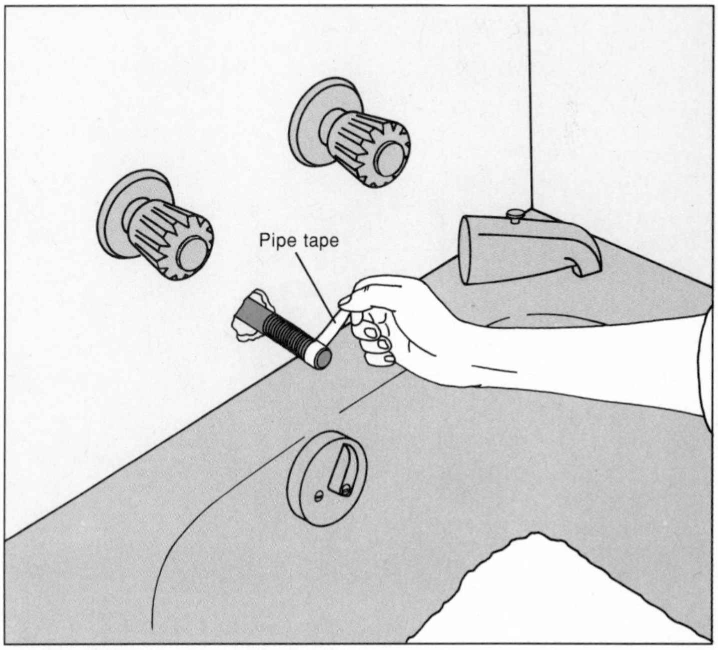

4 Mounting the tub spout. Clean the threads of the nipple with steel wool or a wire brush. Apply pipe tape to the threads (unless the directions on the new spout advise otherwise) and silicone sealant to the base of the spout. Thread the spout onto the nipple and tighten it by hand, then wrap the tub spout with masking tape and tighten with a pipe wrench.

REPLACING A SLIP-FIT TUB SPOUT

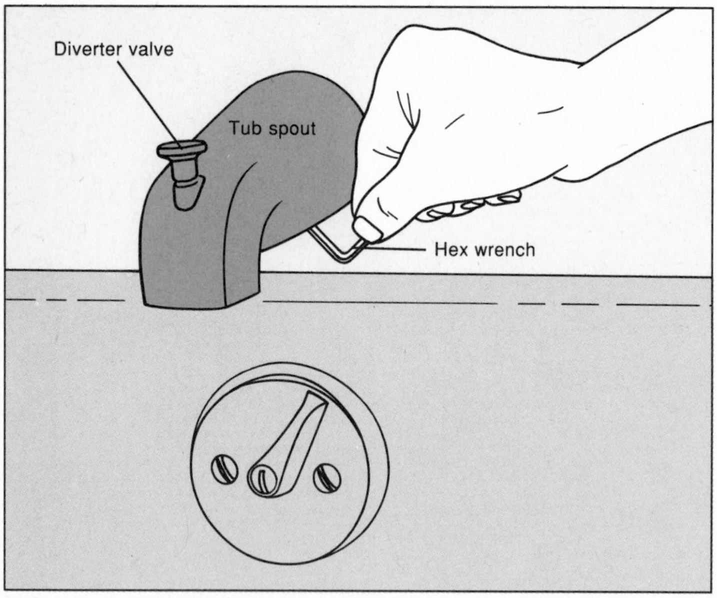

1 Removing the old spout. Loosen the clamp screw on the underside of the spout with a hex wrench *(above)*. Grasp the spout firmly and twist it off the copper pipe.

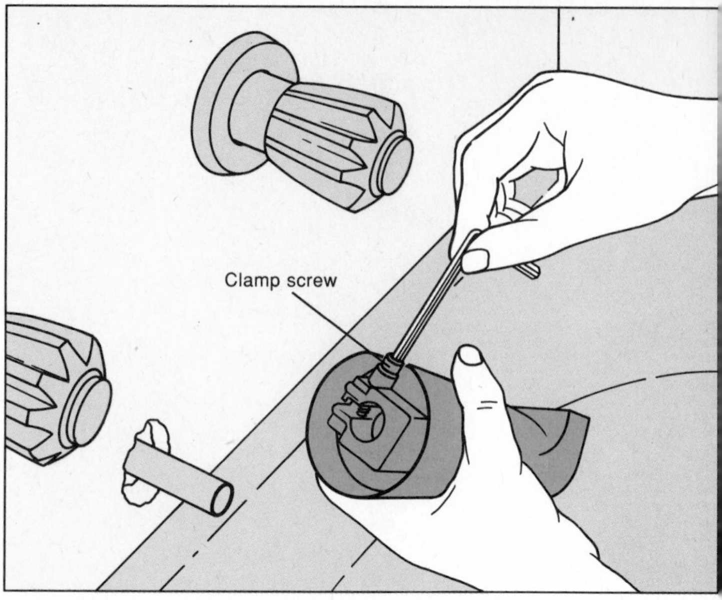

2 Mounting the new spout. Loosen the clamp screw on the new spout with a hex wrench *(above)*, and twist the spout onto the copper pipe. Turn the spout so that the clamp screw faces up, and partially tighten the screw. Twist the spout into position and finish tightening the screw with the hex wrench. Do not overtighten.

REPAIRING PUSH-PULL DIVERTERS

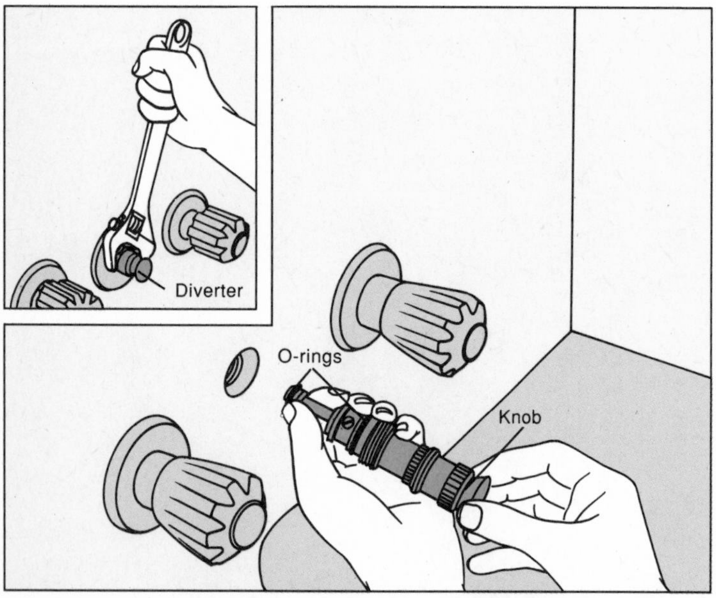

Repairing a push-pull diverter (double-handle faucets). Turn off the water supply and open the faucets. Close the drain and set a bath-mat down to prevent loss of parts and protect the tub. To remove a diverter located between two faucet handles, wrap it in tape and unscrew it with an adjustable wrench *(inset)*, exposing two O-rings. Unscrew the diverter knob by hand *(above)* to reveal another O-ring and a spring. Clean the spring and lubricate it lightly with petroleum jelly or silicone lubricant. Replace any O-rings that appear cracked or worn. If problems persist, replace the entire diverter with an identical part.

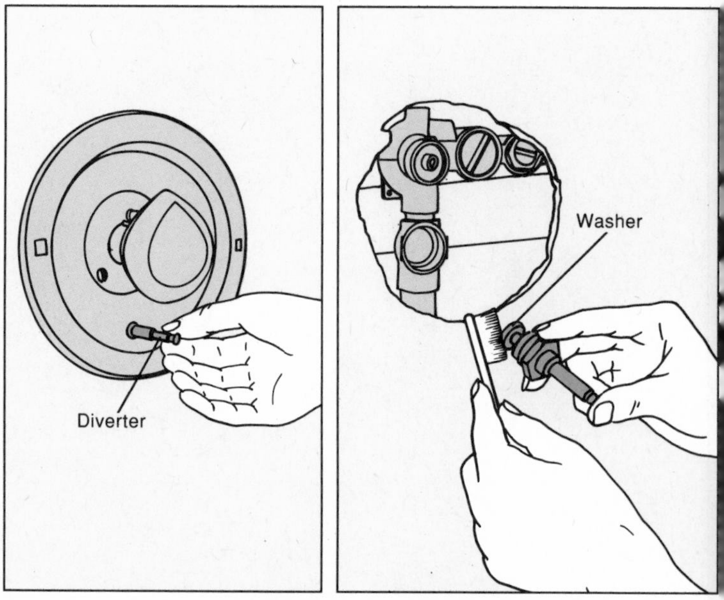

Repairing a diverter (single-lever faucets). Remove the faucet handle and escutcheon, then unscrew the diverter. If water is not being properly diverted from the tub spout to the shower head, clean any sediment off the washer with vinegar and an old toothbrush *(above, right)*. If water leaks from around the diverter, or if its parts are worn, replace the entire mechanism with one of the same make.

REPAIRING A SHOWER HEAD

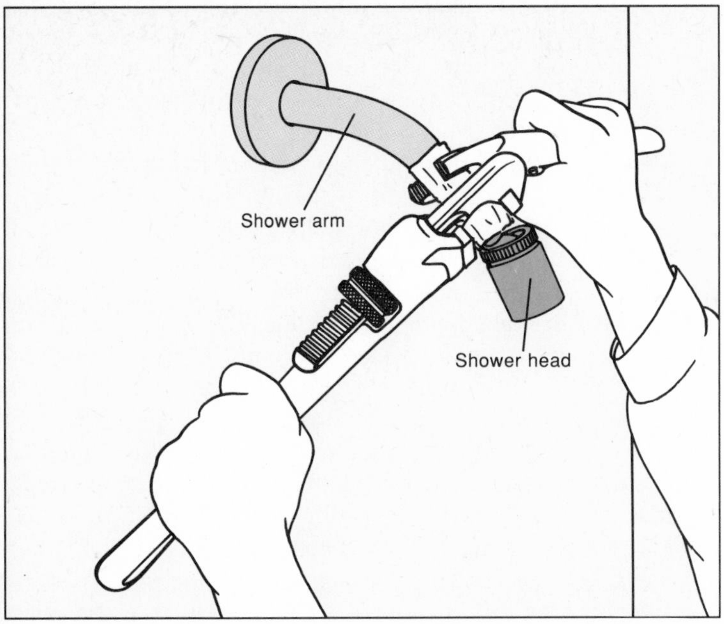

1 Removing the shower head. When the flow of water from a shower head is uneven or insufficient, first disassemble and clean it. Close the drain and cover it with a bath mat to prevent loss of parts and to protect the tub. Wrap the shower head collar in masking tape and turn it counterclockwise with a pipe wrench. Or use a strap wrench, which requires no tape *(page 136)*. For greater leverage, grip the shower arm with one wrench and turn the collar counterclockwise with a second wrench *(above)*. Twist off the loosened shower head by hand.

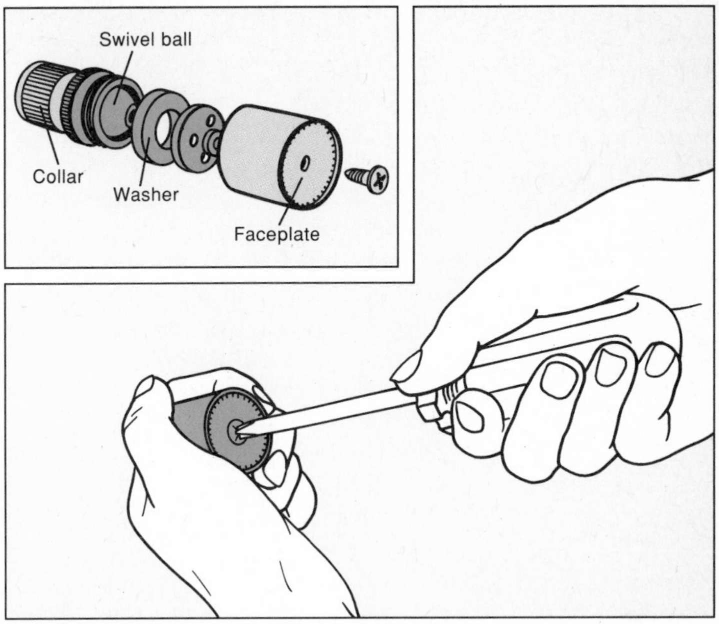

2 Disassembling the shower head. Remove the screw *(above)* or the knob that secures the faceplate to the shower head. Unscrew the collar from the shower head to reveal the swivel ball, and pry out the washer. To clean the disassembled parts shown in the inset, go to step 3. Go to step 4 to replace a shower head whose parts appear badly worn or corroded.

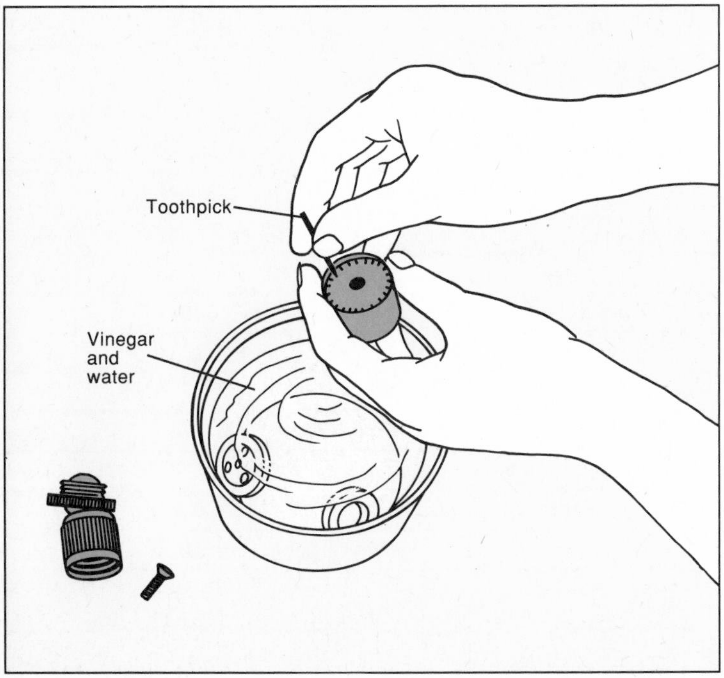

3 Cleaning the shower head. Soak the entire shower head or its disassembled parts overnight in vinegar. Scrub with steel wool or an old toothbrush, and clear the spray holes with a needle or toothpick *(above)*. If possible, buy replacements for the worn parts of an expensive shower head rather than buy a new one. Lubricate the swivel ball with petroleum jelly or silicone lubricant and reassemble the parts in the reverse order.

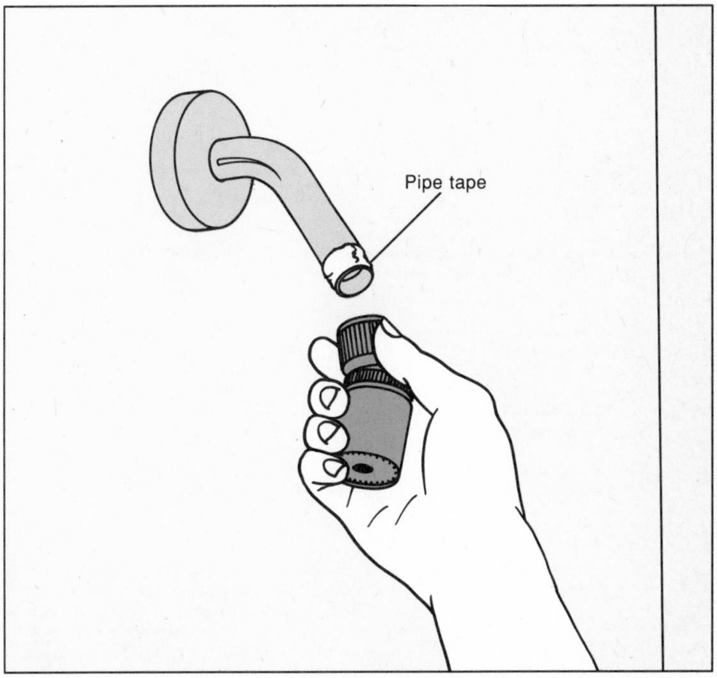

4 Installing the shower head. When replacing an old shower head with a new one, consider water-saving heads, heads with adjustable spray, and heads with plastic parts that collect less sediment. Clean the pipe threads of the shower arm with steel wool or a wire brush. Apply pipe tape to the threads to seal the joints. Hand-tighten the head on the shower arm *(above)*. Turn on the water to test the shower; if it leaks, tighten it an additional half-turn.

MAIN DRAINS

The home drainage system is called the drain-waste-vent (DWV) system because of its three major functions: drainage of "gray water," waste removal and venting of sewer gas, all of which are interconnected. The DWV system is completely separate from the supply system and, because it must maintain modern sanitation standards, it is closely regulated by plumbing codes. Consult local authorities regarding these codes before working on your home drainage system.

Under each sink, bathtub and toilet is a P- or S-shaped trap containing water, which acts as a seal to prevent waste and gases from rising out of the drainage system and into the home. Beyond the traps, gravity carries the waste along horizontal branch drains to the main soil stack. Most fixtures and appliances are clustered around this main drainpipe *(page 14)*. Local plumbing codes regulate how many fixtures can empty into a stack, so a house may have more than one soil stack.

The portion of the soil stack above the branch drains is called the vent, a single pipe rising above the roof line. Air pressure in the vent prevents fixtures from being siphoned dry each time they are used. A common problem is snow or leaves blocking the vent, creating a vacuum that stops waste water from draining properly farther down the system. In the harsh winters of some parts of the United States and Canada, snow and ice would soon block the standard vent; codes there call for oversize pipes, often fitted with cages or screens.

At the bottom of the soil stack, the pipe makes a 90-degree turn to become the main drain, which slants to a public sewer or private septic system. The main cleanout, which provides access to the main drain, is located near this turn. In warmer climates, or in houses without basements, the cleanout may be outside, above the point where the main drain leaves the foundation. Another access point to the main drain is the house trap, often found on older homes close to the foundation wall.

Two or more clogged or sluggish drains point to blockage somewhere in the branch drains, soil stack, main drain or sewer line. Often the hardest step in dealing with such a problem is gaining access to the drainpipe. Opening a rusted cleanout plug or house trap may be so difficult that it is easier to work down from the vent stack on the roof. (However, if the roof is high or steeply pitched, it may be best to hire a plumber for this job.) When you can open the cleanout or house trap, a garden hose, especially with a rubber expansion nozzle, or a manual auger will usually clear the drain. Power augers are available from plumbing supply or rental companies, but you should consider this as a last resort, and only if you have experience with heavy power tools. Sometimes a blockage affecting the entire drainage system is caused by a crack in the sewer pipe. Repairing the pipe is a formidable job, but an experienced amateur plumber can do this too, after informing the local authorities.

TROUBLESHOOTING GUIDE

SYMPTOM	POSSIBLE CAUSE	PROCEDURE
Drain sluggish or backed up	Trap or branch drain clogged at fixture	Clear trap or drain at fixtures: Kitchen sinks *(p. 36)* Bathroom sinks *(p. 46)* Toilets *(p. 56)* Bathtubs and showers *(p. 68)*
Two or more drains backed up	Drainage system blocked below fixtures	Unblock main drain from main cleanout *(p. 85)* ⬓◒ Unblock house trap *(p. 88)* ⬓◒ Unblock soil stack from roof vent *(p. 88)* ⬓●
Odors or gurgling from drains	Vent blocked	Unblock vent from roof *(p. 88)* ⬓●
	Septic tank full	Have septic system serviced professionally *(p. 92)*
Basement floor drain backed up	Main drain blocked	Unblock main drain from main cleanout *(p. 85)* ⬓◒
	House trap or sewer line blocked	Unblock house trap or sewer line *(p. 88)* ⬓◒
All drains sluggish or backed up	Main drain blocked	Unblock main drain from main cleanout *(p. 85)* ⬓◒
	House trap or sewer line blocked	Unblock house trap or sewer line *(p. 88)* ⬓◒
	Leak in sewer line	Replace damaged section of sewer line *(p. 90)* ■●▲
	Septic tank full	Have septic system serviced professionally *(p. 92)*
	Main drain or sewer line has insufficient slope	Call for service

DEGREE OF DIFFICULTY: □ **Easy** ⬓ **Moderate** ■ **Complex**
ESTIMATED TIME: ○ **Less than 1 hour** ◒ **1 to 3 hours** ● **Over 3 hours** ▲ **Special tool required**

ACCESS TO THE MAIN CLEANOUT

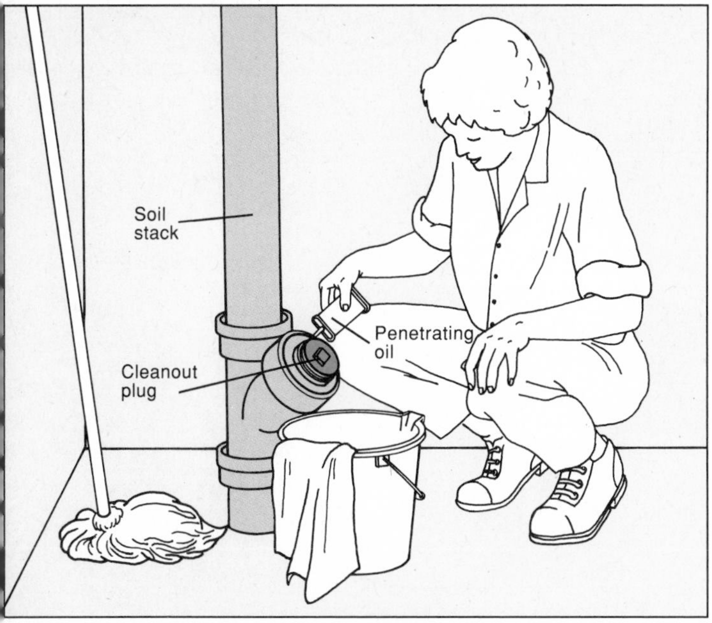

1 Loosening the cleanout plug. The main cleanout is usually near the bottom of the soil stack, in the basement or crawlspace, close to where the stack makes a 90-degree turn to leave the house. If the main cleanout is hard to find, locate the vent on the roof and follow it down to the corresponding spot in the basement, crawlspace or lawn. If the cleanout is difficult to reach, search for a house trap *(page 88)*. Before opening the cleanout, either shut off the main water supply or do not use any water in the house. Allow waste to drain out of the system, preferably overnight, so that water does not pour out after the cleanout plug is off. Have a mop, rags and a bucket ready. If the cleanout plug is metal, it may be stubborn to remove. Apply penetrating oil to the threads, as shown, waiting overnight if possible before trying to remove it.

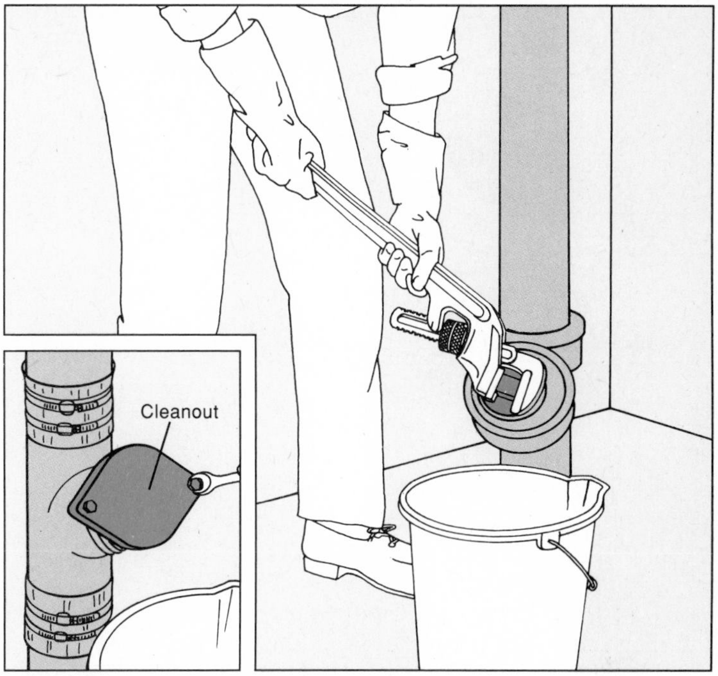

2 Removing the cleanout plug. Loosen, but do not remove, the cleanout plug with a large pipe wrench *(above)*. If the nut has been rounded off from previous repairs, first file it square with a metal file. If it still does not loosen, go to step 3. Let some of the trapped water ooze out into the bucket, tighten the nut and mop up, repeating the process until no water remains. Clear the blockage with an auger or hose *(page 86)*.

Some cleanouts have a metal cover secured by two brass bolts. Apply penetrating oil, wait a few minutes and use an open-end wrench *(inset)* or adjustable wrench to loosen the bolts.

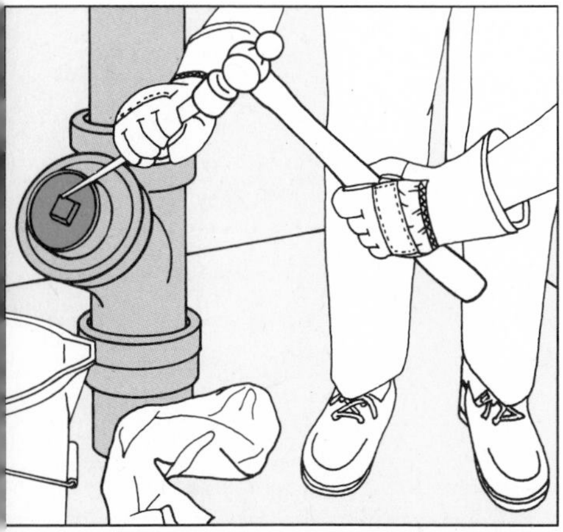

3 Loosening the plug with a hammer and chisel. If the plug remains stubborn, place a cold chisel on one edge of the nut and tap it firmly counterclockwise with a ball-peen hammer, as shown, then move to the next face. Continue hammering until the plug is loose enough to turn with a wrench. If the nut remains stubborn, apply heat with a propane torch *(next step)*.

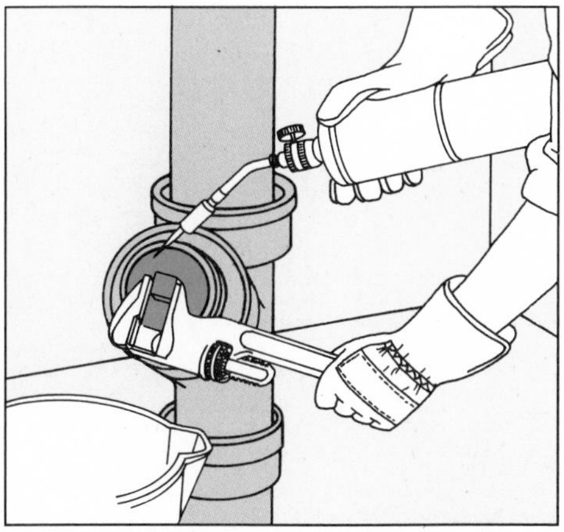

4 Applying heat to the plug. If the plug will not move because it has rusted in place, try burning off the rust with a propane torch. The heat needed to loosen the plug dissipates very quickly, so hold the wrench with one hand while you play the torch over the rusted fitting with the other hand *(above)*.

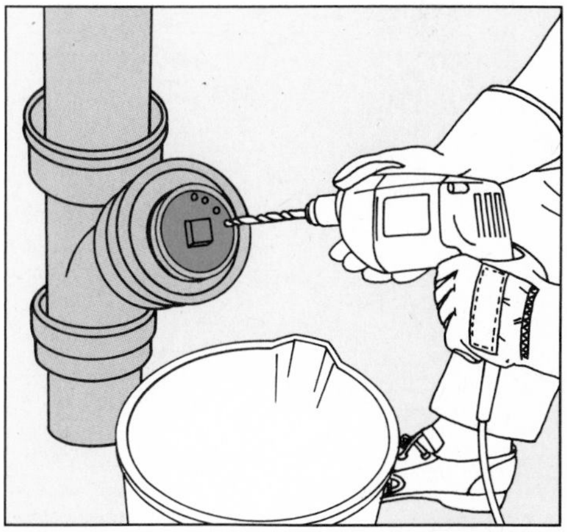

5 Breaking the plug. As a last resort, you may have to destroy a rusted cleanout plug and replace it with a new one. Using a power drill with a 3/8-inch metal bit, drill a ring of holes within 1/4 inch of the edge of the plug *(above)*. With a hammer and chisel, knock out the center of the plug, then the pieces between the holes. Do not let the pieces fall down the drain. Replace the plug after you have serviced the drain.

UNBLOCKING THE MAIN DRAIN

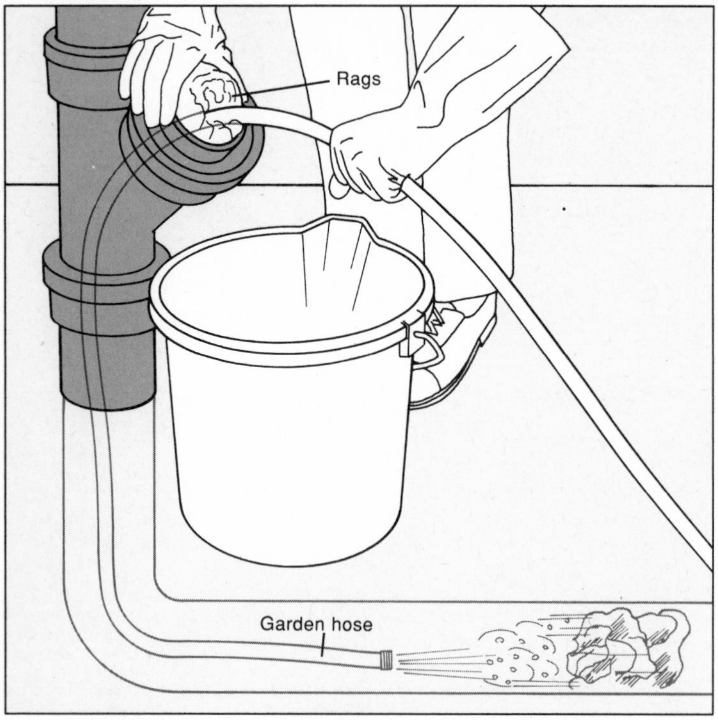

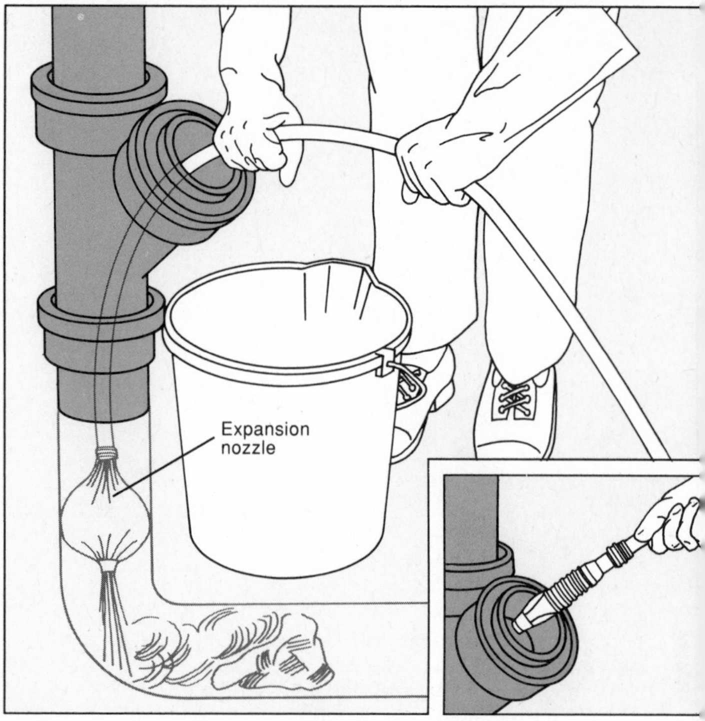

1 Using a hose to break up small blockages. Insert a garden hose as far as possible into the cleanout, or until you hit the blockage. Pack large, wet rags around the hose to create a seal. Hold the rags and hose firmly with both hands *(above)*, while a helper turns the water on full pressure. (Do not use water elsewhere in the house while doing this.) If the blockage seems to have been cleared, test the drain *(step 4)*. If not, add an expansion nozzle to the hose *(step 2)* or use an auger *(step 3)*.

2 Using an expansion nozzle. Choose a nozzle made to fit a 3- to 6-inch drainpipe, available at a plumbing supply or hardware store, and screw it to the end of the hose *(inset)*. Push the hose as far as possible into the drain and turn on the water. The nozzle will expand to seal the pipe *(above)*, forcing the water to blast the line clear if the blockage is not too large. If the blockage seems to have been cleared, turn off the water, detach the hose from the faucet, and wait a few minutes for the nozzle to deflate before removing it.

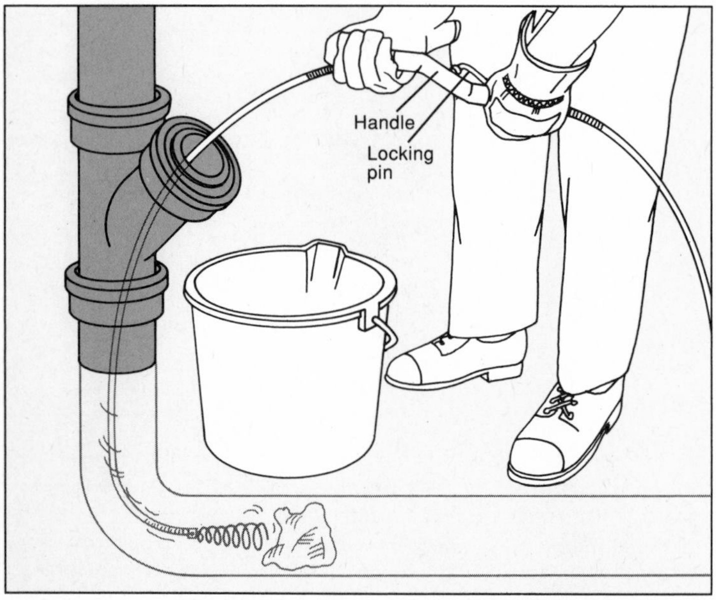

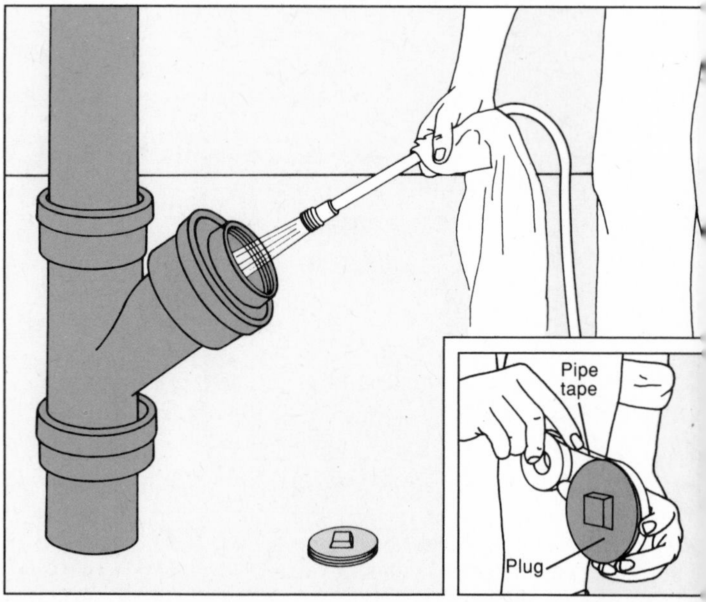

3 Using a manual auger. Use a drain auger 1/4 inch in diameter and 25 to 100 feet long. Uncoil it on the floor, allowing enough room to push and pull. Feed the auger as far as possible into the drain, lock the handle in place, then wind it clockwise to break up the blockage *(above)*. When the auger moves freely, test the drain *(step 4)*. If the blockage remains intact or out of reach, you can rent a power auger *(page 87)* or work on the house trap *(page 88)*.

4 Testing the drain. Test a cleanout located near the floor by sending water through it with a hose *(above)*. If the cleanout is near the ceiling, replace the plug, then turn on a faucet upstairs. When the water runs freely through the drain, rub petroleum jelly or wind pipe tape on the threads of the plug *(inset)* and replace and tighten it.

UNBLOCKING THE MAIN DRAIN WITH A POWER AUGER

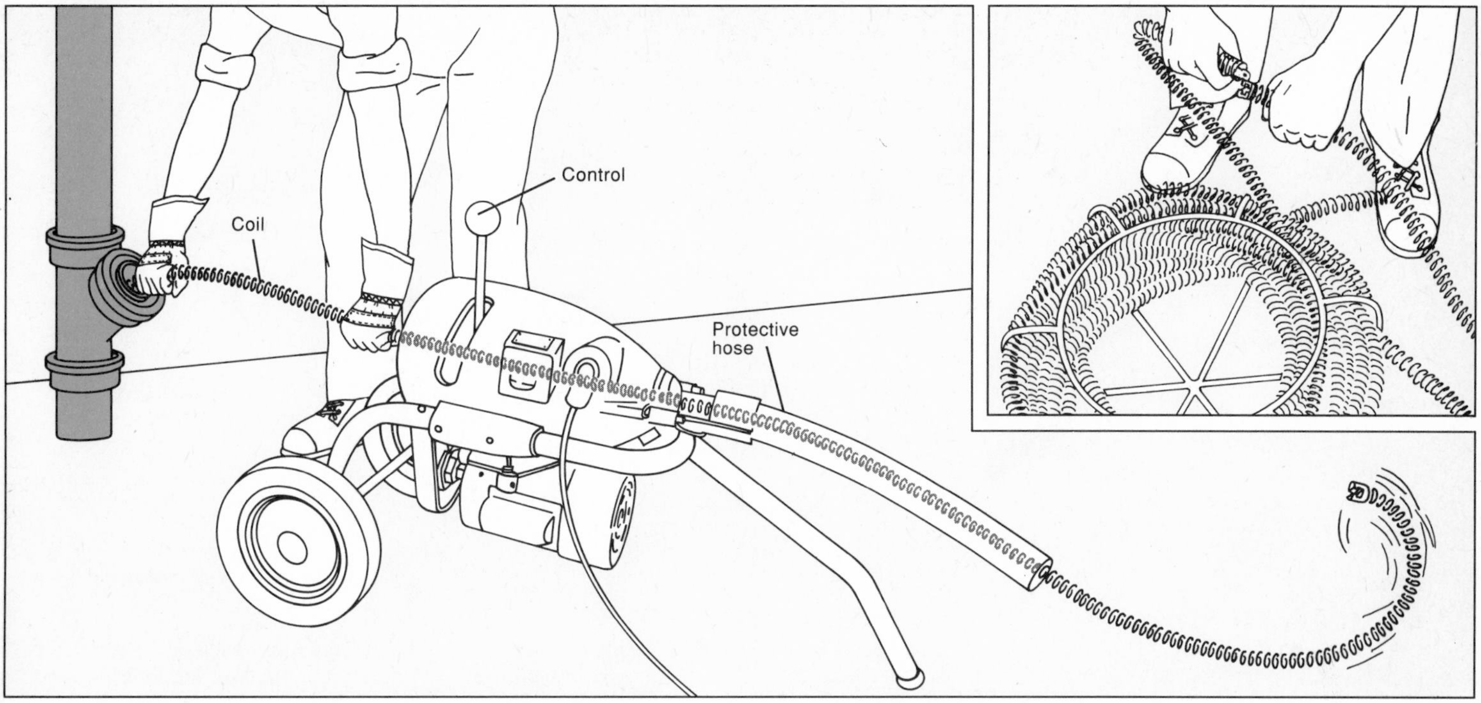

1 Feeding the auger into the drain. Rent a power auger designed for a 3- or 4-inch drain, with a single length of coil rolled on a drum and several cutting heads. Power augers are heavy-duty tools that must be used with great care to protect pipes; follow the manufacturer's directions carefully. (Some augers have automatic feed, for example.) Wear work gloves, heavy boots and eye protection, since the auger may flail about as you feed it into the drain. Attach an all-purpose head, which breaks up loose blockages such as grease, fabric, hair and sand, to a length of coil. With the motor off, feed the coil into the drain by hand *(above)*. If the first length of coil does not reach the blockage, clip a second length of coil to the end of the first *(inset)* and continue feeding it down the drain.

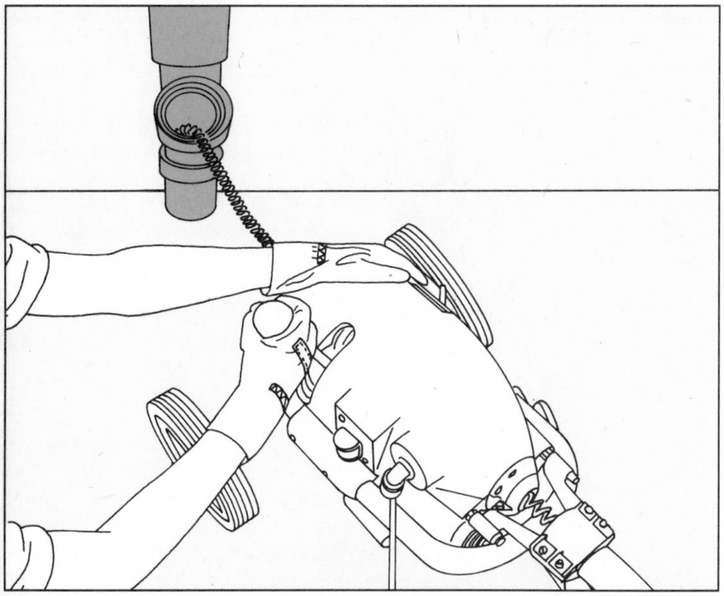

2 Activating the auger. Depending on the model, turn the control to the drive position *(above)* or press the foot pedal to activate the auger. Keep hands and feet clear of all moving parts. As its cutting head turns, the auger will begin to break up the blockage. If it jams, pull the auger out a few inches, then advance it again. When the auger spins freely, turn the control to the idle position, feed a few more inches of coil into the drain, then activate the auger again. Repeat this procedure until the blockage is cleared, then turn off the auger and pull out the coils by hand. If the blockage remains, switch to a heavier-duty head *(next step)*.

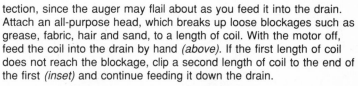

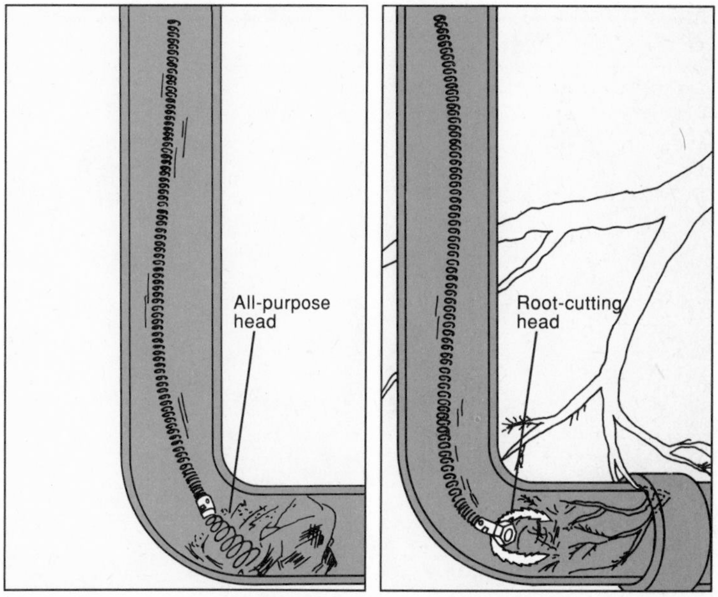

3 Changing the cutting head. When the all-purpose cutting head *(above, left)* fails to break up the blockage, or you suspect that tree roots are the cause, replace the head with a root-cutting head *(above, right)*. The spinning blades will scrape the sides of the pipe to remove tree roots, rust and mineral deposits. **Caution:** Do not turn on the auger until it reaches the blockage. When the drain is clear, turn off the auger before removing it. Do not use sharp or saw-tooth heads for plastic drainpipe.

UNBLOCKING THE MAIN DRAIN AT THE HOUSE TRAP

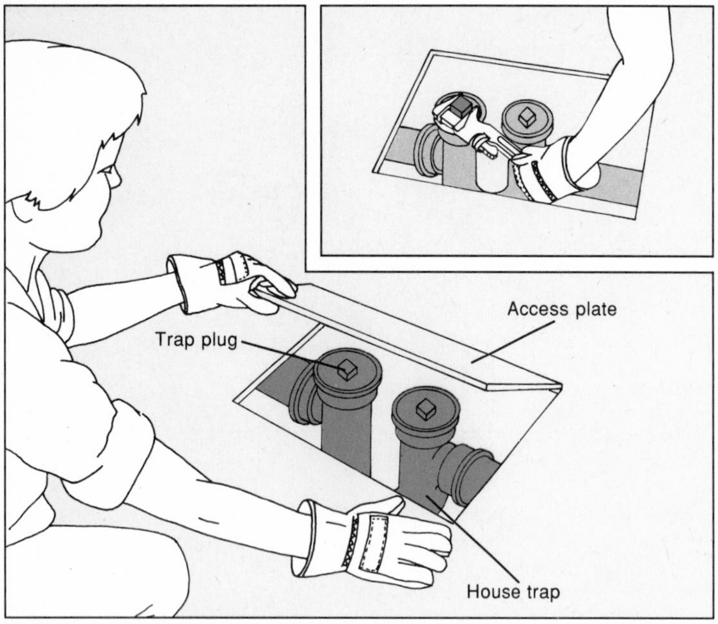

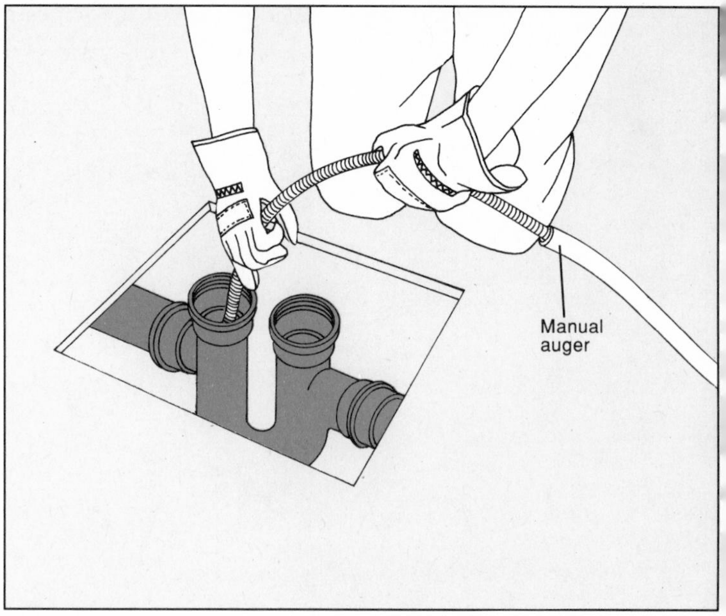

1 Working on the house trap. If your home has a house trap, it will be located on the horizontal main drain—generally laid just beneath the basement floor and sometimes covered by an access plate *(above)*. A house without a basement may have a trap on the main sewer line outside, with access at ground level. To clear a clogged trap, loosen the plug closest to the outside sewer line *(inset)*. Shine a flashlight into the open trap and poke in a sewer rod to release any blockage. If the blockage is not within reach or the sewer rod is ineffective, use an auger *(step 2)*.

2 Using an auger. Use a manual auger designed for a 3- or 4-inch drainpipe. Feed the auger into the the trap, turn the pin to lock the handle in place and rotate the coil clockwise. If the clog is not in the trap but in the main drain between the trap and the main cleanout, remove the second plug and auger toward the cleanout.

SERVICING THE ROOF VENT

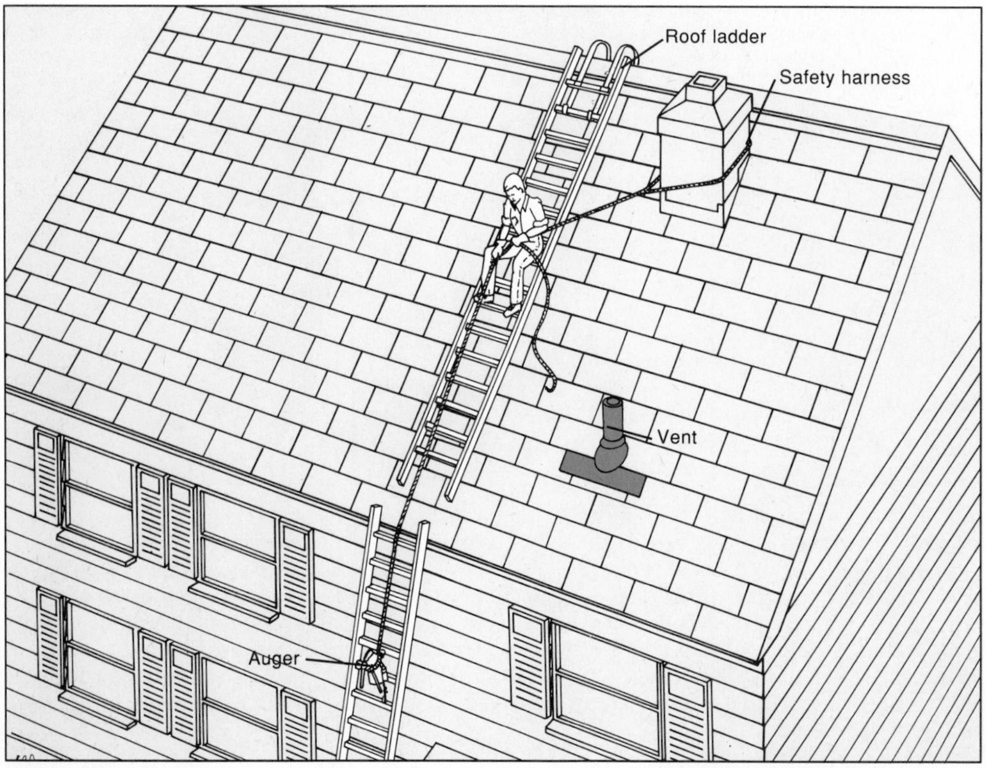

1 Preparing to work on the roof. When it is impossible to remove a blockage from the branch drains, main cleanout or house trap, or when all the drains are slow or smell bad, work from the roof to clear the vent and soil stack. Work in good weather and wear rubber-soled shoes to avoid slipping. If the roof is pitched, rent a ladder that hooks over the peak of the house. For extra security, also rent a safety belt and attach its free end to a fixed point on the roof, such as a chimney, especially if you will be working with a power auger. Work kneeling or sitting if the angle of the roof allows, and have a helper on the ground. **Caution:** If you are not comfortable about working on your roof or with heavy power tools, call for professional service.

SERVICING THE ROOF VENT (continued)

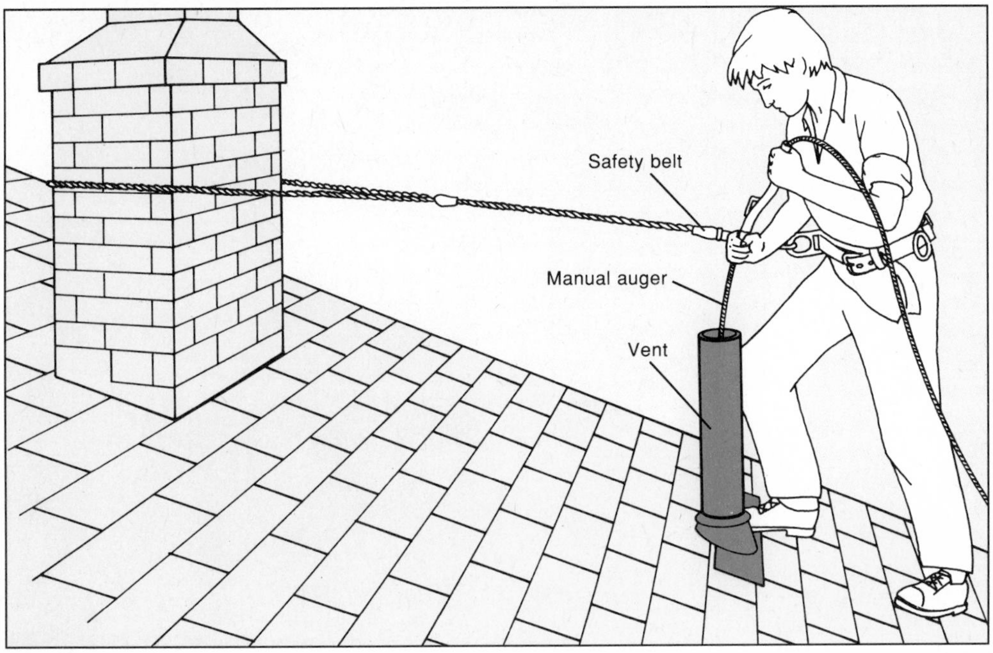

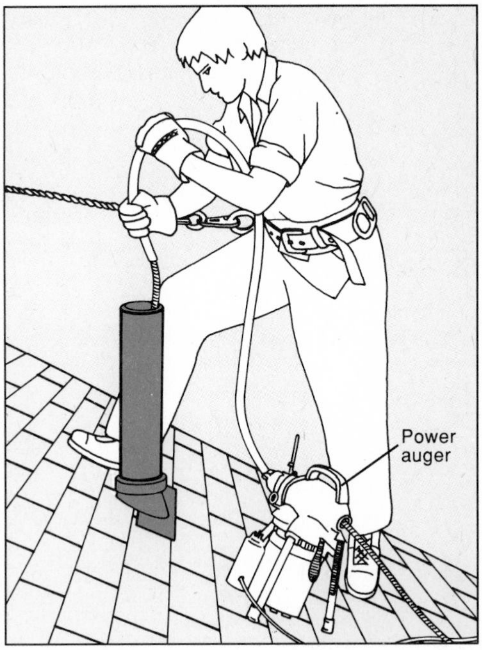

2 Using an auger. Use a manual auger at least as long as the distance from the roof to the bottom of the soil stack. Feed the auger into the vent until it reaches the blockage, lock the handle in place and rotate the auger clockwise *(above, left)*. Feed in more coil as needed and stop occasionally to pull out any debris. Next, remove the auger, insert a garden hose, and flush the vent. The water will drain freely through the vent if the blockage has been cleared. If the problem was drains smelling bad, wait a few days to determine if the smell has gone away, then protect the vent with a cage *(step 3)*. If the water backs up from the vent, the blockage remains. If you are experienced with power tools, consider renting a lightweight power auger *(page 87)*. Feed in the coil until the cutter head meets the blockage *(above, right)*, then turn on the auger. Stop the auger frequently and try to fish out the blockage rather than push it farther down the drain. Test with a hose (but have someone inside the house watch for overflowing fixtures). If the blockage remains, call for service.

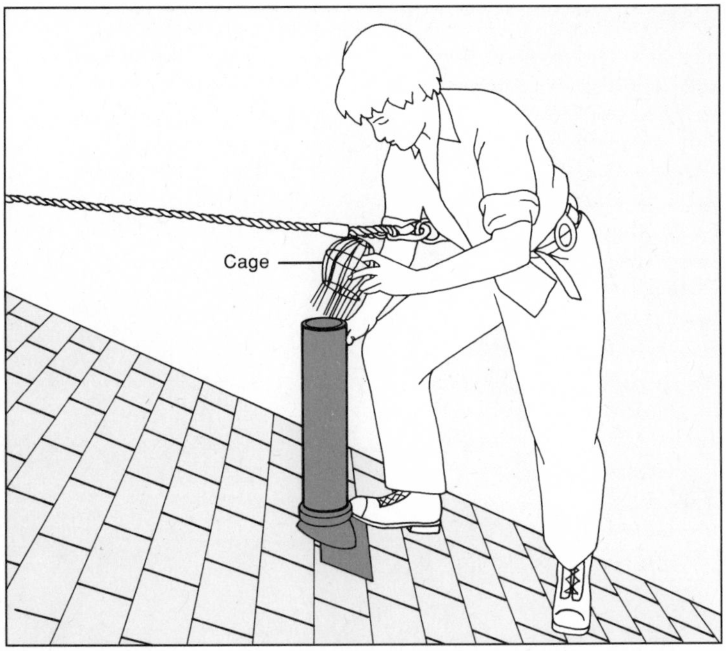

3 Protecting the vent. To prevent objects from falling into the vent, causing sluggish or smelly drains, add a cage to the end of the pipe. Cages are especially useful on flat roofs, where the vent is shorter and more exposed. Buy a metal cage the width of the vent at a hardware or plumbing supply store and insert it into the top of the vent *(above)*. Or cut a piece of metal screening slightly larger than the opening, bend it over the pipe and wire it to the vent.

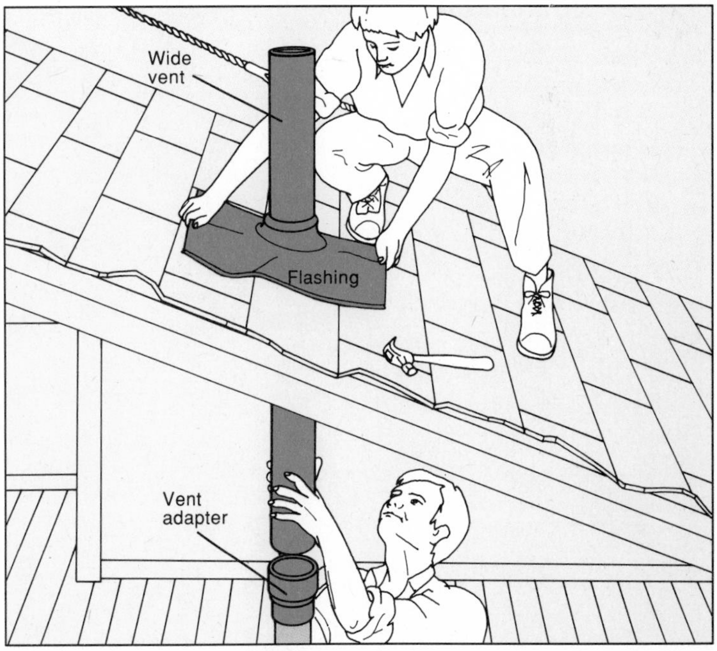

Expanding the top of the vent stack. Increasing the diameter of the vent will help keep it from being blocked by ice in colder climates. As much a roofing project as a plumbing repair, it is a professional job that involves removing the flashing and adjacent shingles, removing a section of the old vent from inside the house, enlarging the hole in the roof, adding a wider vent and adapter *(above)*, and applying new flashing and shingles. The entire assembly must be properly sealed to prevent leaking.

REPAIRING THE SEWER LINE

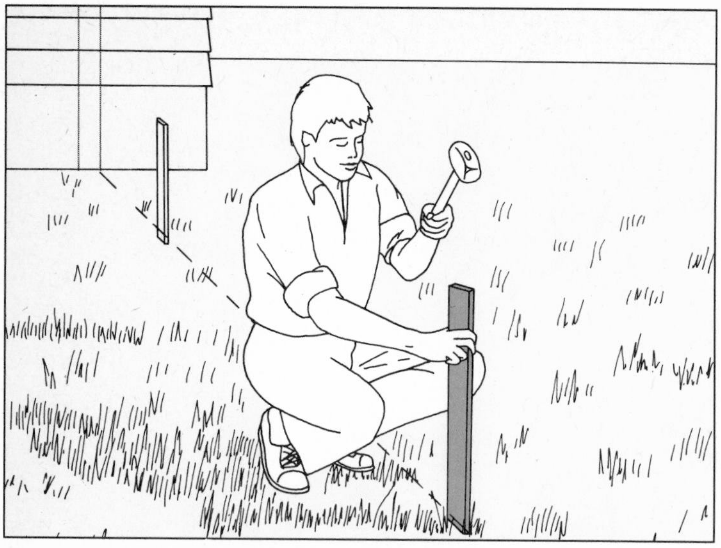

1 Marking the sewer line. Soil can crush or roots can grow into a cracked main sewer line, causing the entire plumbing system to back up. Even without a blockage, a soggy, smelly lawn may indicate a leaking pipe below. Excavating your yard to locate and repair a sewer line—and landscaping the damage afterward—is a major job, but you can save money by doing much of the work yourself.

Begin by tracing the path of the sewer line from the main drain to the sewer or septic tank. Ask your local authorities for permission to work on the sewer pipe, and for information about its location and depth. Next, stake out the path of the sewer line *(above)* and hire a contractor with a backhoe to dig a trench to within a foot of the pipe.

2 Reinforcing the trench. If the soil is loose or sandy, make the trench safe to work in by shoring its walls. (Clay soil generally does not require shoring; gravel soil requires shoring at the bottom only.) Lower sheets of 3/8-inch exterior-grade plywood into the trench to line the walls *(above)*. If the soil is especially loose, brace the trench *(next step)* for extra support.

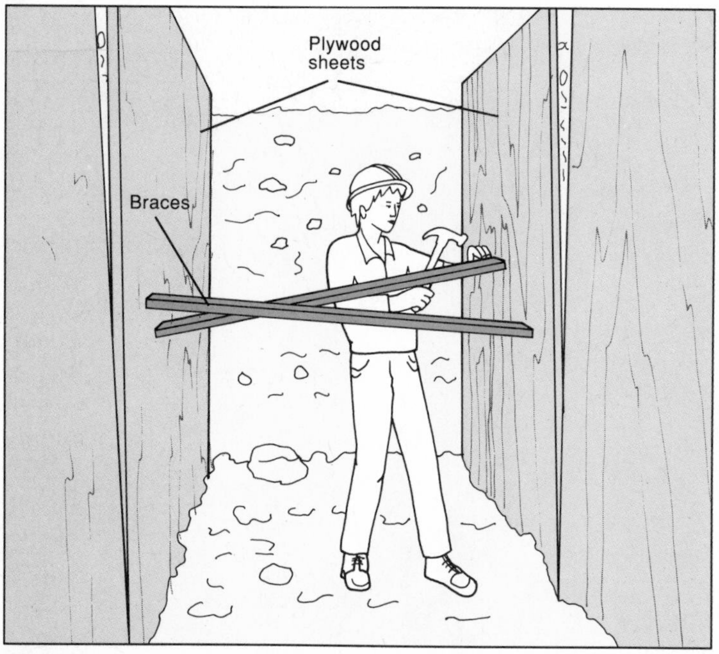

3 Bracing the trench. Cut enough 2 x 4s, slightly longer than the width of the trench, to brace it every three or four feet. Nail each of the supports to the plywood sheets between waist and shoulder height *(above)*. Leave enough space between braces to dig with a shovel. **Caution:** Plumbing supply lines or electrical cables may run parallel to the sewer line.

4 Exposing the sewer line. To locate the leak, look for an area of wet soil in the base of the trench. Turn off the main water supply and be sure that nobody uses any fixtures during the repair. Dig around the broken section of pipe so that it is completely exposed *(above)*. If the sewer pipe is plastic, use plastic replacement pipe. If it is cast iron, replace it with plastic if the plumbing code permits, cast iron if it does not. Both are attached with hubless fittings—neoprene sleeves held in place by stainless steel clamps.

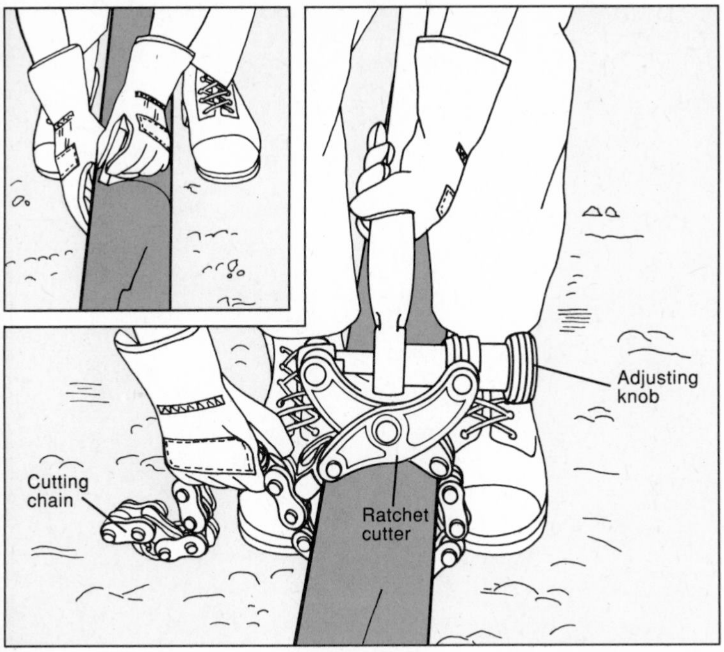

5 Preparing to cut the pipe. With a piece of chalk, mark the section of pipe to be replaced *(inset)*. Cut plastic sewer pipe with a short-handled saw at the chalk lines. Then push loosely wadded paper towels into the pipe ends to block harmful sewer gas. If the sewer pipe is cast iron, you must rent a ratchet cutter *(above)* to remove the broken section. Wrap the cutting chain around the chalk line and secure it to the handle. Set the dial to CUT and turn the adjusting knob clockwise to tighten the chain around the pipe.

6 Cutting cast-iron pipe. Wearing safety goggles, pump the handle of the ratchet back and forth *(above)* to cut through the pipe. Loosen the adjusting knob, unlock the chain, slide the ratchet cutter to the other chalk mark and repeat the procedure. Remove the damaged section of pipe and wad paper towels into both of the standing pipes to block harmful sewer gas.

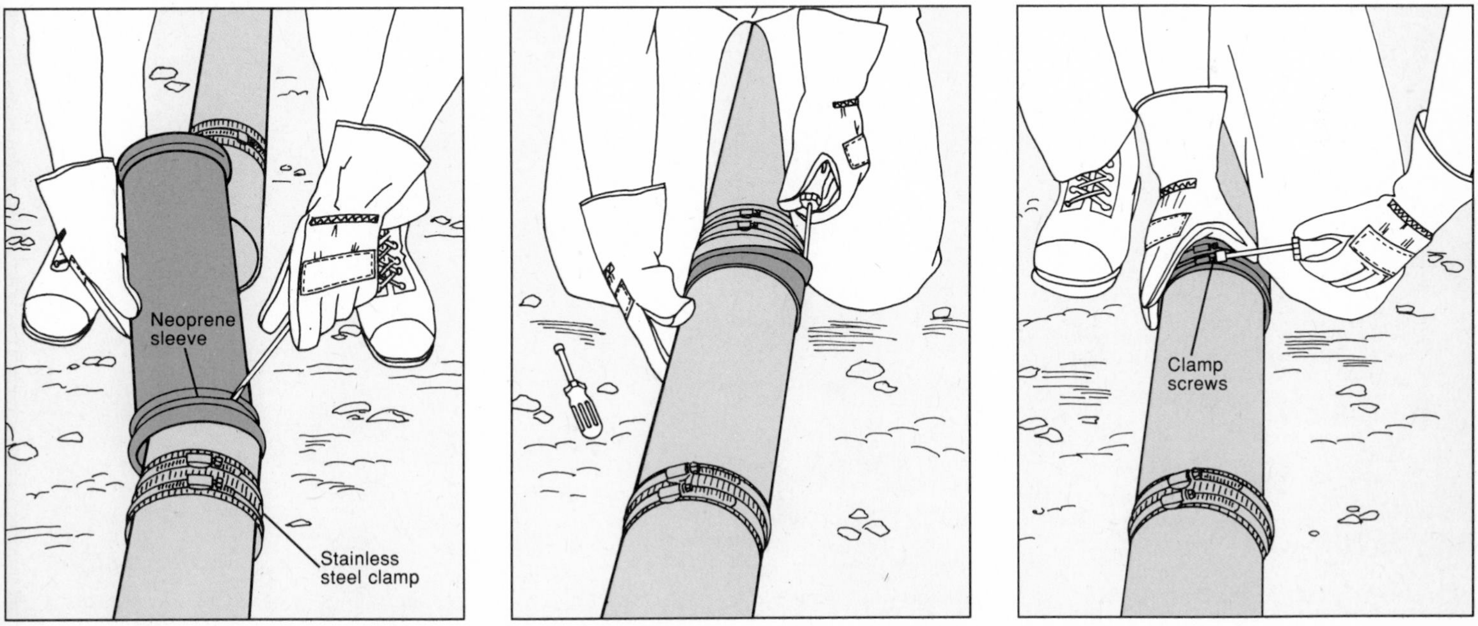

7 Cutting and fitting the new pipe. Cut the replacement pipe, whether cast iron or plastic, about 1/4 inch shorter than the old piece. Slide a loosened clamp onto each standing pipe. Force both ends of the replacement pipe halfway into neoprene sleeves, then fold the lips of the sleeves back over the pipe. Fit the section between the standing pipes *(above, left)* and fold the sleeves in place with the help of a screwdriver *(above, center)*. Position a clamp over each joint, then tighten the clamp screw securely with a socket wrench or nut driver *(above, right)*. Have the repair examined if required by local plumbing codes. Pack earth firmly around the pipe so that it slopes away from the house. Remove the plywood and supports, but wait a day or two to be sure there are no leaks before filling in the trench.

SEPTIC SYSTEMS

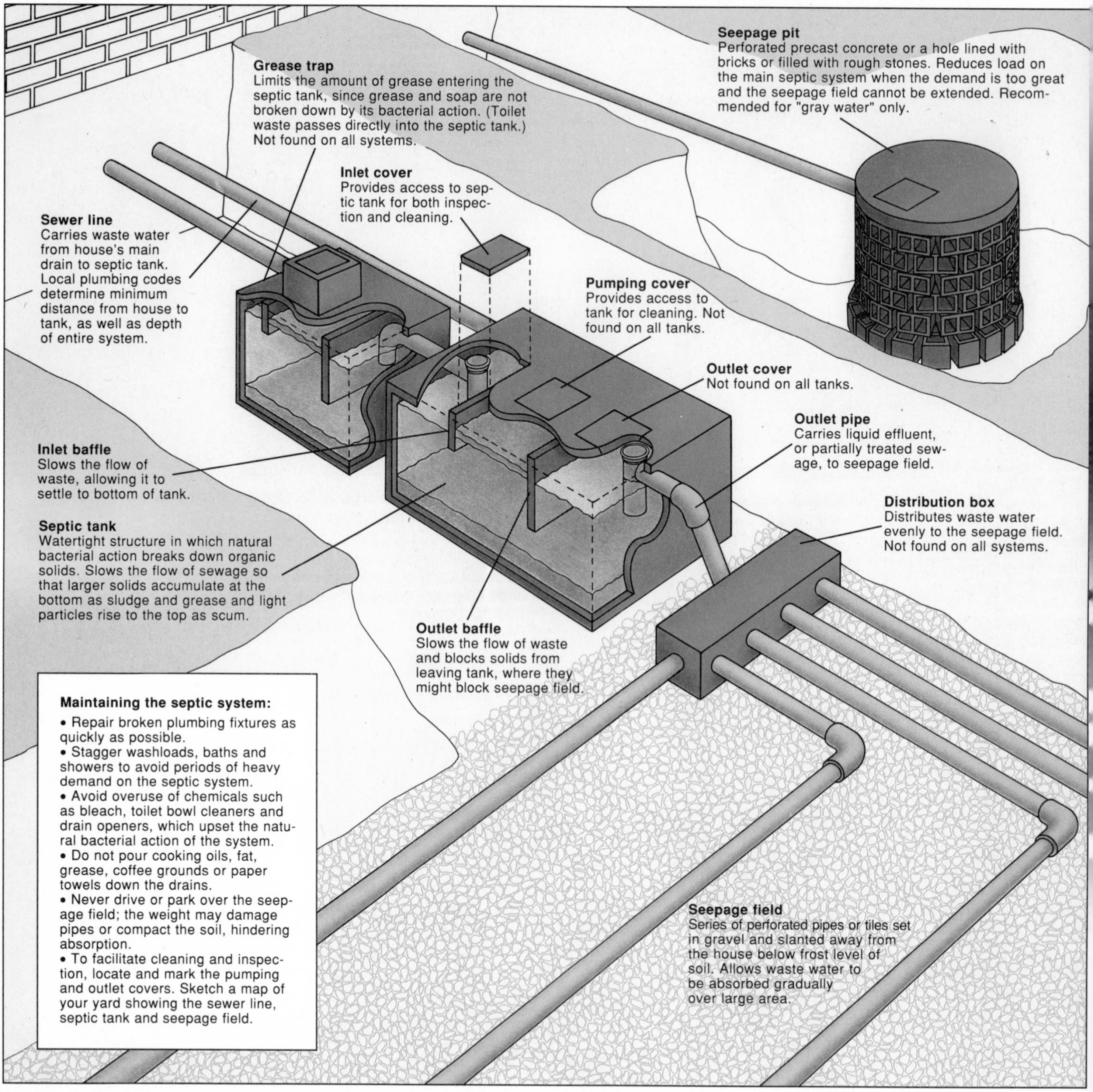

Grease trap
Limits the amount of grease entering the septic tank, since grease and soap are not broken down by its bacterial action. (Toilet waste passes directly into the septic tank.) Not found on all systems.
Seepage pit
Perforated precast concrete or a hole lined with bricks or filled with rough stones. Reduces load on the main septic system when the demand is too great and the seepage field cannot be extended. Recommended for "gray water" only.
Inlet cover
Provides access to septic tank for both inspection and cleaning.
Sewer line
Carries waste water from house's main drain to septic tank. Local plumbing codes determine minimum distance from house to tank, as well as depth of entire system.
Pumping cover
Provides access to tank for cleaning. Not found on all tanks.
Outlet cover
Not found on all tanks.
Outlet pipe
Carries liquid effluent, or partially treated sewage, to seepage field.
Inlet baffle
Slows the flow of waste, allowing it to settle to bottom of tank.
Distribution box
Distributes waste water evenly to the seepage field. Not found on all systems.
Septic tank
Watertight structure in which natural bacterial action breaks down organic solids. Slows the flow of sewage so that larger solids accumulate at the bottom as sludge and grease and light particles rise to the top as scum.
Outlet baffle
Slows the flow of waste and blocks solids from leaving tank, where they might block seepage field.
Seepage field
Series of perforated pipes or tiles set in gravel and slanted away from the house below frost level of soil. Allows waste water to be absorbed gradually over large area.

Maintaining the septic system:

- Repair broken plumbing fixtures as quickly as possible.
- Stagger washloads, baths and showers to avoid periods of heavy demand on the septic system.
- Avoid overuse of chemicals such as bleach, toilet bowl cleaners and drain openers, which upset the natural bacterial action of the system.
- Do not pour cooking oils, fat, grease, coffee grounds or paper towels down the drains.
- Never drive or park over the seepage field; the weight may damage pipes or compact the soil, hindering absorption.
- To facilitate cleaning and inspection, locate and mark the pumping and outlet covers. Sketch a map of your yard showing the sewer line, septic tank and seepage field.

Servicing a private septic system. Household sewage—99 percent of which is water—moves from the house through the sewer line to a septic tank for treatment. There it is divided into four components: sludge and scum, which are trapped in the tank; gas, which escapes back through the sewer pipe and out the house vent; and liquid sewage, which flows to a seepage field beneath the lawn.

If your system is sluggish or overloaded, first check that indoor plumbing fixtures are not at fault. A leaking faucet or running toilet may be sending hundreds of gallons of water into the system unnecessarily. Have the tank inspected and cleaned at least as often as required by the local health department, usually every two to five years.

When part of the lawn remains soggy even in dry weather, or household drains are sluggish or smell bad, have a professional service the septic system. The tank may need to be pumped out, the grease trap may require cleaning, or the outlet pipe could be blocked. More serious problems include compacted soil, broken pipes or baffles, or a too-small seepage field. One answer to increased demand is to add a seepage pit to back up the main system. (When a seepage pit is installed without its own septic tank, it can treat only uncontaminated water.) If the soil is no longer absorbent, it may be necessary to dig up and replace the seepage field and its gravel bed. In all cases, refer to local plumbing codes and leave the job to a professional.

PIPES AND FITTINGS

New methods and materials have put the art of pipefitting well within reach of the homeowner, who can simplify repairs by selecting the easiest material for the job at hand.

Copper, today's standard for supply pipes, is durable, carries water efficiently, and is relatively easy to handle. One disadvantage is its high cost; another is the danger of lead pollution from the solder used to join copper pipes and fittings. (Therefore consider low-lead or silver solder). Copper comes in rigid, 10-foot lengths and in flexible coils up to 60 feet long. Two thicknesses are available for home use: medium-thick Type L (color-coded green) and low-cost but thin-walled Type M (color-coded red), suitable for most repairs.

Galvanized steel, the strongest plumbing material, is best suited for pipes that will be exposed to damage or vibration. It is prone to corrosion, however, and the heavy pipe must be joined with threaded fittings. Steel pipe normally comes in 21-foot lengths; most plumbing supply stores will cut and thread a shorter piece for you. Threaded brass pipe is available in 12-foot lengths but, because of its cost, is used for nipples and extensions. When joining copper pipe to steel, use a special dielectric union to prevent an electrolytic reaction that occurs between the two metals, causing corrosion.

If your local code permits their use, two types of plastic—chlorinated polyvinyl chloride (CPVC) and polybutylene (PB)—are light, inexpensive and easy to use. Flexible PB is joined with screw-together compression fittings; rigid CPVC is joined with solvent cement. PB pipe is especially useful for long runs around beams and corners, and behind walls. When installing plastic pipe, allow enough room—about 1/2 inch per 10-foot length—for thermal expansion.

Cast-iron, the heaviest and most durable plumbing material, is used only for drains and vents. Older pipes have hub-and-spigot joints sealed with ropelike caulking and molten lead. Today, cast iron is joined with metal-clamped neoprene gaskets called hubless fittings. Polyvinyl chloride (PVC) and acrylonitrile butadiene styrene (ABS) are also used for drainage where codes permit. Both are available in 10-foot lengths, and can be joined with solvent cement or hubless fittings.

TROUBLESHOOTING GUIDE

SYMPTOM	POSSIBLE CAUSE	PROCEDURE
Copper pipe leaks	Pinhole leak in pipe	Patch hole temporarily with tape or clamp *(p. 11)* □○
	Joint loose	Unsolder and resolder joint *(p. 140)* ⬓◒
	Loose joint caused by water hammer	Resolder joint *(p. 140)* ⬓◒ and install shock absorber *(p. 113)* ⬓◒
	Pipe cracked or corroded	Replace damaged section with copper *(p. 96)* ⬓◒, with PB *(p. 98)* □○, or with CPVC *(p. 99)* ⬓◒
Steel or brass pipe leaks	Pinhole leak in pipe	Patch hole temporarily with tape or clamp *(p. 11)* □○
	Fitting loose	Tighten fitting *(p. 101)* □○
	Pipe cracked or corroded	Replace damaged section with steel or brass (p. 101) ⬓◒, with copper *(p. 102)* ⬓◒, with PB *(p. 103)* □○, or with CPVC *(p. 104)* ⬓◒
Plastic pipe leaks (CPVC or PB)	Pinhole leak in pipe	Patch hole temporarily with tape or clamp *(p. 11)* □○
	Compression fitting loose	Tighten or replace fitting *(p. 98)* □○
	Pipe cracked	Replace damaged area with CPVC *(p. 105)* ⬓◒ or PB *(p. 106)* □○
Cast-iron drainpipe leaks	Lead in joint displaced	Tamp down old lead or add cold caulking compound *(p. 107)* □◒
	Pipe cracked or corroded	Replace damaged section with cast iron or ABS *(p. 108)* ⬓◒▲
Plastic drainpipe leaks (PVC or ABS)	Pipe cracked	Replace damaged section with hubless fittings *(p. 108)* ⬓◒ or solvent cement *(p. 109)* ⬓◒
Valve leaks	Valve dirty or corroded	Disassemble and clean valve *(p. 110)* □○
	Packing or washer worn	Replace packing or washer *(p. 110)* □○
	Valve worn	Replace valve *(p. 110)* ⬓◒
Valve won't close	Gate or washer blocked	Disassemble and clean valve *(p. 110)* □○
	Valve worn	Replace valve *(p. 110)* ⬓◒
No shutoff valve at fixture	Water supply to fixture controlled by main shutoff valve	Install shutoff valve at fixture *(p. 111)* ⬓◒
Pipe or faucet noisy	Water hammer	Install shock absorber *(p. 113)* ⬓◒

DEGREE OF DIFFICULTY: □ **Easy** ⬓ **Moderate** ■ **Complex**
ESTIMATED TIME: ○ **Less than 1 hour** ◒ **1 to 3 hours** ● **Over 3 hours** ▲ **Special tool required**

TYPES OF FITTINGS

SUPPLY FITTINGS

Tee
Joins a 90-degree branch run to a straight run of pipe.

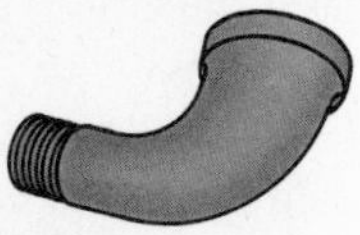

Street elbow
Attaches to another fitting to change pipe direction.

Elbow
Changes the direction of a pipe run; Available in 45- and 90-degree bends.

Nipple
Joins threaded fittings that are close together. Often used to complete a run.

Coupling
Joins lengths of pipe running in the same direction. Pipe ends bottom out against a shoulder inside standard coupling; slip coupling has no shoulder.

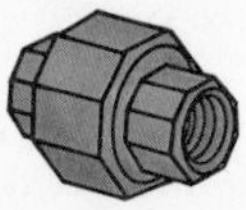

Union
Joins threaded pipe, usually galvanized steel, so that the run can be disassembled.

Plug
Closes an unused opening in a pipe or fitting. Also used to temporarily plug an open pipe during a repair if the water must be turned on.

TRANSITION FITTINGS

Dielectric union
Joins copper to steel pipe to prevent electrolytic corrosion between the different metals. The threaded end screws onto the steel pipe; the brass end is soldered onto the copper pipe.

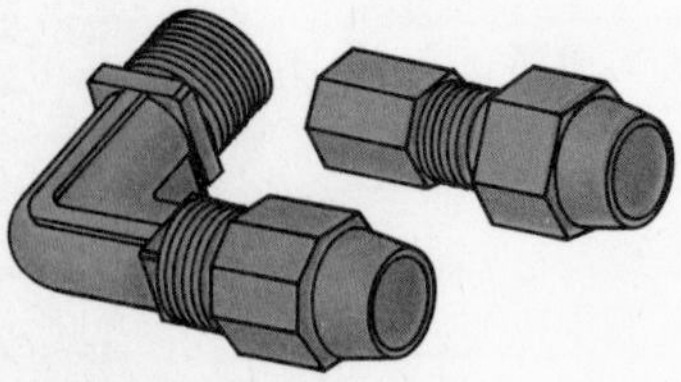

Threaded adapters
Used to join PB, copper or CPVC to threaded pipe. When buying adapters, specify whether they will be used for hot or cold water supply pipes.

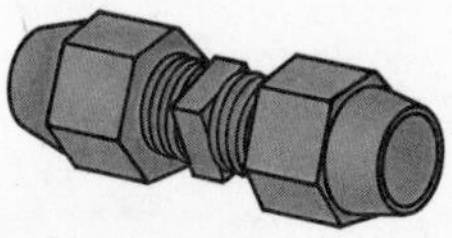

Compression fitting
Joins PB, copper or CPVC supply lines without cement or solder.

CPVC-to-unthreaded pipe adapter
Compression side of adapter clamps to copper or PB pipe; socket end is cemented to CPVC pipe.

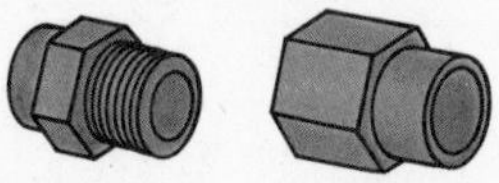

CPVC-to-threaded pipe adapter
Threaded side of adapter screws onto steel or brass pipe; socket end is cemented to CPVC pipe.

DRAIN FITTINGS

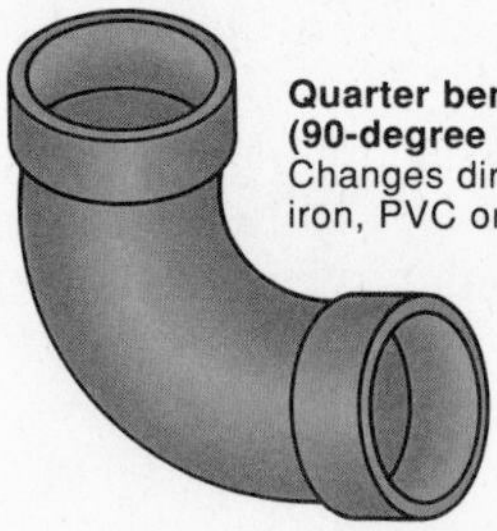

Quarter bend (90-degree elbow)
Changes direction of cast-iron, PVC or ABS drainpipe.

Hub
Joins straight lengths of cast-iron, PVC or ABS drainpipe. Old cast-iron hub-and-spigot joints are sealed with oakum and lead. Plastic joints are commonly called couplings.

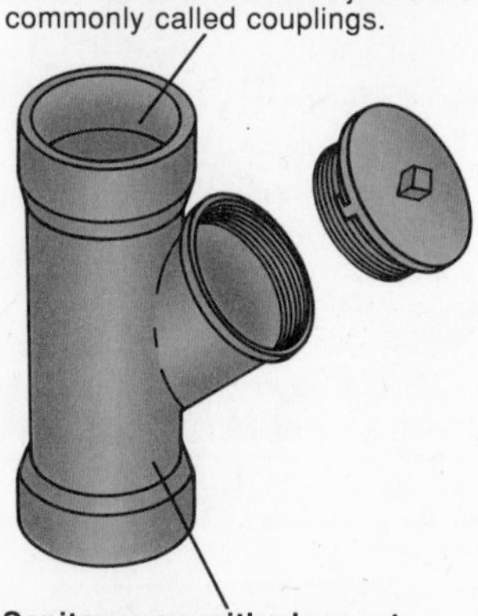

Sanitary wye with cleanout
Joins two lengths of drainpipe; threaded cleanout allows access to the pipe for cleaning and augering.

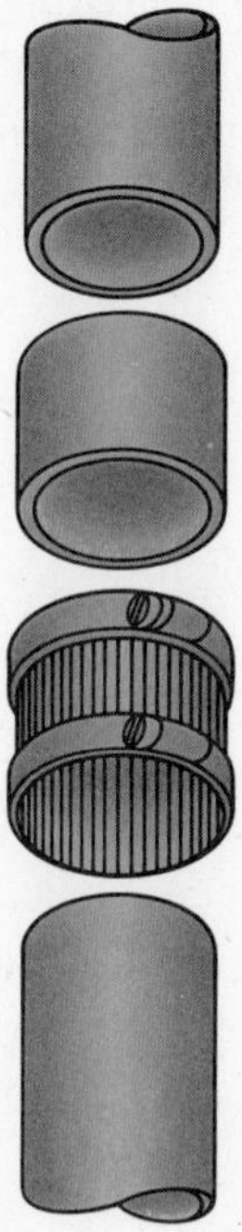

Hubless fitting
Joins hubless cast-iron, ABS or PVC drainpipe without caulking or cement. A tight-fitting neoprene sleeve is held in place over the joint by stainless steel collar and clamps.

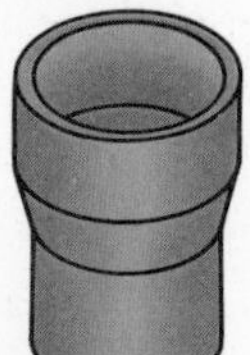

Reducer
Connects drainpipes of different diameters.

Wye branch
Joins a branch pipe at a 45-degree angle to a straight pipe run.

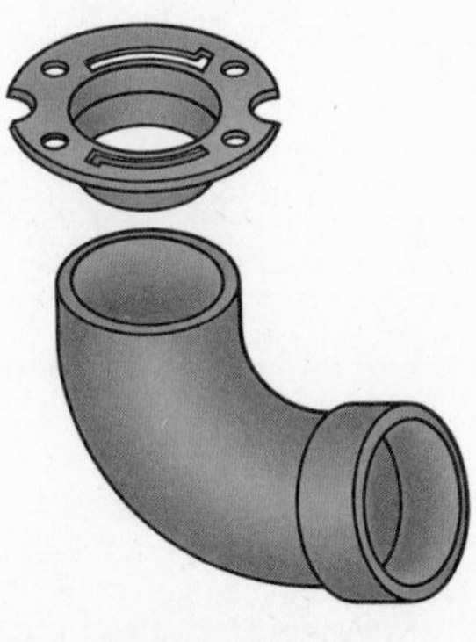

Closet flange and bend
Connects the toilet to its branch drain, and helps anchor the toilet to the bathroom floor.

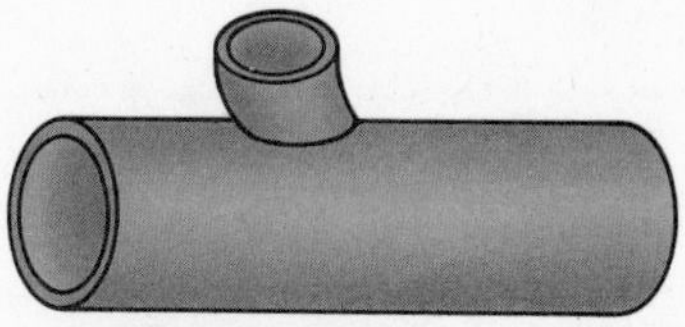

Reducing tee-wye
Joins a branch pipe at a 90-degree angle to a larger-diameter pipe run.

MEASURING PIPES AND FITTINGS

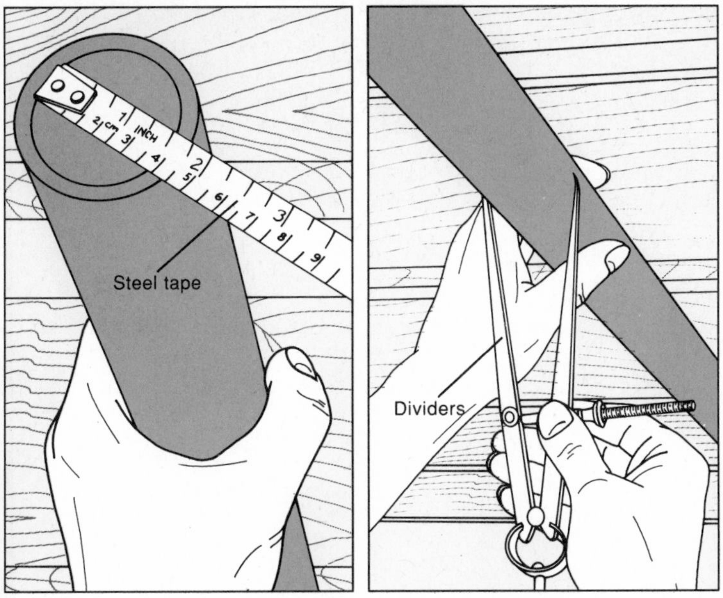

Finding the pipe diameter. Hold a ruler or steel tape across the widest part of a cut pipe and measure from one inner wall to the other *(above, left)* to determine the inside diameter (ID), or nominal size of the pipe. If the pipe is part of a run, first determine the outside diameter (OD) using dividers *(above, right)*, calipers or a C-clamp, then measure the result with a ruler. Take several readings and average them, then refer to the chart at right to find the inside diameter (ID). Always choose replacement pipe and fittings based on inside diameter.

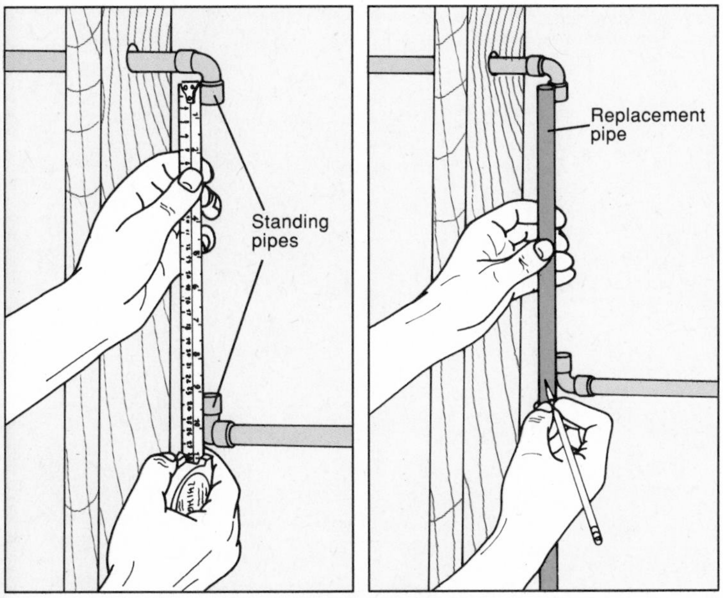

Measuring the replacement pipe. After cutting out or unthreading the damaged pipe, buy a replacement pipe that is several inches longer than the gap. Place the new fittings on the ends of the standing pipes. Hold a steel tape where the new pipe will reach inside the fitting, and measure to the same point on the opposite fitting *(above left)*. Or, hold the replacement pipe up to the gap and mark the exact length—including the depth of the fittings—with a pencil *(above right)*.

CALCULATING PIPE DIMENSIONS

TYPE OF PIPE	OUTSIDE DIAMETER (OD)	INSIDE DIAMETER (ID)	DEPTH OF FITTING SOCKET
COPPER	3/8 in.	1/4 in.	5/16 in.
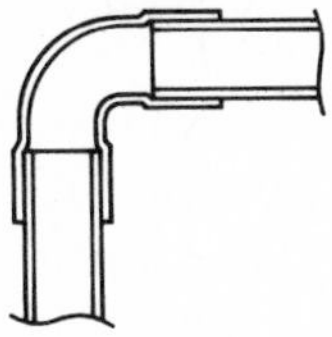	1/2 in.	3/8 in.	3/8 in.
	5/8 in.	1/2 in.	1/2 in.
	7/8 in.	3/4 in.	3/4 in.
	1 1/8 in.	1 in.	15/16 in.
	1 3/8 in.	1 1/4 in.	1 in.
	1 5/8 in.	1 1/2 in.	1 1/8 in.
THREADED	3/8 in.	1/8 in.	1/4 in.
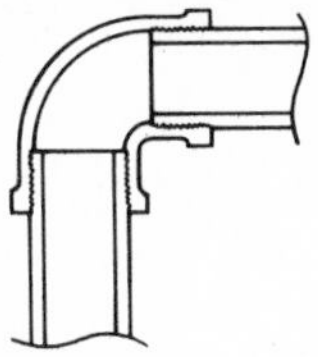	1/2 in.	1/4 in.	3/8 in.
	5/8 in.	3/8 in.	3/8 in.
	3/4 in.	1/2 in.	1/2 in.
	1 in.	3/4 in.	9/16 in.
	1 1/4 in.	1 in.	11/16 in.
	1 1/2 in.	1 1/4 in.	11/16 in.
	1 3/4 in.	1 1/2 in.	11/16 in.
	2 1/4 in.	2 in.	3/4 in.
PLASTIC	7/8 in.	1/2 in.	1/2 in.
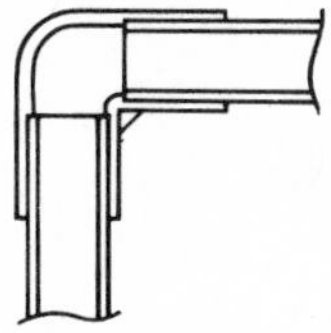	1 1/8 in.	3/4 in.	5/8 in.
	1 3/8 in.	1 in.	3/4 in.
	1 5/8 in.	1 1/4 in.	11/16 in.
	1 7/8 in.	1 1/2 in.	11/16 in.
	2 3/8 in.	2 in.	3/4 in.
	3 3/8 in.	3 in.	1 1/2 in.
	4 3/8 in.	4 in.	1 3/4 in.
CAST IRON	2 1/4 in.	2 in.	2 1/2 in.
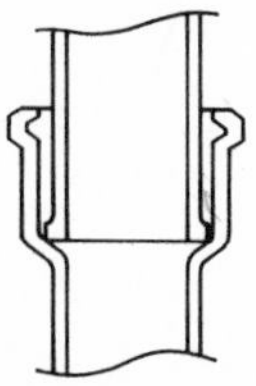	3 1/4 in.	3 in.	2 3/4 in.
	4 1/4 in.	4 in.	3 in.
	5 1/4 in.	5 in.	3 in.
	6 1/4 in.	6 in.	3 in.

Reading the chart. Always choose replacement pipe with the same inside diameter (ID) as the old pipe. For example: You have removed a damaged section of copper pipe with a 7/8-inch outside diameter (OD) and a 3/4-inch ID and have chosen to replace it with CPVC pipe. Buy CPVC pipe with a 3/4-inch ID (its nominal size) and 3/4-inch fittings, (which have a socket depth of 5/8 inch). When you cut the replacement pipe, be sure to account for this 1 1/4 inches.

REPAIRING COPPER PIPE

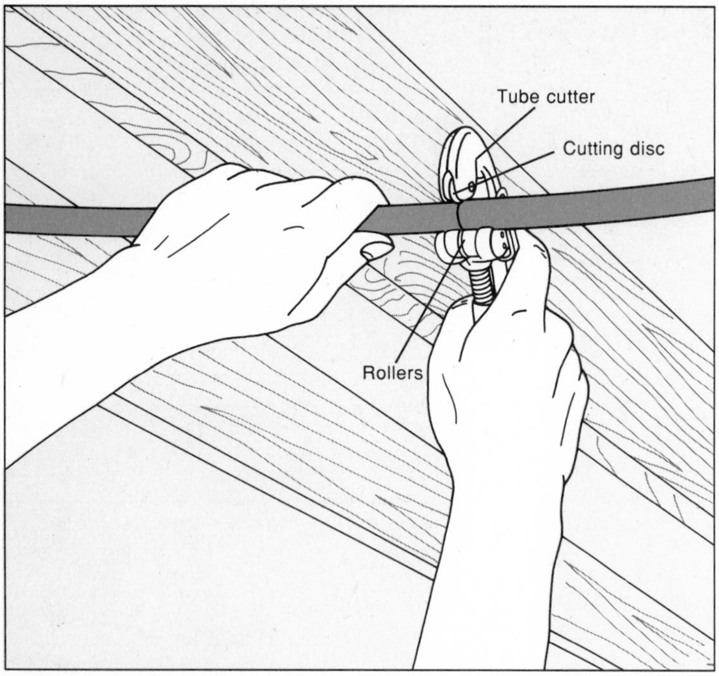

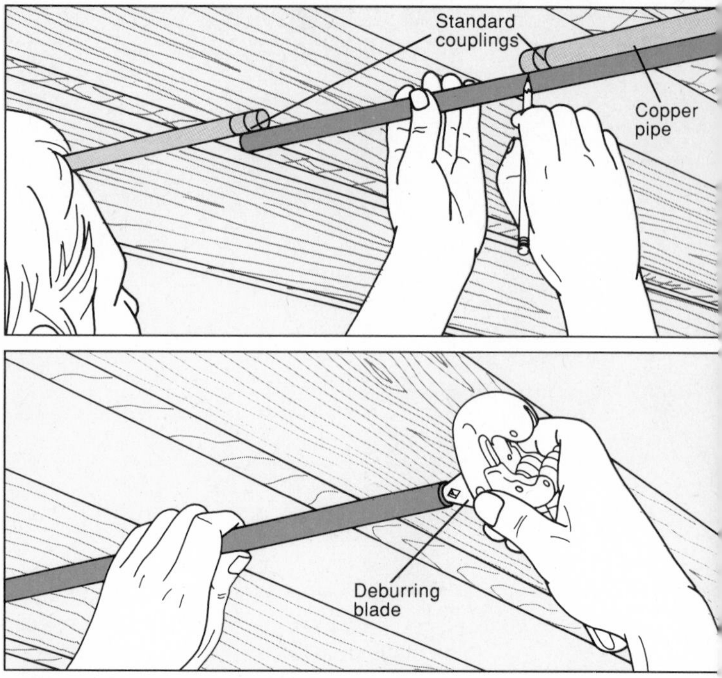

1 Cutting copper pipe. Close the main shutoff valve and drain the supply lines. If the broken pipe is hard to reach, cut it with a hacksaw. If it is accessible, use a tube cutter. Fit the jaws of the cutter around the pipe at one end of the defective section. Screw the knob clockwise until the rollers and the cutting disc grip the pipe firmly *(above)*. Give the knob another quarter-turn so that the disc bites into the pipe. Rotate the cutter once around the pipe, then tighten the knob and rotate again. Continue tightening and turning the tube cutter until the pipe is severed. Loosen the knob, slide the cutter down the pipe, and cut through the other side of the broken section.

2 Measuring and deburring copper pipe. If you are using standard couplings, fit them on the ends of the old pipe, hold the new pipe against the gap *(above, top)* and mark it at the coupling ridges. (If you are using slip couplings, mark the new pipe flush with the old pipe.) Cut the replacement pipe to length with the tube cutter. Use the triangular blade attached to the cutter (or a round metal file) to ream out the burrs inside the old and new pipes *(above, bottom)*. Next, prepare the joints for soldering *(step 4)* or bend the pipe first if necessary *(step 3)*.

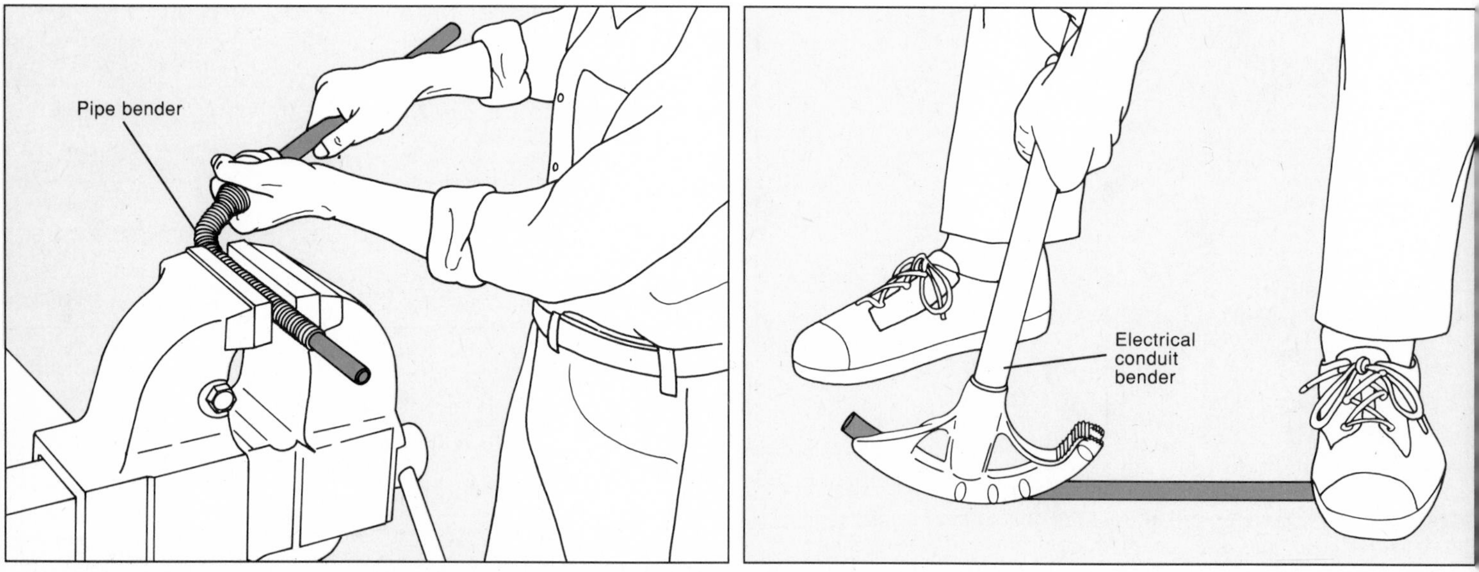

3 Bending copper pipe. To prevent kinks in a replacement pipe that must be bent, slip a coiled-spring pipe bender onto the pipe using a clockwise, twisting motion. If the pipe is flexible, bend it with your hands or form it over your knee. Overbend the pipe slightly, then ease it back to the correct angle. If the pipe is rigid, clamp the pipe bender in a vise and bend the pipe with both hands *(above, left)*. Or, for greater leverage, slip a rigid pipe into the slot of an electrical conduit bender and push down on the handle *(above, right)*.

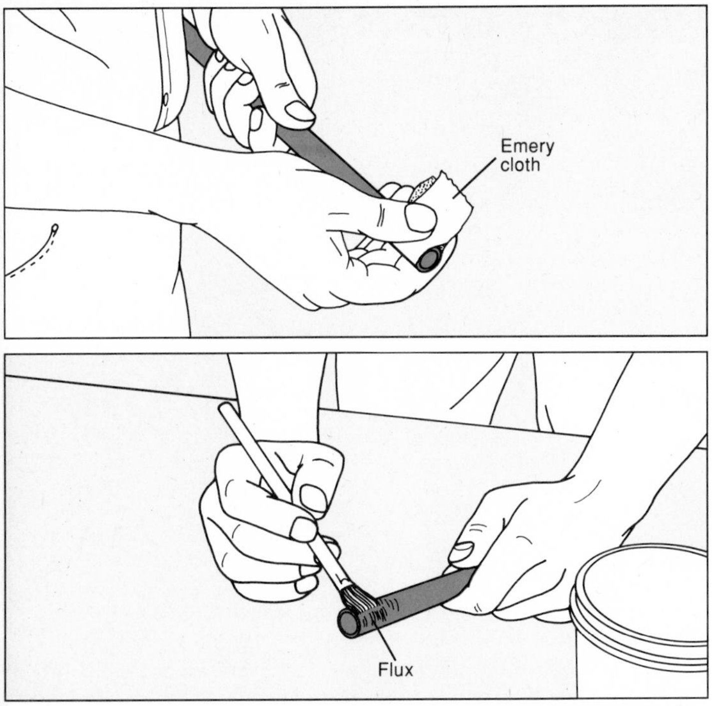

4 Preparing the joints. Rub the inside of the couplings and the ends of the old and new pipes with emery cloth or fine steel wool until they are brightly burnished *(above, top)*. Remove any grit left on the surfaces with a clean, dry cloth. With a small brush, spread a thick, even coat of soldering flux on all of the pipe surfaces *(above, bottom)*. Brush a small amount of flux inside the couplings.

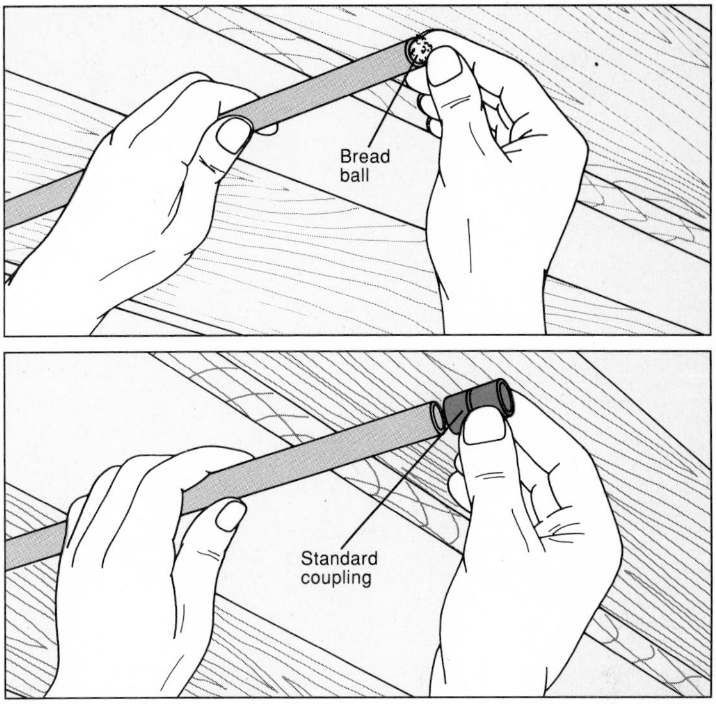

5 Fitting the couplings. Fit a coupling onto each end of the old pipes and give them a quarter-turn to evenly spread the flux *(above, bottom)*. If water drips from the standing pipes, the joint cannot be properly soldered. An old plumber's trick is to ball up a piece of white bread and plug the pipe before soldering to absorb moisture *(above, top)*. The bread will later dissolve and drain away.

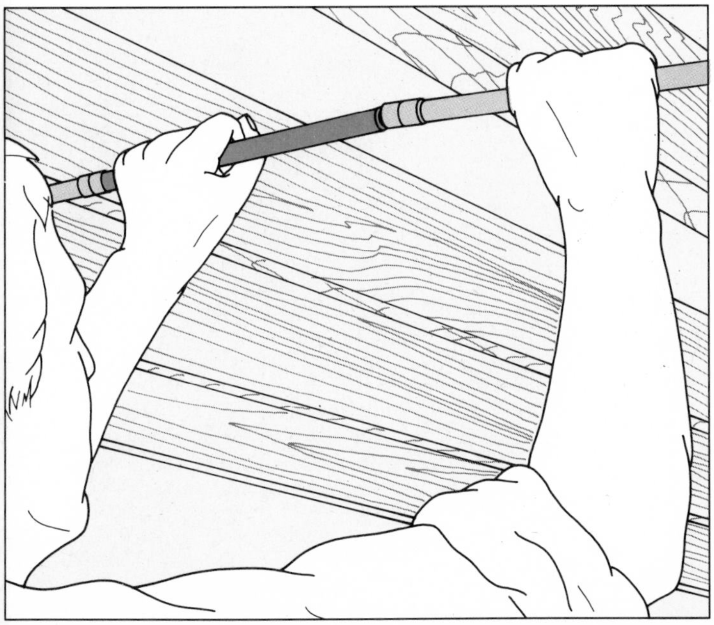

6 Fitting the replacement pipe. If you are using standard couplings, fit one end of the new pipe into the coupling until it bottoms out. Gently pull the pipes toward you until you can slip the other end into the second coupling *(above)*. If you are using slip couplings, hold the new pipe in place, then slide the couplings over the joints. Give the new pipe a quarter-turn to evenly distribute the flux.

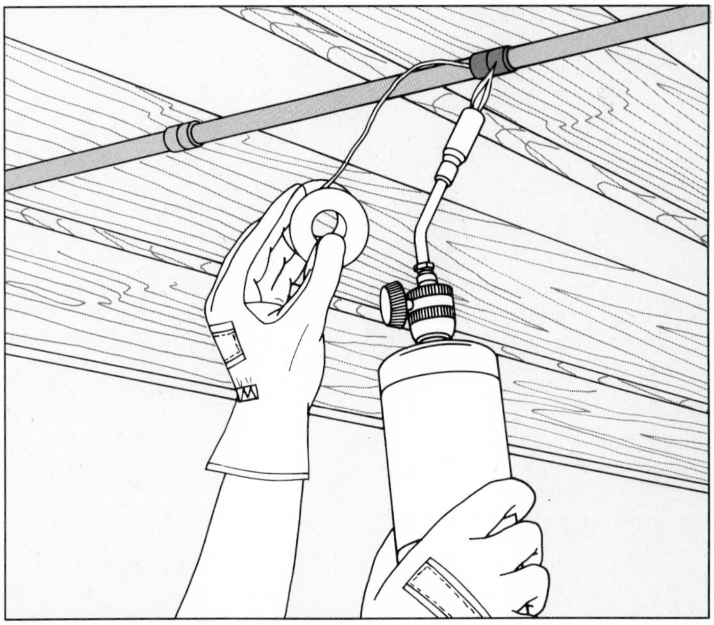

7 Soldering the joint. Wear safety glasses and work gloves, and protect flammable materials with a fireproof shield. Light the propane torch and play the flame over the fitting and nearby pipe, heating them as evenly as possible. Touch the tip of the wire solder to the joint *(above)* until it melts into the fitting, but do not let the flame touch the solder. When the joint is properly heated, the flux inside draws molten solder into the fitting to seal the connection. Feed solder into the joint until a bead of metal appears around the edge.

REPAIRING COPPER PIPE WITH POLYBUTYLENE (PB) PIPE

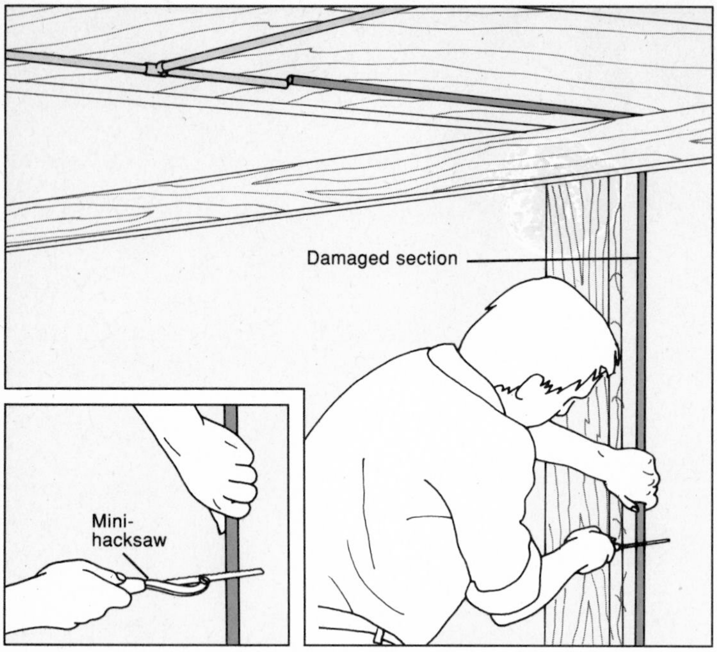

1 Removing the damaged section. Close the main shutoff valve and drain the supply lines. If the damaged pipe is in an awkward location *(above)*, cut it with a mini-hacksaw *(inset)*, then file the ends square with a metal file. (Otherwise, use a tube cutter.) Deburr both ends of the standing pipe *(page 96)*.

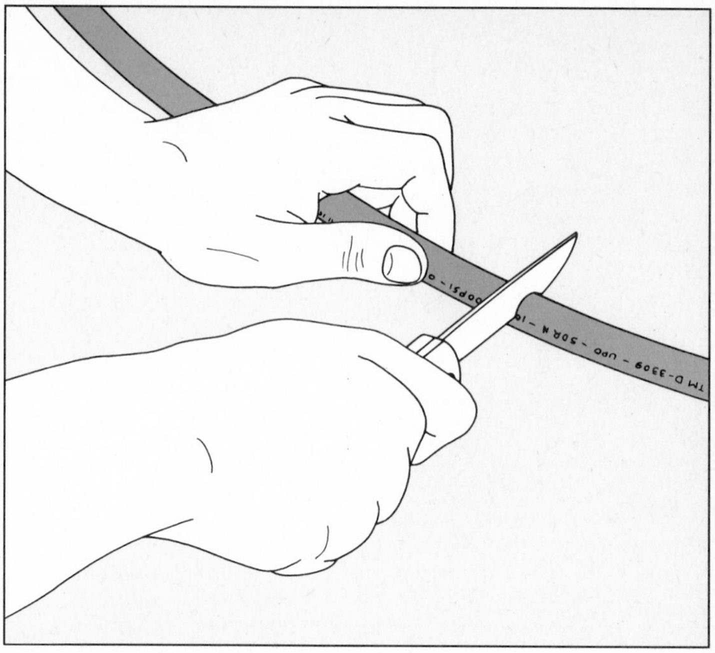

2 Cutting flexible polybutylene (PB) pipe. Measure the gap in the copper pipe and transfer that measurement to the PB pipe. Hold the PB pipe against a flat surface and press down squarely with a sharp knife *(above)*. Once the knife bites into the surface it will easily slice through the pipe. PB pipe can also be cut using a plastic tube cutter with a guillotine blade.

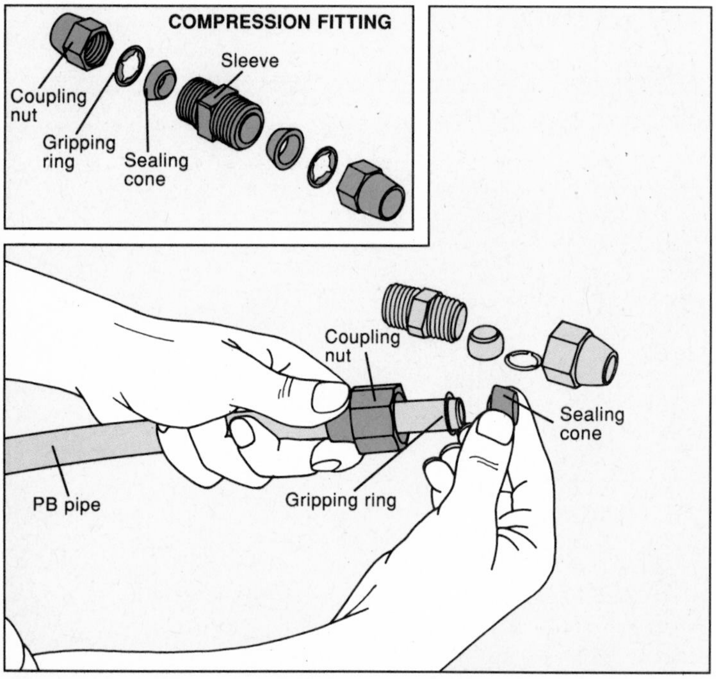

3 Fitting the adapters. Disassemble a compression fitting *(inset)*, used to join unthreaded pipe without solder or cement. Slide a coupling nut onto one end of the PB pipe, then a metal gripping ring and plastic sealing cone *(above)*. The end of the PB pipe should protrude 1/4 to 1/2 inch from the cone. Other compression fittings may simply slide onto the two pipes to be joined.

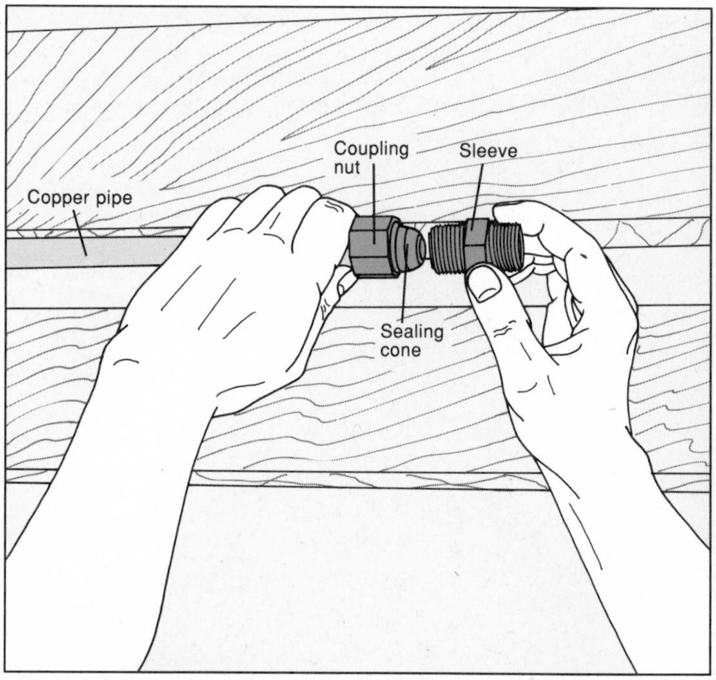

4 Adapting copper to PB. Fit the second nut, gripping ring and sealing cone onto the end of the copper pipe. With one hand, hold the sleeve against the sealing cone on the copper pipe *(above)*. With the other hand, screw the nut to the sleeve *(above)* and tighten by hand. Then fit the PB pipe, with its half of the fitting, onto the other end of the sleeve and tighten by hand.

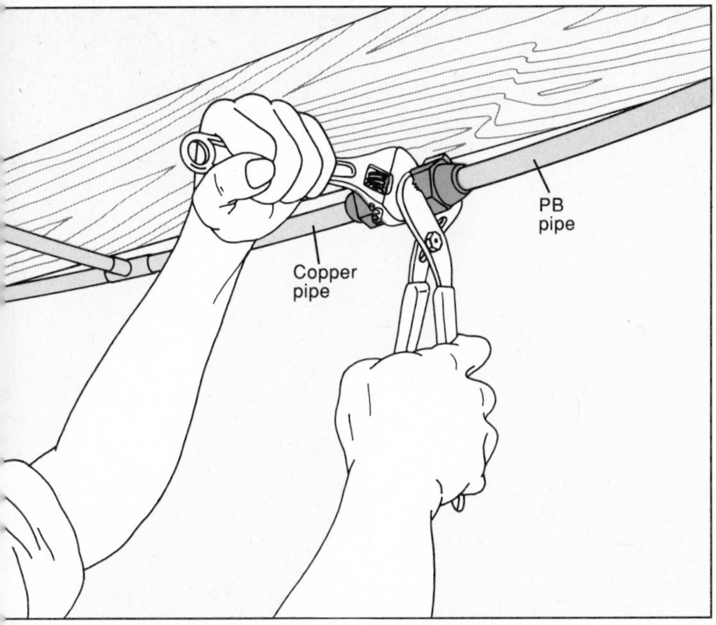

5 Tightening the fitting. Use two adjustable wrenches or channel-joint pliers to tighten the compression fitting. Grip the sleeve nut with one wrench and tighten the coupling nut one-half turn (but no more) with the second wrench *(above)*. Overtightening will displace the sealing cone or strip the plastic threads.

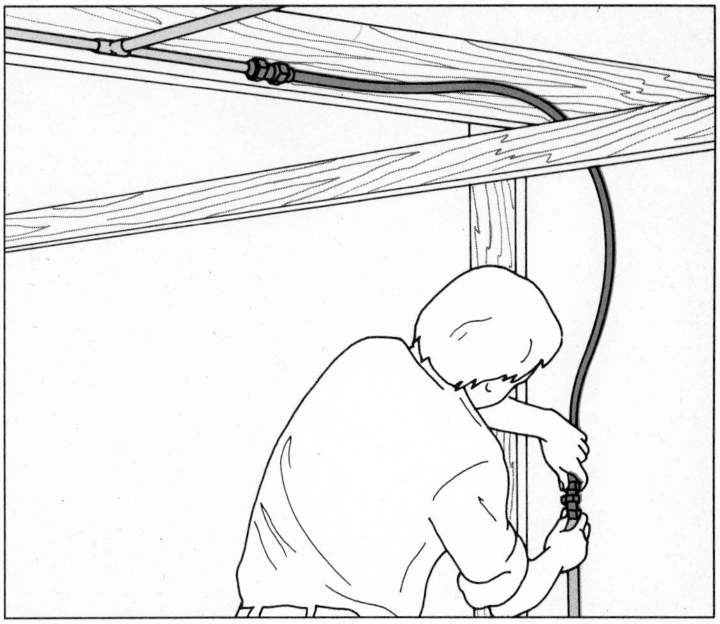

6 Completing the repair. Connect the other end of the PB pipe to the copper pipe *(above)* with a second compression fitting, repeating steps 3, 4 and 5. Turn on the water; if a joint leaks, reassemble the fitting rather than tightening it further. The problem is likely a sealing cone that is not properly seated inside the fitting.

REPAIRING COPPER PIPE WITH CPVC PIPE

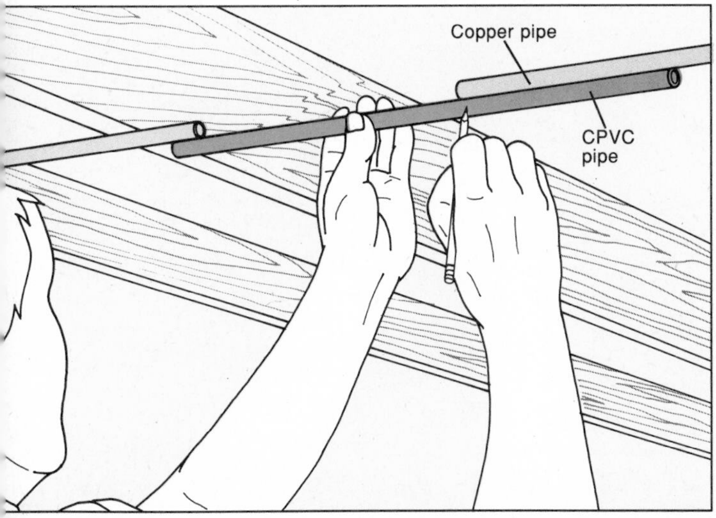

1 Removing the damaged section. Prepare to be without water for as long as it takes the CPVC solvent cement to cure (two hours at temperatures above 60°F). Close the main shutoff valve and drain the supply lines. Cut the broken section of copper pipe with a tube cutter or hacksaw and deburr the ends of the standing pipes (*page 96*). Hold the CPVC pipe against the gap and mark it with a pencil *(above)*.

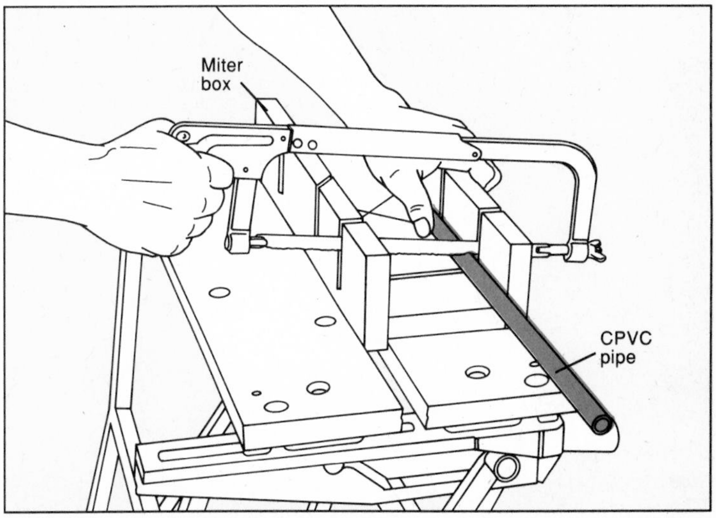

2 Cutting the replacement pipe. Place the CPVC pipe in a miter box, lining up the pencil mark on the pipe with the slotted saw guides *(above)*. Brace the pipe with one hand and saw through the mark with a hacksaw. If you are cutting several pipes, a plastic tube cutter will speed the job.

REPAIRING COPPER PIPE WITH CPVC PIPE (continued)

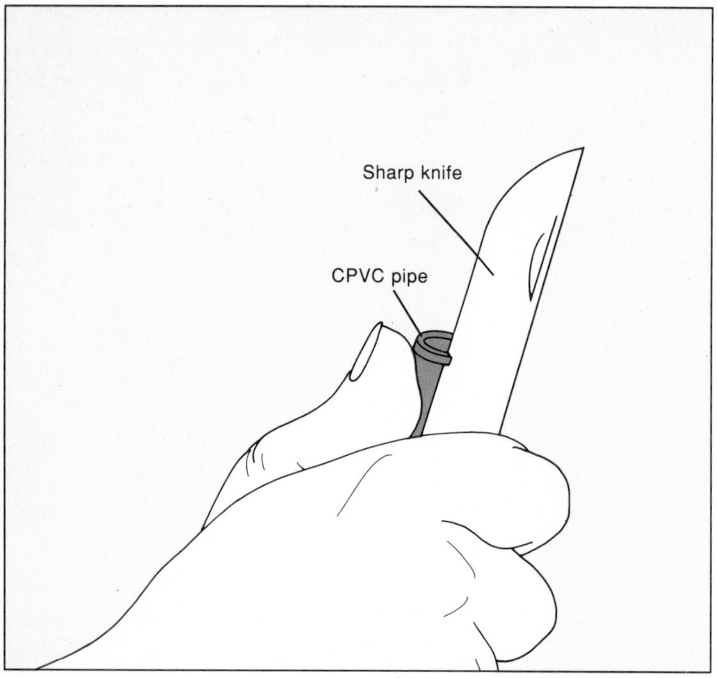

3 Deburring and beveling CPVC. The ends of a sawed-off CPVC pipe must be deburred (but not a pipe cut with a plastic tube cutter) and beveled. Using a sharp knife *(above)*, trim the inside edge to aid water flow and the outside edge to improve the welding action of the solvent.

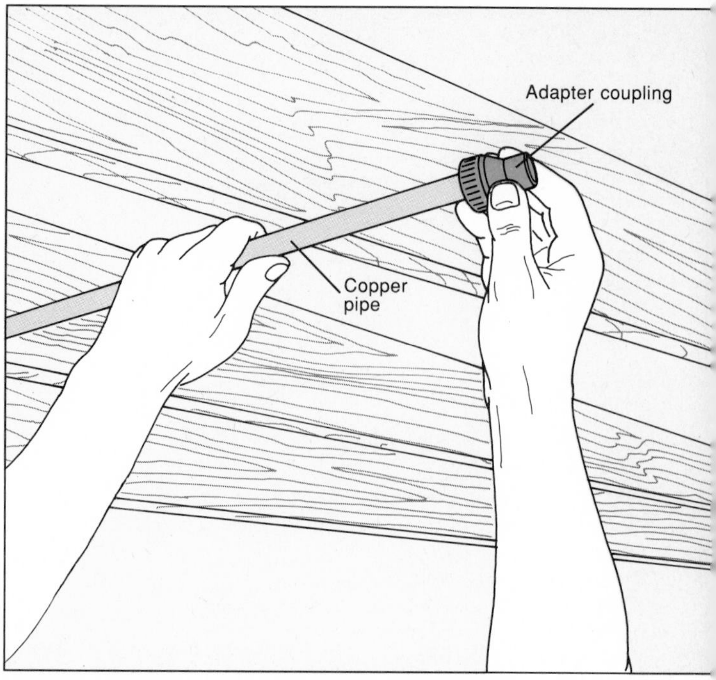

4 Fitting the adapter couplings. Loosen two adapter couplings and push them onto the copper pipes until the pipe ends bottom out inside the coupling sockets. To make the job easier, file the outside edges of the copper pipes, then lubricate them with petroleum jelly. If you mark the pipes about 1 1/2 inches from the ends, you will be able to tell if they are properly seated. Hand-tighten both couplings.

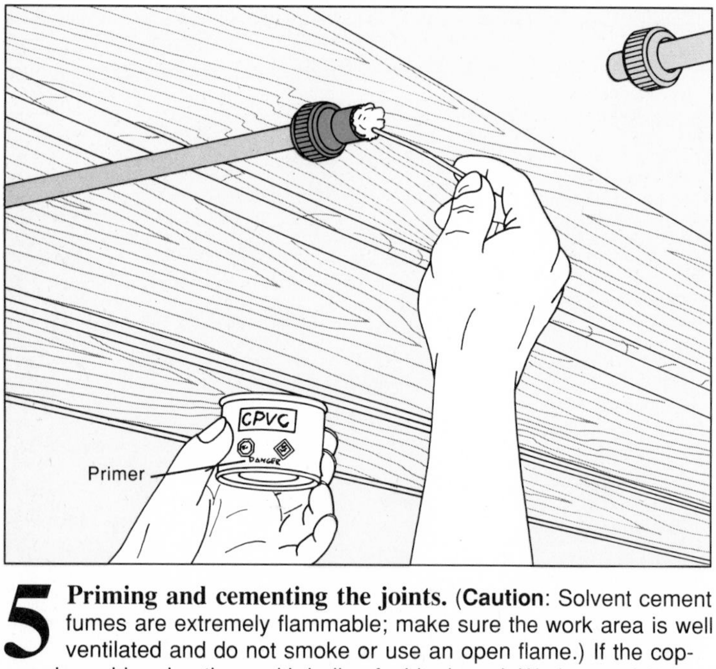

5 Priming and cementing the joints. (**Caution**: Solvent cement fumes are extremely flammable; make sure the work area is well ventilated and do not smoke or use an open flame.) If the copper pipes drip, plug them with balls of white bread. Work as quickly as possible with the cement, which sets in less than 30 seconds. With an applicator or clean cloth, apply a coat of primer to the ends of the CPVC pipes and the adapter couplings *(above)*. With another applicator, apply a liberal coat of CPVC cement as deep as the coupling sockets to the outside ends of the CPVC pipe. Then apply a light coat of cement inside the coupling sockets.

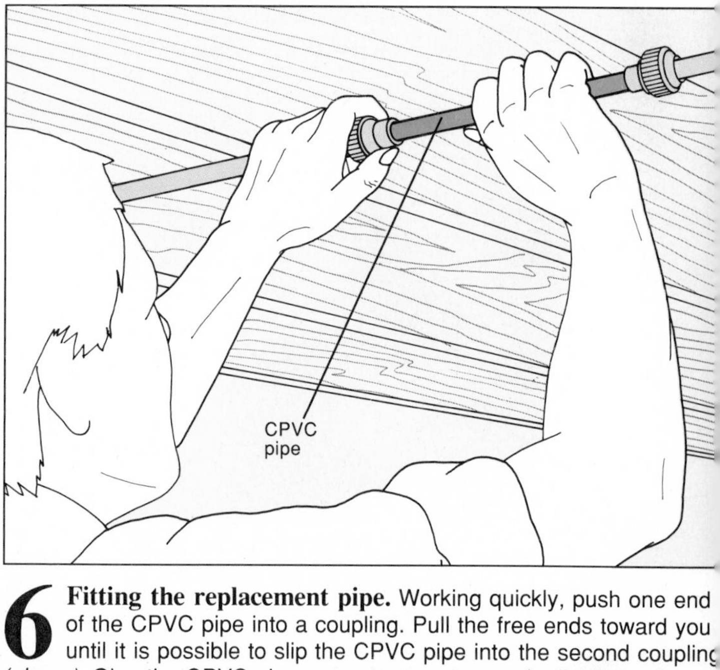

6 Fitting the replacement pipe. Working quickly, push one end of the CPVC pipe into a coupling. Pull the free ends toward you until it is possible to slip the CPVC pipe into the second coupling *(above)*. Give the CPVC pipe a quarter-turn to evenly distribute the cement inside the sockets. Wipe away any excess cement with a clean, dry cloth.

REPAIRING GALVANIZED STEEL PIPE

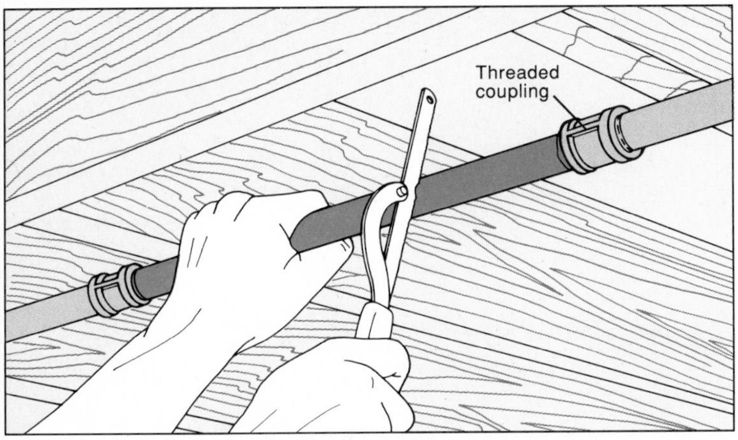

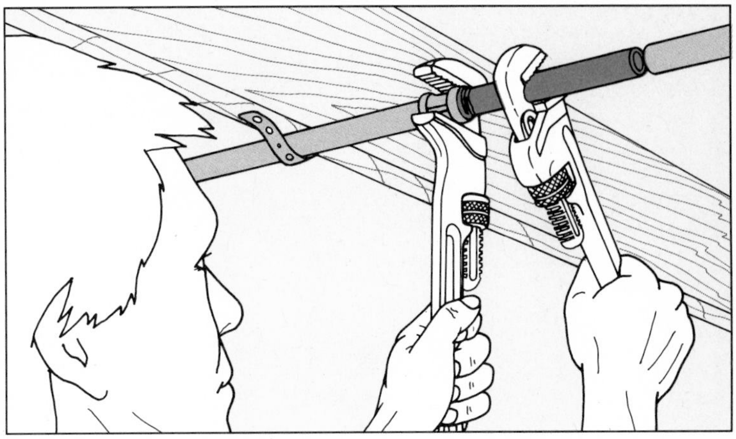

1 **Removing the defective run.** Threaded joints make galvanized pipes easy to assemble but complicated to remove. Once it is in place, a threaded pipe cannot be unscrewed as one piece, since loosenening it at one fitting will tighten it at the other. Close the main shutoff valve and drain the supply lines. Look for a union near the damaged section and, if there is one, unscrew the pipe from it. Unscrew the other end from its coupling. If there is no union, cut through the pipe with a fine-tooth hacksaw or mini-hacksaw *(above, left)* and unthread the two pieces from their couplings. To unthread pipe from a union or coupling, grip the coupling with one wrench and turn the pipe with another so that the rest of the run will not be twisted or strained. The jaws of the wrenches should face the direction in which force is applied *(above, right).* Since galvanized pipe is expensive, you may want to save any long, undamaged sections of cut pipe and have them rethreaded by a plumbing supplier.

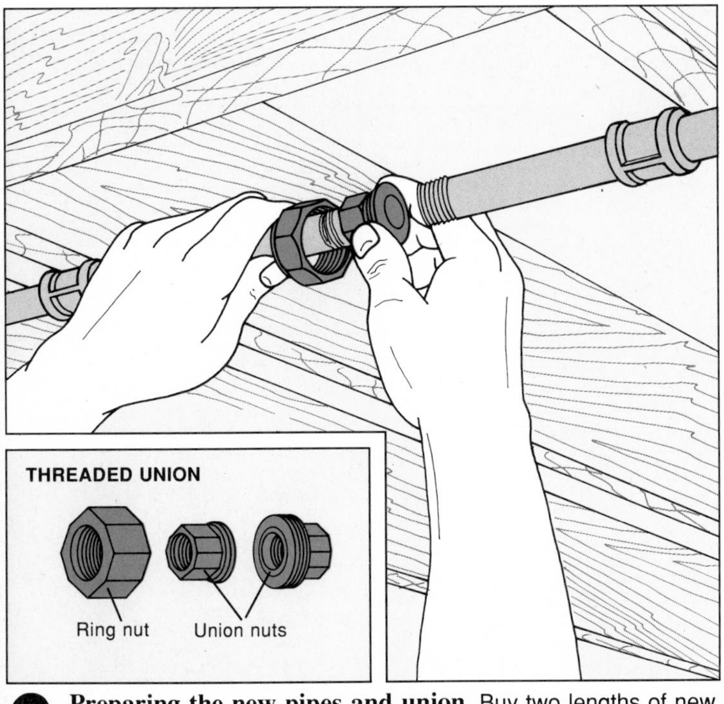

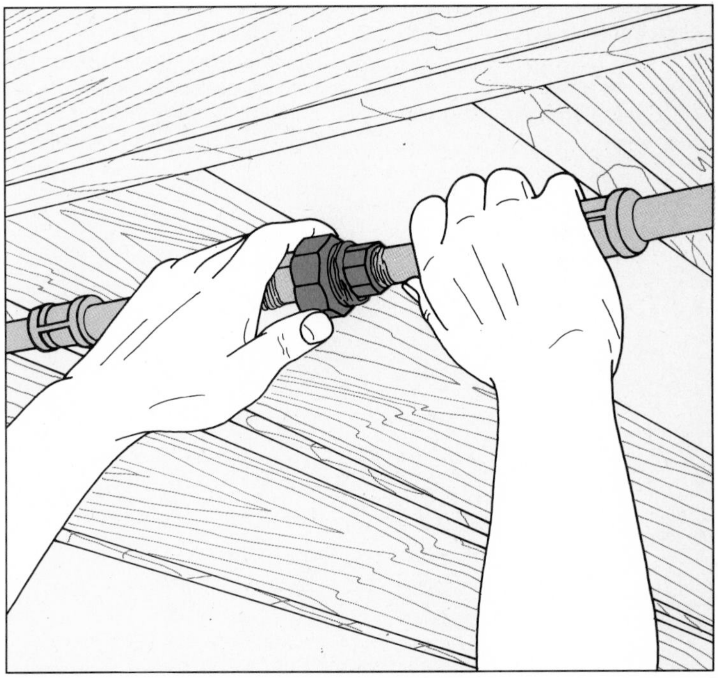

2 **Preparing the new pipes and union.** Buy two lengths of new pipe and a union whose combined length, when threaded together, is the same as the broken section. Wind 1 1/2 turns of pipe tape clockwise over the threads of the new pipe, tightly enough so that they show through. Thread one of the pipes into the nearest coupling using the same double-wrench technique described in step 1. Then thread the other pipe into its fitting, leaving a slight gap in the run. Disassemble a threaded union *(inset).* Slide the union's ring nut onto one of the pipes, then screw union nuts to each pipe end *(above).*

3 **Connecting the union.** Slide the ring nut to the center of the union and screw it onto the exposed threads of the union nuts *(above)*, joining the two pipes. Grip the exposed union nut with one wrench and tighten the ring nut with a second wrench.

REPAIRING STEEL PIPE WITH COPPER PIPE

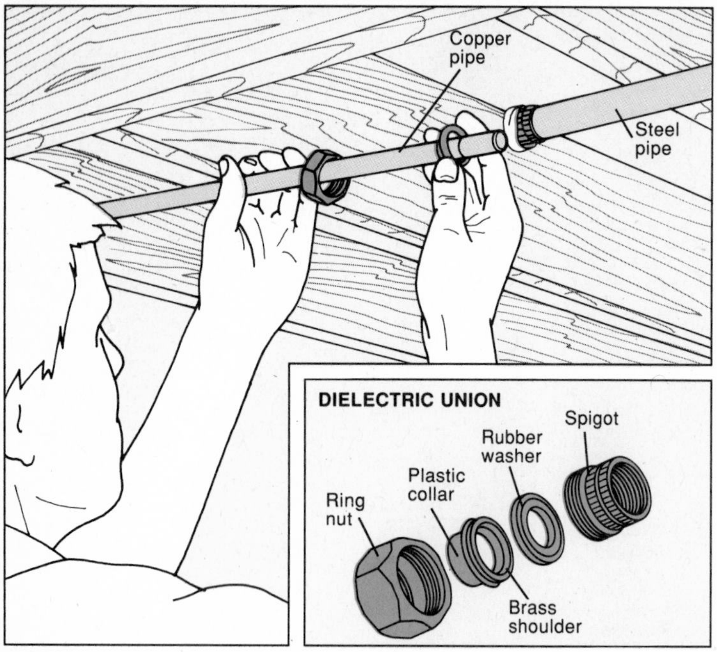

1 Working with dielectric unions. Close the main shutoff valve and drain the supply lines. Unthread or cut out the damaged section of galvanized pipe *(page 101)*, and measure the amount of replacement pipe you will need *(page 95)*. Special dielectric unions *(inset)* are required to join copper pipe to galvanized; otherwise, the electrolytic reaction between the two metals will encourage corrosion. Screw the steel spigot of one dielectric union onto a steel pipe, and apply pipe tape to its exposed threads. Slip the ring nut and plastic collar past the end of the copper pipe *(above)*. Fit the second dielectric union onto the other pipe ends in the same way *(above)*.

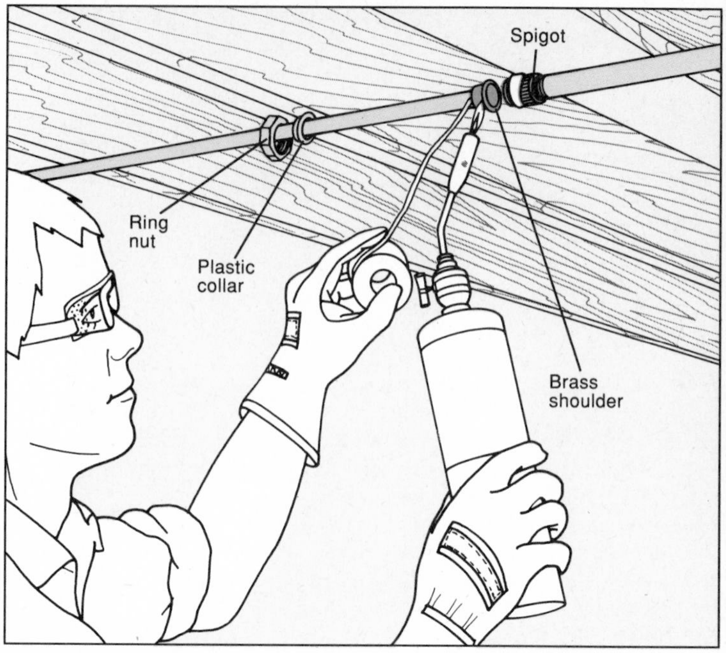

2 Soldering the brass fittings. Wear safety glasses and work gloves, and protect flammable materials with a fireproof shield. With an emery cloth, burnish the brass shoulders of the two dielectric unions and both ends of the copper replacement pipe. Apply flux to the inside of the brass shoulders and the outside of the copper pipes. Slip the shoulders onto the ends of the pipes and give them a quarter-turn to evenly distribute the flux. Solder the shoulders to the pipes *(page 140)* and allow them to cool before completing the repair.

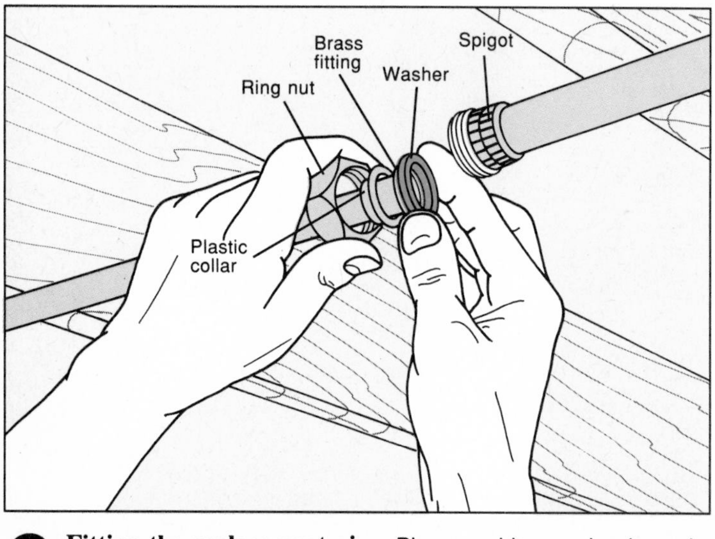

3 Fitting the replacement pipe. Place a rubber washer in each of the soldered brass shoulders, then fit the copper replacement pipe between the galvanized standing pipes *(above)*.

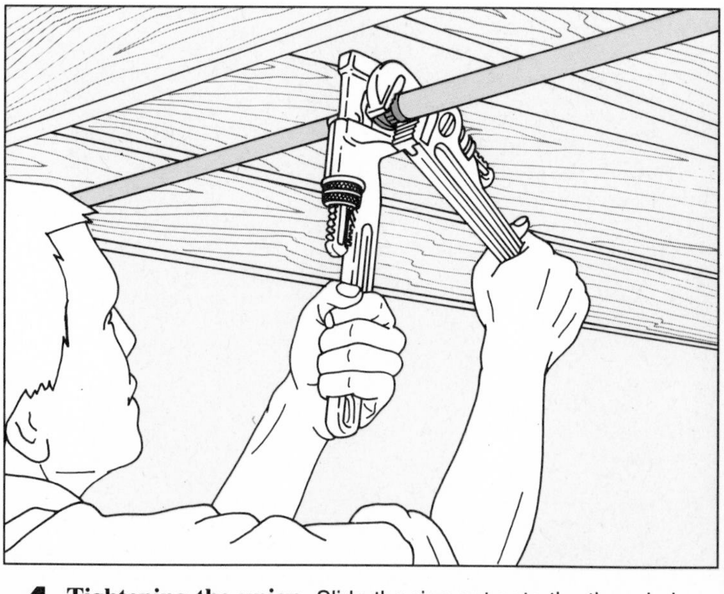

4 Tightening the union. Slide the ring nut onto the threaded spigot and tighten by hand. Grip the spigot with one wrench and tighten the ring nut one-half turn with a second wrench *(above)*. Repeat this procedure to assemble and tighten the other union. Note that if the defective pipe was part of your home's electrical grounding system, the circuit will have been broken by the dielectric unions. Have an electrician install a grounding jumper to maintain continuity.

REPAIRING STEEL PIPE WITH POLYBUTYLENE (PB) PIPE

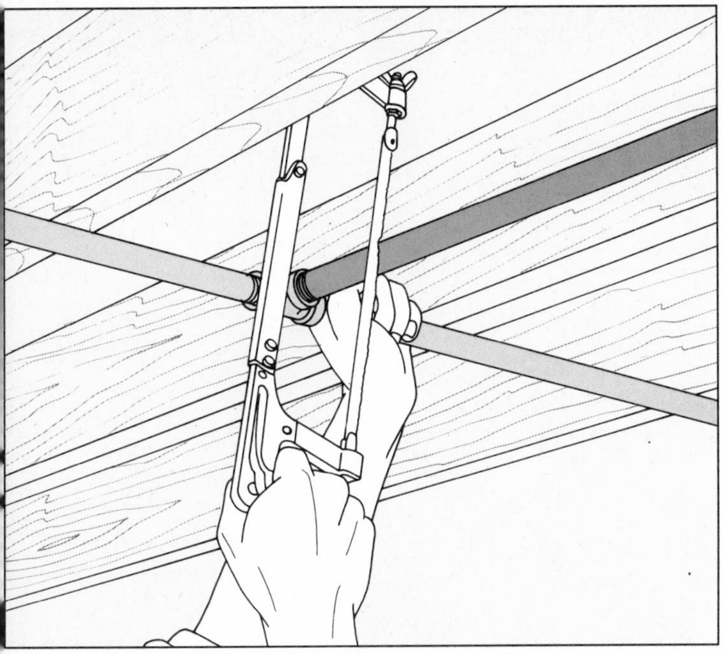

1 Removing the broken pipe. Close the main shutoff valve and drain the supply lines. Unthread or cut through the broken pipe and remove it from the nearest fittings *(above)*. If the fittings are in good condition, simply clean them. If they are corroded, replace them with identical parts.

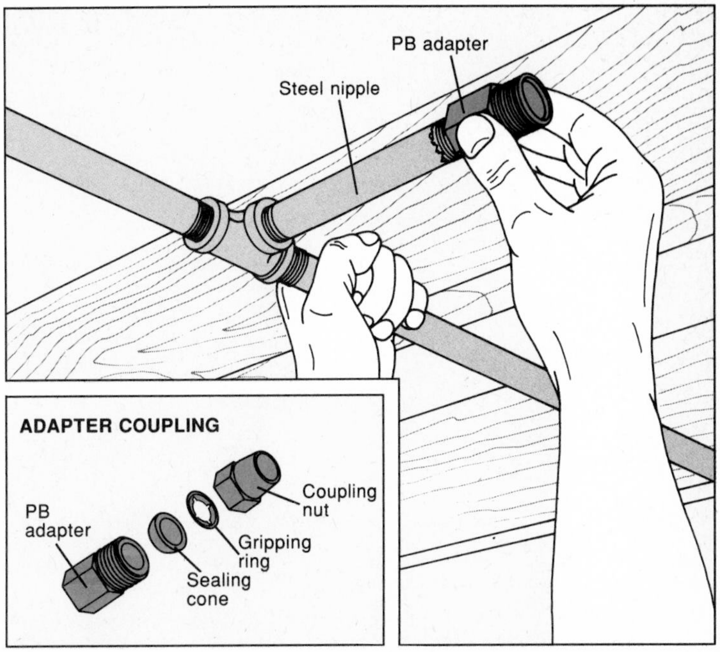

2 Attaching the adapter couplings. Apply pipe tape to both ends of a short steel nipple, then screw it into one of the steel fittings. Grip the fitting with one wrench and tighten the nipple one-half turn with a second wrench (but do not overtighten). Thread a nipple to the other steel fitting, then screw PB adapters onto both nipples by hand *(above)*.

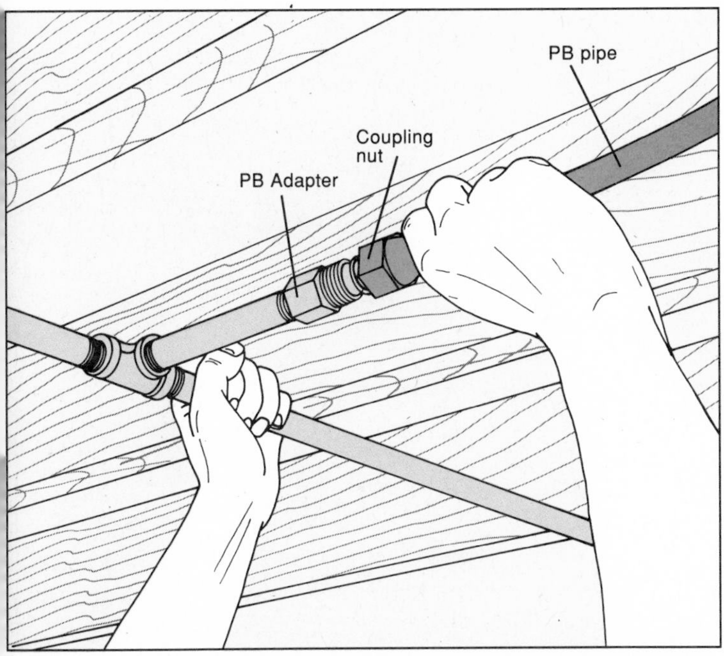

3 Fitting PB pipe to the adapter. Fit the coupling nut, gripping ring and sealing cone onto one end of the PB pipe. Hold the pipes together *(above)* and screw the coupling nut to the PB adapter. Hold the other end of the PB pipe to the second adapter, bending it if necessary, and mark the pipe. Cut the pipe with a sharp utility knife. Fit the second coupling nut, gripping ring and sealing cone assembly onto the end of the PB pipe, hold it against the adapter, and screw the coupling nut to the adapter by hand.

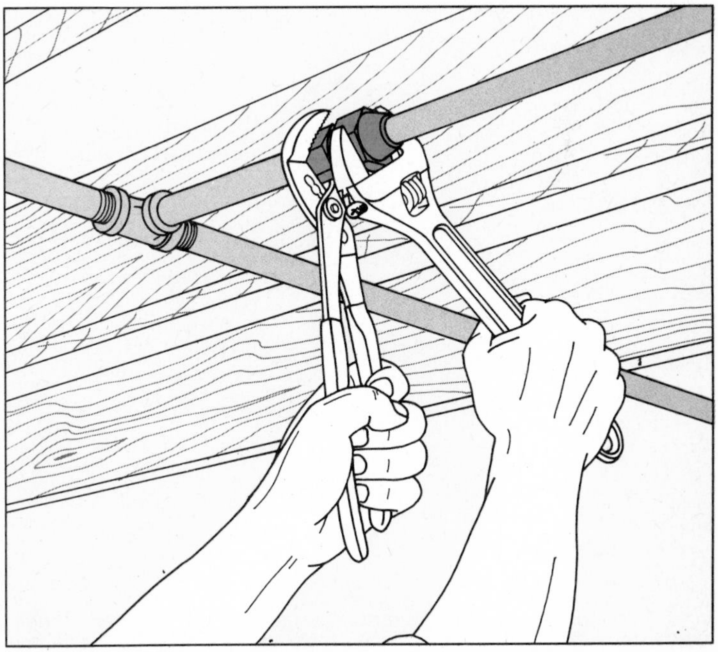

4 Completing the connection. Grip the adapter with one wrench and tighten the coupling nut one-half turn with a second wrench *(above)*. Then tighten the other PB adapter and coupling nut assembly. Turn on the water. If there are any leaks from the joints, tighten another half-turn. If the damaged pipe was part of your home's electrical grounding system, have an electrician install a grounding jumper to maintain continuity.

REPAIRING STEEL PIPE WITH CPVC PIPE

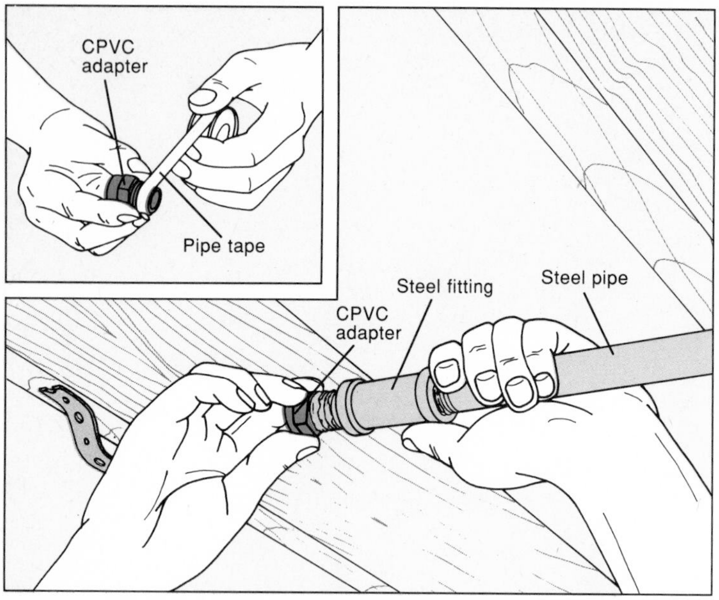

1 Adding the CPVC adapter. Prepare to be without water for as long as it takes the CPVC solvent cement to cure (two hours at temperatures above 60°F). Close the main shutoff valve and drain the supply lines. Cut and unthread the damaged steel pipe from its fittings *(page 101)*. Wrap pipe tape around the threads of two CPVC adapters *(inset)* and screw them into the steel fittings by hand *(above)*. With an adjustable wrench, tighten the adapters just beyond hand-tight.

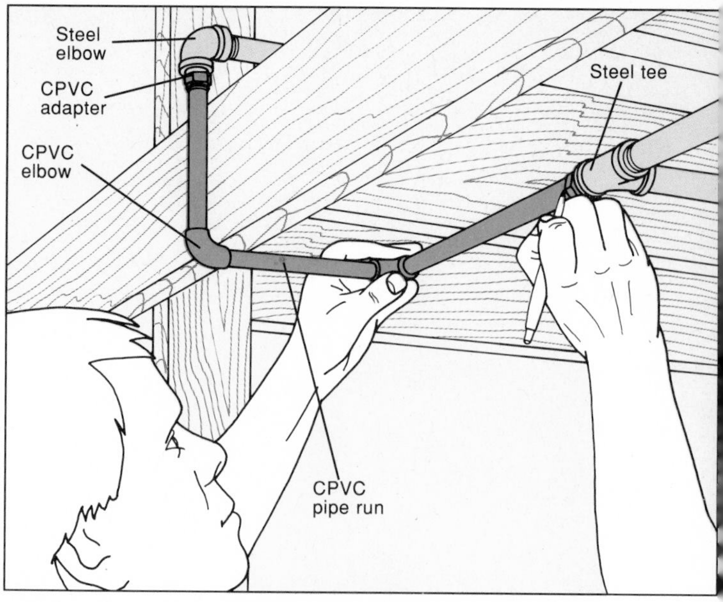

2 Measuring and test-fitting CPVC pipe. Measure the length of CPVC replacement pipe needed (in this example, a short run) and cut the pipes square with a hacksaw in a miter box *(page 99)*. Push the end of one pipe into an adapter socket as far as it will go. To join CPVC sections for a long run of pipe, push an elbow or fitting onto the other end of the pipe. Measure and cut the next length, push it into the elbow or coupling, and continue dry-fitting the new pipe run. At each connection, draw a line across the fitting and adjacent pipe *(above)* as a guide for reassembly and gluing.

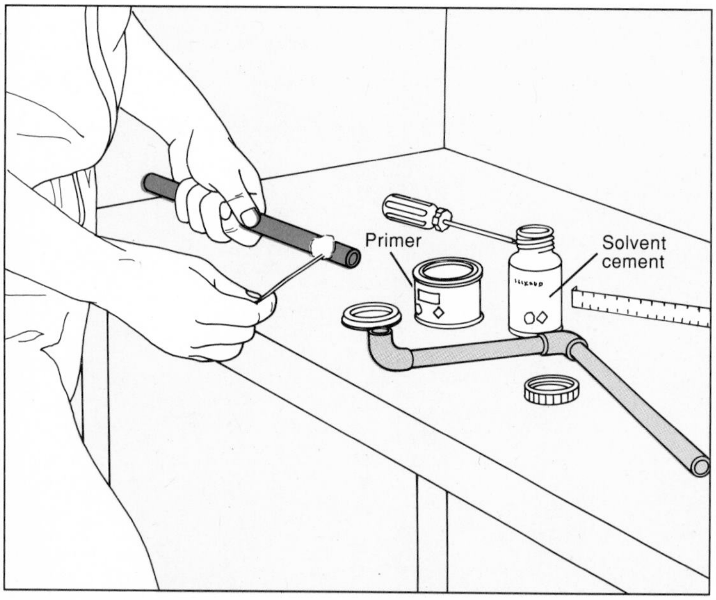

3 Cementing the CPVC pipe. (**Caution:** Solvent cement fumes are extremely flammable; make sure the work area is well ventilated and do not smoke.) Remove and disassemble the CPVC pipes and fittings, keeping track of their proper order. Using a sharp knife, bevel and deburr all pipe ends. Clean the first pipe and fitting in the run with CPVC primer, then apply CPVC solvent cement to both *(above)*. Push the pipe into the fitting, give it a quarter-turn to spread the cement, and align the pencil marks. Hold the pipe and fitting together for about 10 seconds. Continue gluing and assembling the run until the last fitting is in place, but do not attach the last pipe yet.

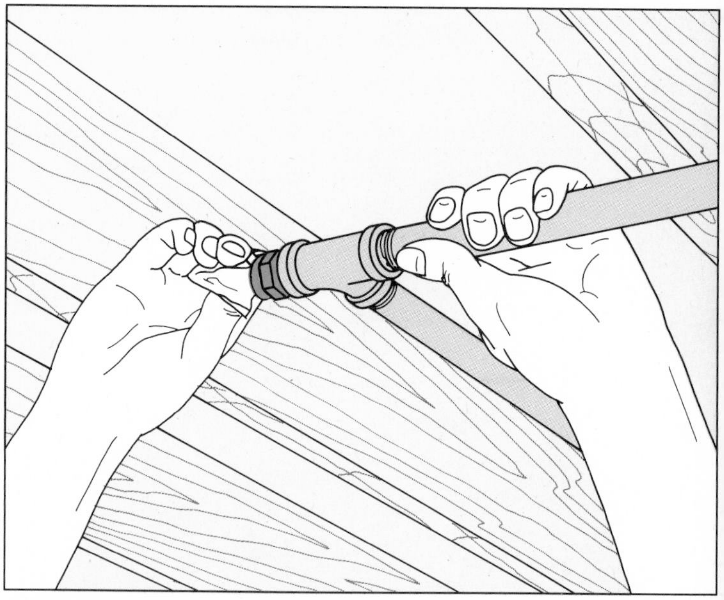

4 Preparing the adapters. Clean the CPVC adapters with primer *(above)*. If the sockets are wet, dry them with a clean cloth and plug the steel pipes with balls of white bread.

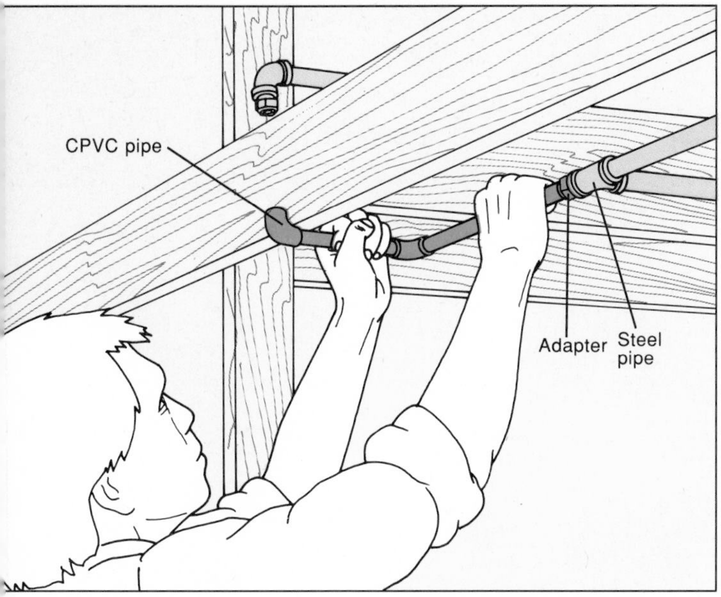

5 Fitting the CPVC run. Coat one of the adapter sockets with solvent cement, then apply cement to the end of the assembled run of CPVC pipe. Push the pipe into the adapter socket as far as it will go, then give the pipe a quarter-turn to spread the cement. Line up the marks on the pipe and socket and hold the pieces together for about 10 seconds *(above)*.

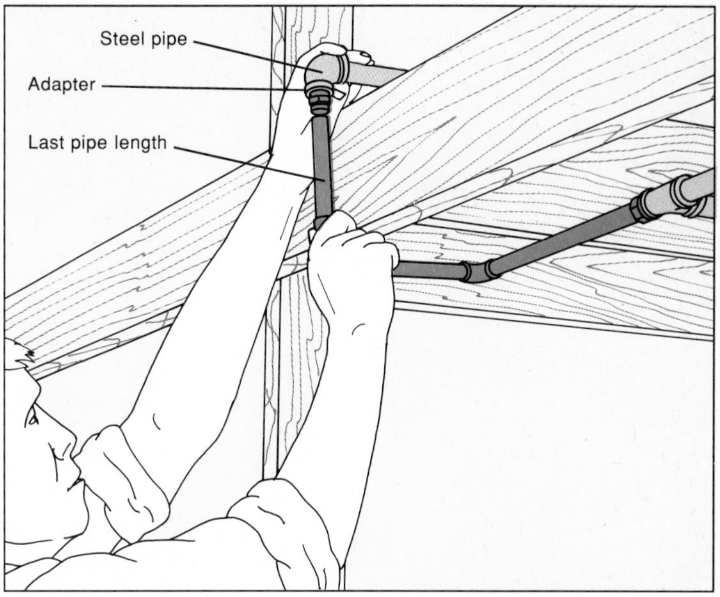

6 Fitting the last pipe. Prime and cement the second adapter, the last fitting on the CPVC run, and both ends of the remaining pipe. Push one end of the pipe into the fitting, gently pull the attached run to make room *(above)*, then push the other end into the adapter socket as far as it will go. Give the pipe a quarter-turn to spread the cement and hold the pieces together, as shown, for about 10 seconds. If the replaced section of steel pipe was part of your home's electrical grounding system, have an electrician install a grounding jumper to maintain continuity.

REPAIRING CPVC PIPE

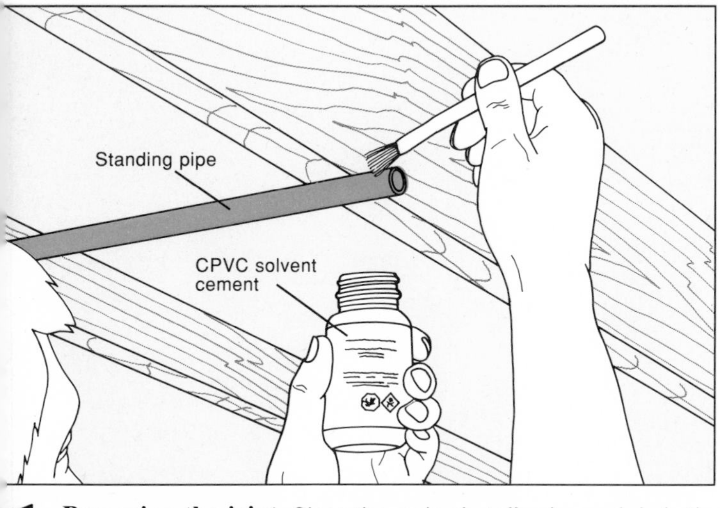

1 Preparing the joint. Close the main shutoff valve and drain the supply lines. Cut out the damaged section of pipe with a hacksaw or plastic tube cutter. Hold the replacement pipe against the gap and mark it with a pencil, then cut it squarely with a tube cutter or hacksaw and miter box. Using a sharp knife, deburr and bevel the ends of the pipes. Clean the ends of the standing pipes and two coupling sockets with CPVC primer. Apply a liberal coat of CPVC solvent cement to the couplings and pipe ends *(above)* to the depth of the socket. Push the couplings onto the pipes, give them a quarter-turn to spread the cement, and hold the pieces together for about 10 seconds. (**Caution:** Solvent cement fumes are extremely flammable.)

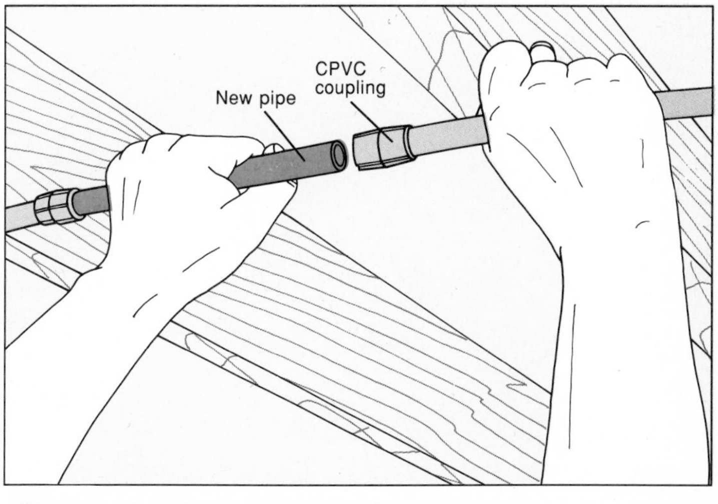

2 Inserting the replacement pipe. Clean and prime the exposed coupling sockets and both ends of the replacement pipe, then apply cement. Working quickly, push one end of the replacement pipe into a coupling, then gently bend the pipes toward you until there is enough room to slip it into the other coupling *(above)*. Give the pipe a quarter-turn to spread the cement and press the pieces together for about 10 seconds. Wipe off any excess cement around the pipe or fittings with a clean, dry cloth. Do not run water in the pipe until the CPVC has cured (about two hours at temperatures above 60°F).

REPAIRING CPVC PIPE WITH POLYBUTYLENE (PB) PIPE

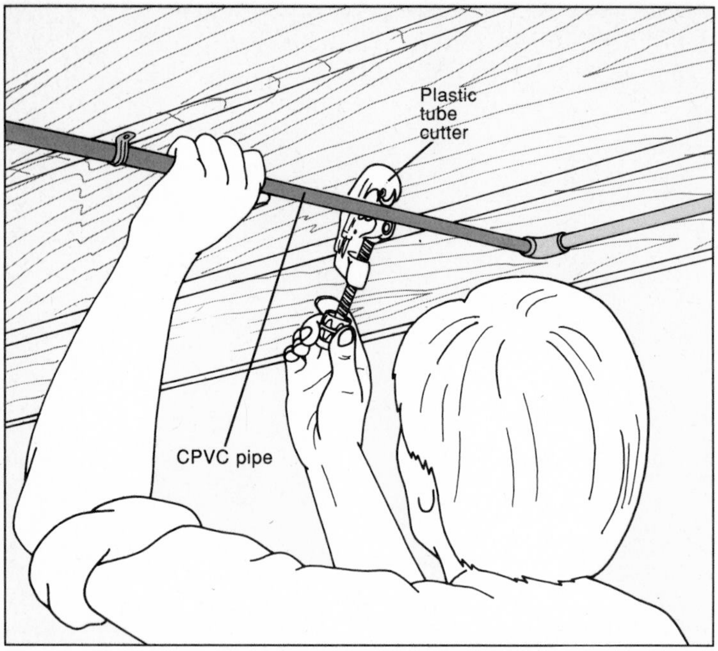

1 Removing the damaged section. Close the main shutoff valve and drain the supply lines. Cut out the damaged section of CPVC pipe with a hacksaw or plastic tube cutter *(above)*. Using a sharp knife, deburr the ends of the standing pipes.

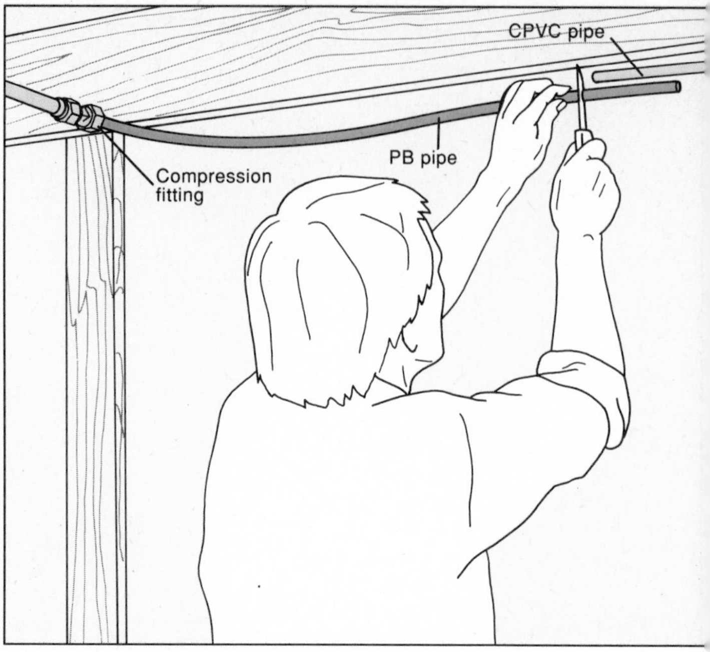

2 Attaching the replacement pipe. Disassemble a compressio fitting *(page 98)* and slide a coupling nut, gripping ring and seal ing cone onto one of the standing CPVC pipes. Slide the other half of the fitting onto the end of the PB replacement pipe. Join the pipe ends and screw the coupling nuts to the connecting sleeve by hand. Grip the sleeve nut with one wrench and tighten each coupling nut with a second wrench (but do not overtighten). Run the other end of the PB pipe to meet the CPVC pipe and cut it to length *(above)*.

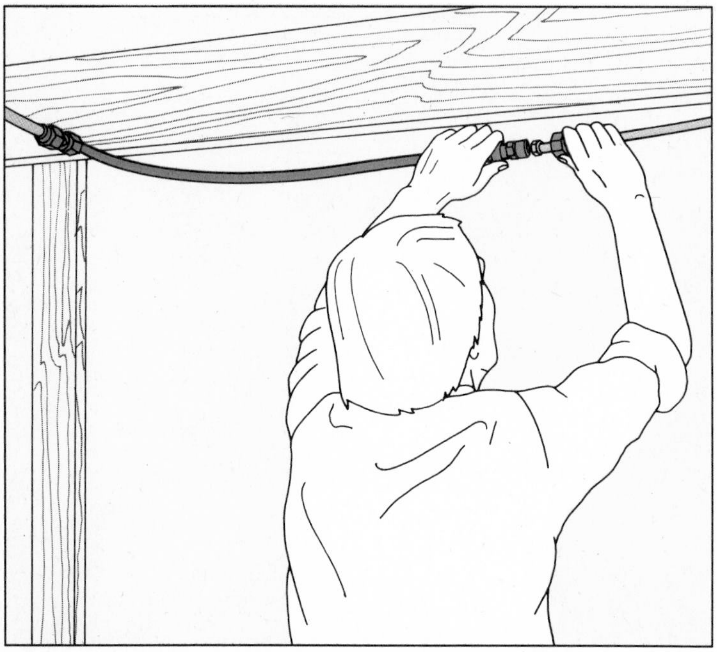

3 Completing the connection. Disassemble a second compression fitting and slide each half onto the free ends of the PB and CPVC pipes, then join them with a connecting sleeve *(above)* as in step 2. Grip the sleeve nut with one wrench and tighten each coupling nut with a second wrench.

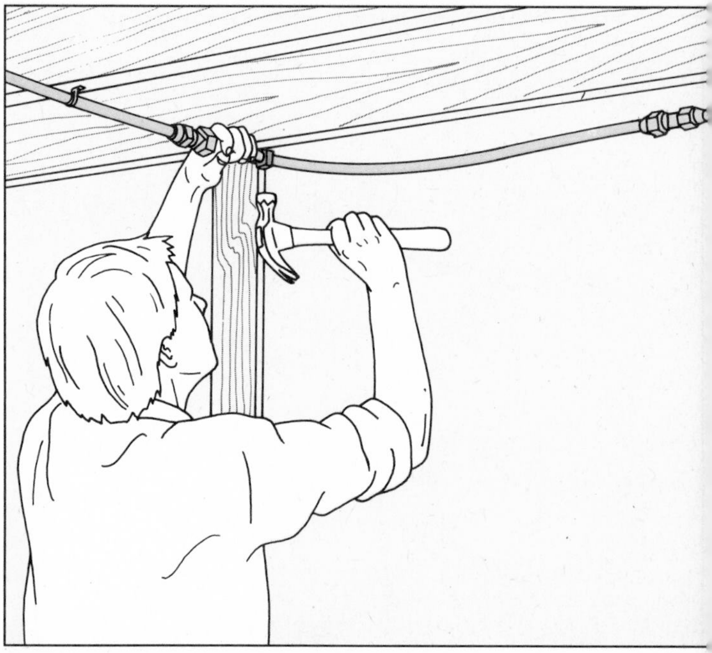

4 Supporting the replacement pipe. In an overhead run, support flexible PB pipe with plastic hangers nailed or screwed to every other joist (every 32 inches), as shown. Support vertical PB pipe every 3 or 4 feet. Turn on the water. If a joint leaks, reassem ble its fitting, making sure that the sealing cone is properly seated.

REPAIRING CAST-IRON JOINTS

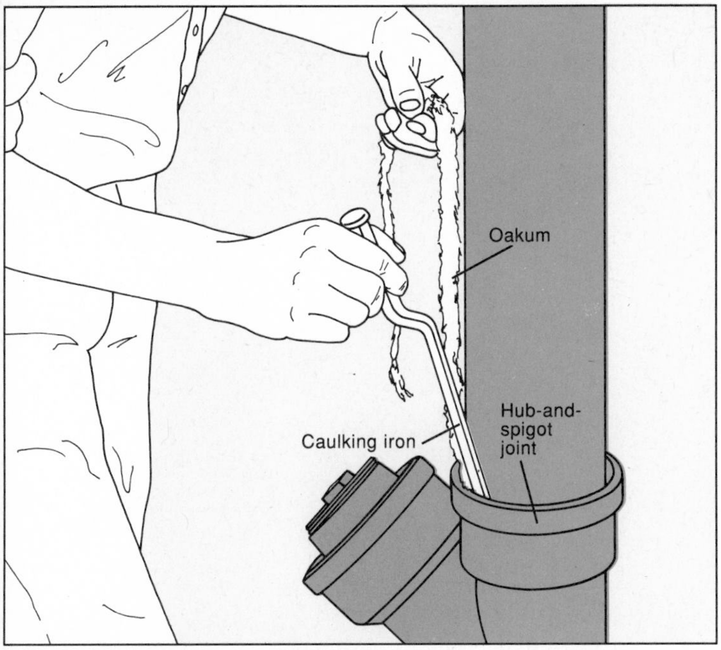

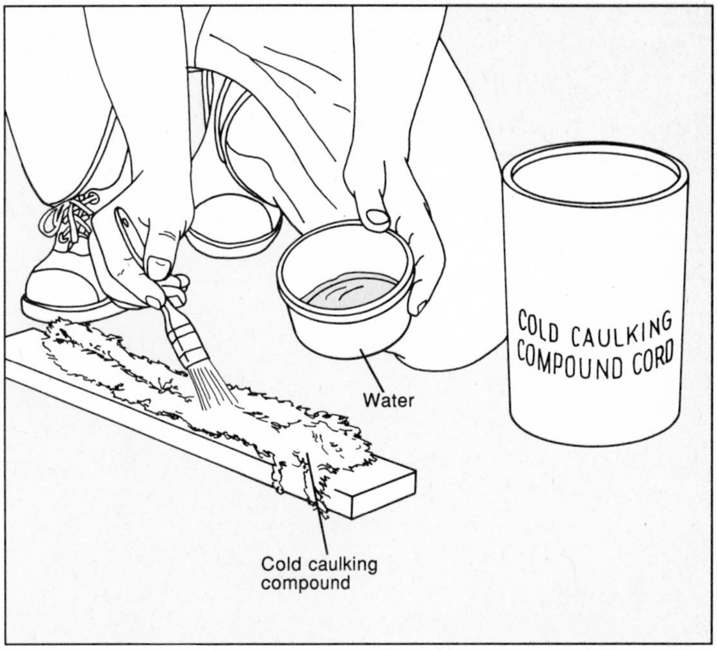

1 Servicing lead-caulked joints. Most cast-iron drainpipe is connected at bell-shaped hub-and-spigot joints, which are sealed with molten lead and a ropelike material called oakum. If water leaks from such a joint, use a hammer and cold chisel to tamp down the lead inside the hub. Since the lead is soft enough to be reshaped over the weak spot, this simple procedure should reseal the joint. If this doesn't work, measure a length of oakum twice the circumference of the pipe, and pack it evenly inside the hub with the end of a caulking tool *(above)* or old screwdriver.

2 Preparing the new caulking. Cold caulking compound and "plastic lead" have replaced molten lead as a caulking for cast-iron joints. Measure a length of the ropelike caulking compound twice the circumference of the pipe and cut it with a sharp utility knife. Place the length on a clean, dry surface and dampen it (do not soak it) with water, using a clean brush *(above)* or a plant sprayer.

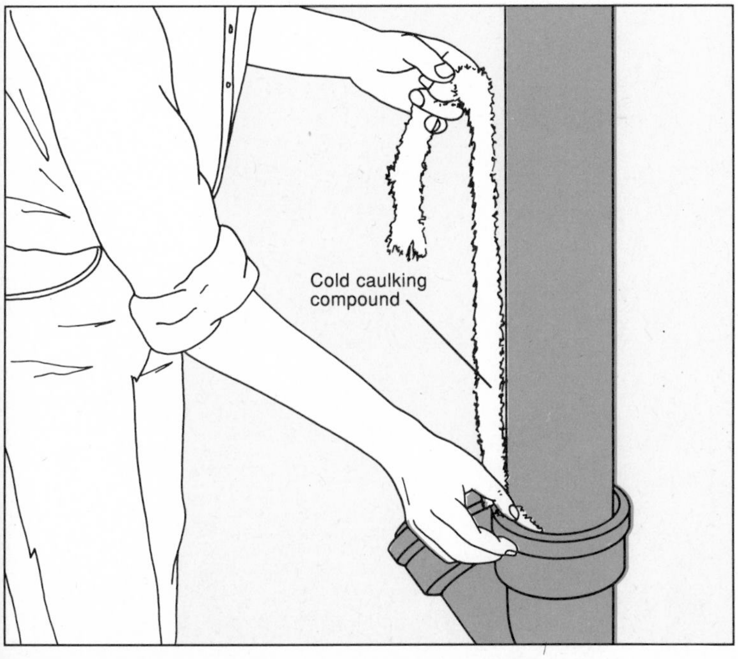

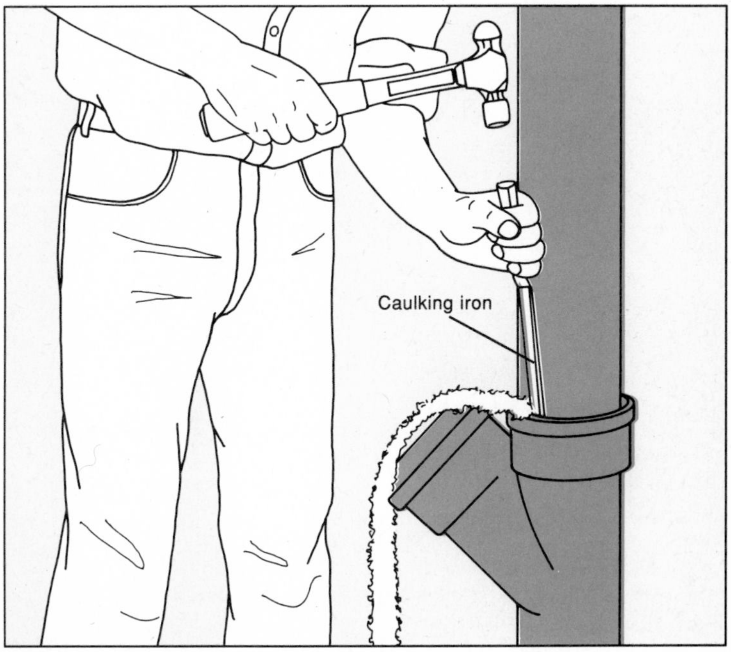

3 Tamping the joint. Work the dampened compound into the joint a few inches at a time *(above, left)*, then tamp it down with a hammer and caulking iron *(above, right)*. Continue adding and tamping the caulking a few inches at a time until the complete length has been packed into the joint. Wait at least six hours for the compound to harden before using the drainpipe. "Plastic lead," an alternative caulking material, is mixed with water and troweled into the joint with a putty knife or an old screwdriver.

REPAIRING CAST-IRON DRAINPIPE

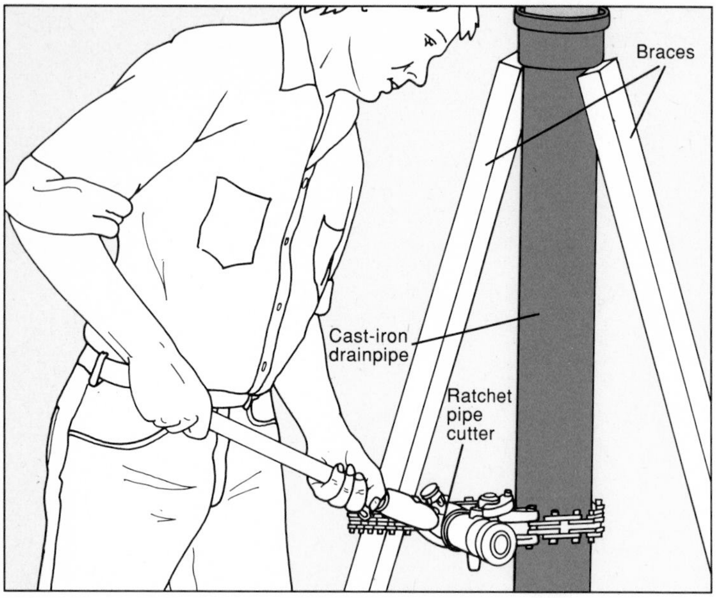

1 Removing the broken pipe. Do not run water in the house during this repair. The easiest way to cut cast-iron pipe is with a ratchet pipe cutter, available at rental stores. Before cutting vertical drainpipe, support it with riser clamps or 2x4s braced against a hub-and-spigot joint above the section to be removed. With chalk, mark the area to be cut, wrap the chain around the pipe and hook it onto the body of the tool. Tighten the knob, turn the dial to CUT and work the handle back and forth until the cutting discs bite through the pipe. If badly corroded pipe crumbles under a pipe cutter, rent an electric saber saw and metal-cutting blade instead.

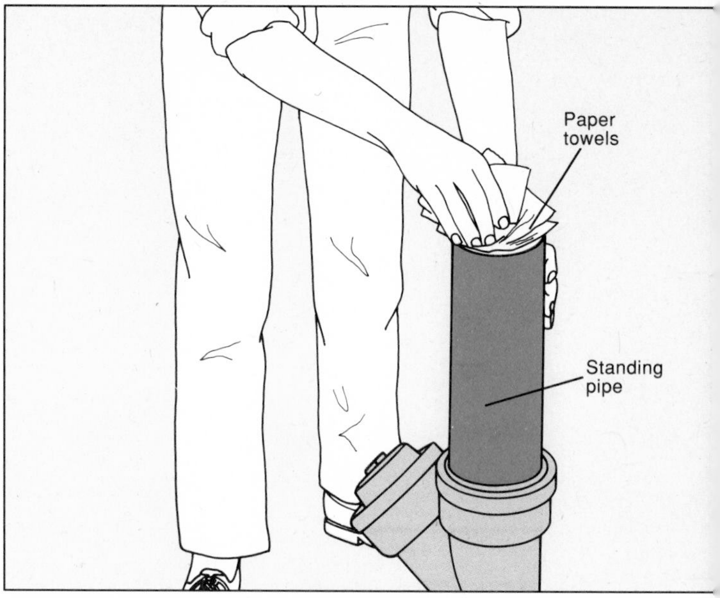

2 Cutting the replacement pipe. Immediately after removing the damaged section, stuff newspapers or paper towels into the standing pipes *(above)* to block dangerous sewer gas. Measure the gap in the pipe and transfer that measurement, less 1/4 inch, to a cast-iron, PVC or ABS replacement pipe. Lay cast-iron pipe across two level 2x4s, spaced to support the pipe ends, and cut it to size with a ratchet cutter *(step 1)* or saber saw. Cut PVC or ABS with a handsaw and miter box. With a sharp knife, deburr and bevel the ends of the plastic pipe.

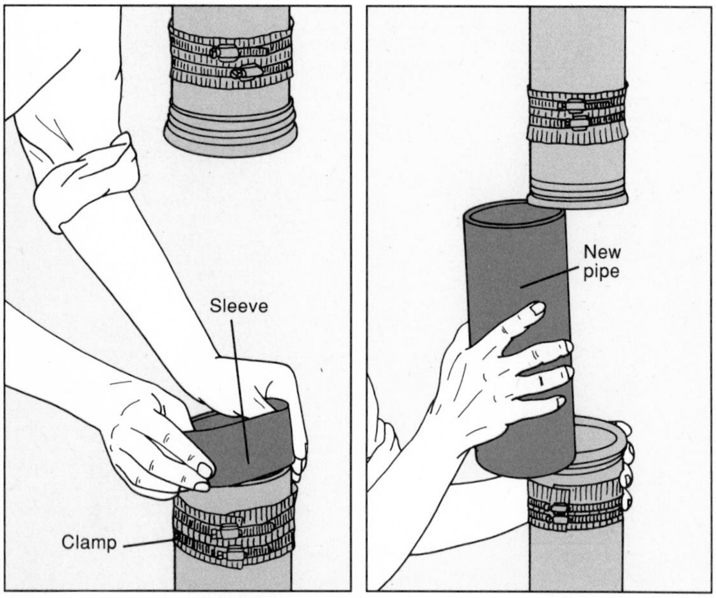

3 Fitting the replacement pipe. The replacement section is joined to the two standing pipes by means of hubless fittings—neoprene sleeves secured by stainless steel clamps. Slide a clamp onto each standing pipe and tighten them temporarily in place. Slip the neoprene sleeves onto each pipe until the ends bottom out inside the sleeves *(above, left)*. Then fold the lip of each sleeve back over the pipe. Work the replacement pipe into the gap between the sleeves *(above, right)* until it is properly seated.

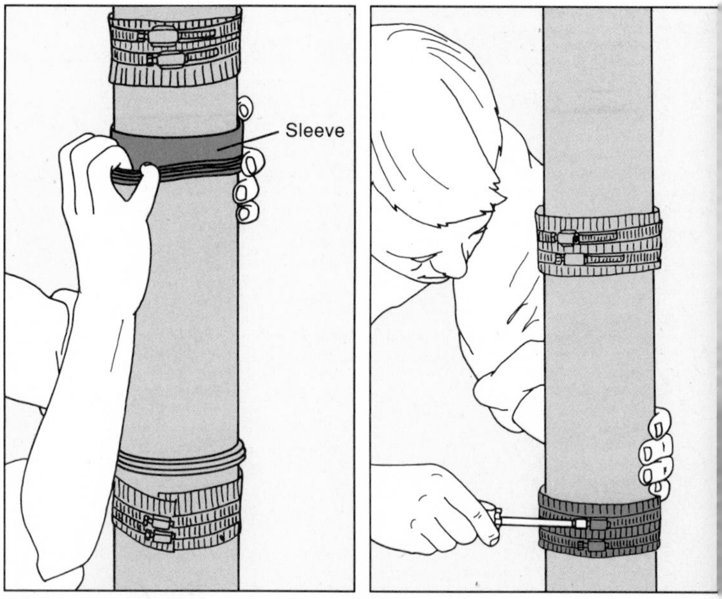

4 Completing the repair. Pull the folded lips of the sleeves over the replacement pipe *(above, left)*. Loosen and slide the clamps over the sleeves, center them over the joints and tighten with a nut driver *(above, right)* or socket wrench. Run water through the drain pipe to test the repair. If a joint leaks, take apart and reassemble the hubless fitting.

REPAIRING PLASTIC DRAINPIPE

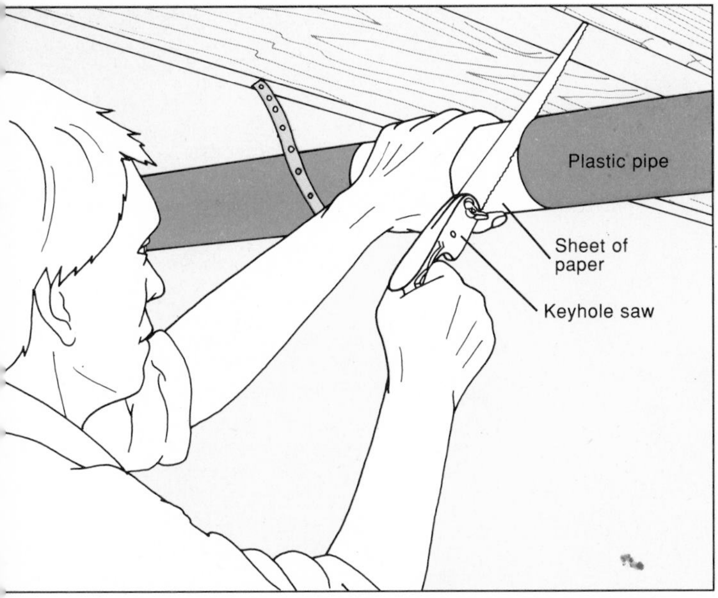

1 Removing the damaged section. Do not run water in the house during this repair. Wrap a sheet of paper around the pipe as a saw guide and cut squarely through the pipe with a keyhole saw *(above).* Repeat at the other end of the damaged section. If the pipe is directly connected to the sewer, stuff newspapers or paper towels into the standing pipes to block sewer gas.

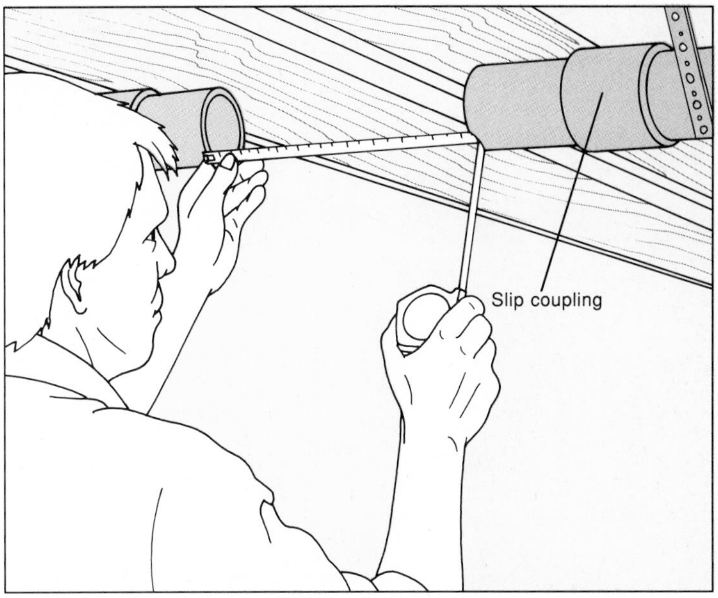

2 Cutting the replacement pipe. Fit a slip coupling over each standing pipe. Carefully measure the gap *(above)* and transfer the measurement to the new pipe. Cut the replacement section to length with a handsaw and miter box, using a paper guide as in step 1 to make a square cut.

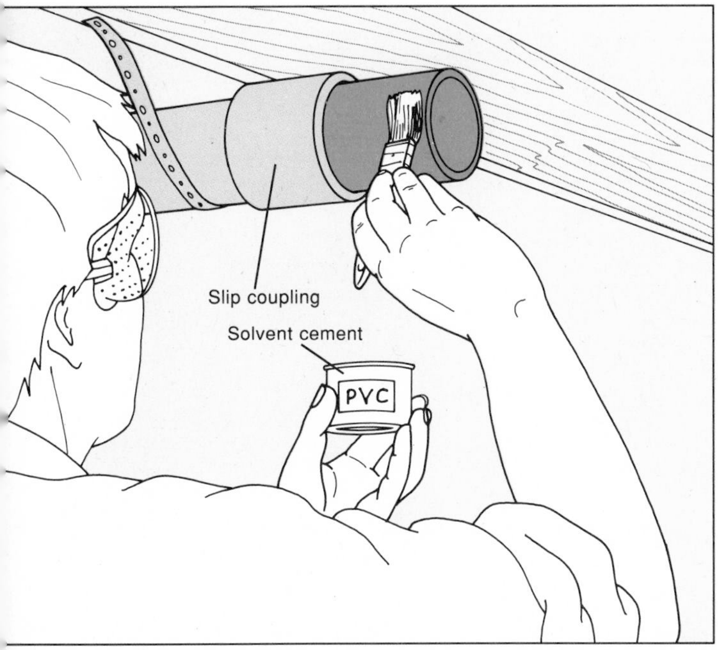

3 Applying solvent cement. With a sharp knife, deburr and bevel the pipe ends to aid water flow and improve the welding action of the solvent cement. Clean the ends of PVC and ABS pipes (but prime only PVC pipes). Apply a thick coat of solvent cement to the pipe ends, half the width of the slip coupling *(above).* To make the job easier, use a 1 1/2-inch brush or applicator for a 3-inch-wide coupling.

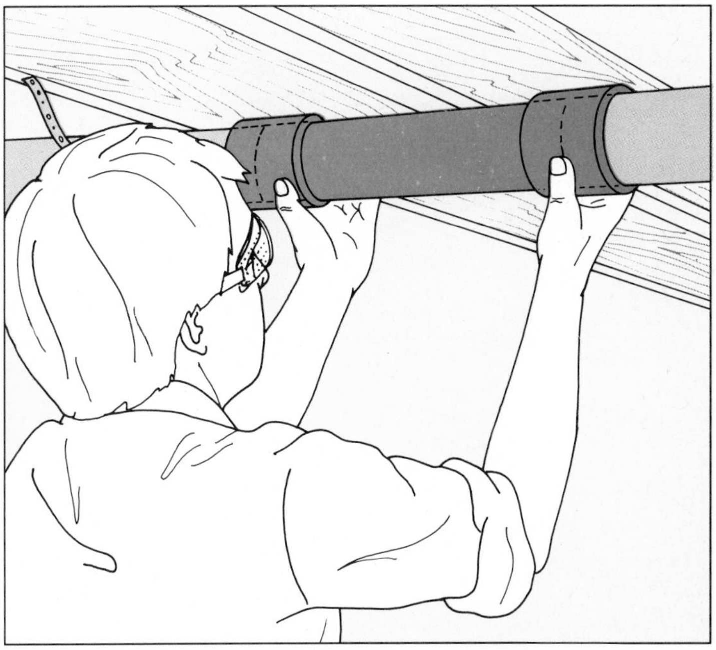

4 Fitting the replacement pipe. Working quickly before the cement sets, lift the new pipe into place and slip one coupling over its joint, then the other. Each coupling must cover an equal amount of old and new pipe *(above).* Give each coupling a quarter-turn to spread the cement, and wipe away excess cement with a clean cloth. Allow the joint to cure (at least two hours), then run water through the drainpipe. If it leaks, the couplings were not properly cemented. Cut out the replacement pipe and fittings and start again.

SHUT-OFF VALVES

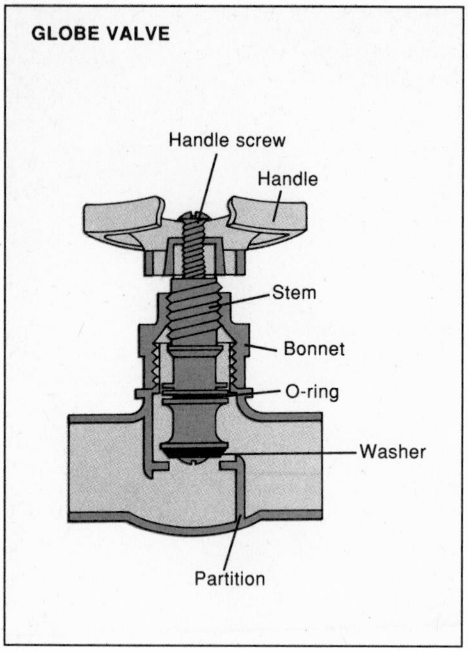

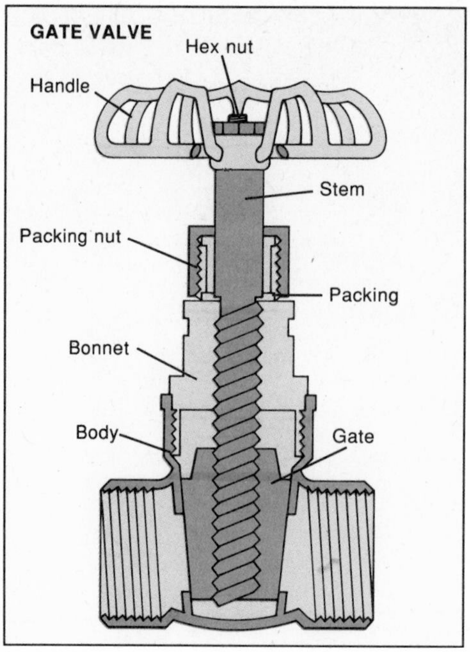

Gate valves *(near left)* are located on the main water supply line. When the handle is turned clockwise to close the valve, a gate lowers to block the flow of water. Since a gate valve does not divert the flow of water from one direction to another, the pressure of the water is the same at both ends of the valve. On a horizontal run, the valve should be installed in an upright position to prevent sediment from collecting in the closing mechanism.

Globe valves *(far left)* are installed on branch lines and fixtures, which may be opened and closed frequently. A partition within the valve body slows the flow of incoming water and therefore reduces water pressure in a supply line. When the handle is turned clockwise to close the valve, a rubber washer or disc presses against a matching seat to block the flow of water. Globe valves must be installed in the proper direction of flow; an arrow is usually stamped on the valve body.

SERVICING VALVES

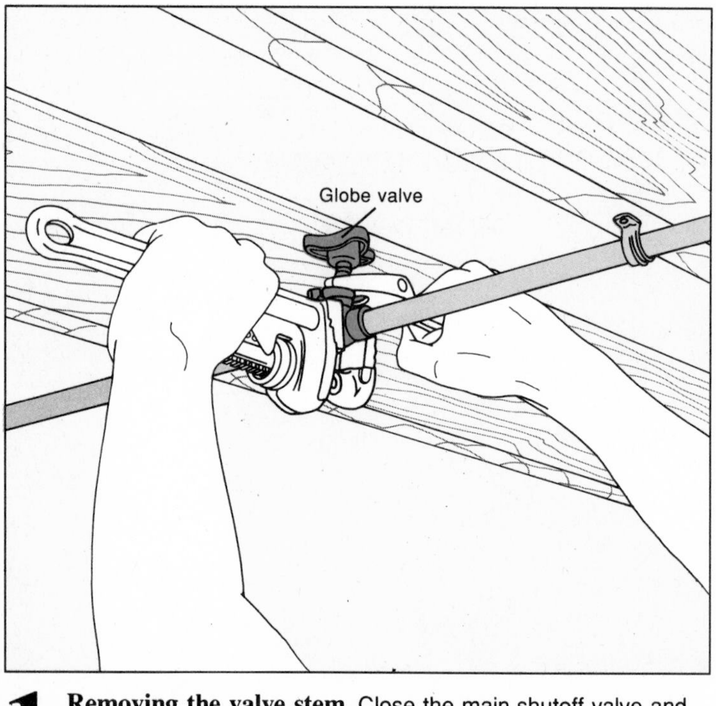

1 Removing the valve stem. Close the main shutoff valve and drain the supply lines. If it is the shutoff valve that is broken, call your local water department to close the curb valve outside the house *(page 14)*. Grip the valve body with a pipe wrench and hold it steady. Fit an adjustable wrench onto the bonnet and turn to looosen it *(above)*. If it is stubborn, apply penetrating oil and wait 15 minutes before trying again. Remove the bonnet and valve stem. If water had been seeping from the handle, service the valve body in place *(step 2)*. If the valve was leaking badly or did not close, remove it from the run to repair or replace *(step 3)*.

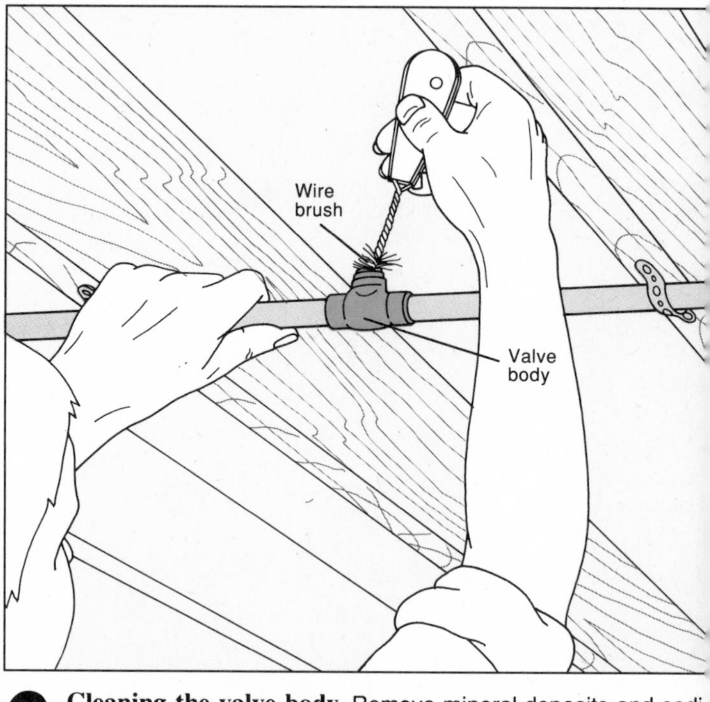

2 Cleaning the valve body. Remove mineral deposits and sediment inside the valve body with a wire brush designed for cleaning copper fittings. Scour the valve opening *(above)*, then bend the wire shaft of the brush with a pair of pliers to reach even farther inside the valve. To cure leaks around the stem and handle, replace the washer, O-ring or packing and reassemble the faucet.

SERVICING VALVES (continued)

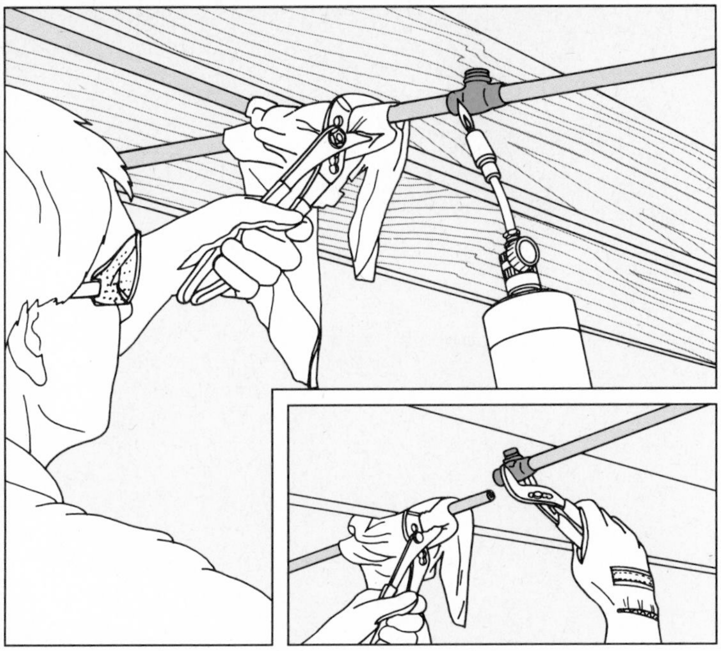

3 Removing the valve. Wear safety glasses and work gloves, and protect flammable materials with a fireproof shield. Open any other valves or faucets on the line, and wrap a rag around adjacent pipes to keep them cool. Play the flame of a propane torch on one of the soldered joints *(above)*. As the solder melts, pull the pipe straight out of the valve without bending it. If necessary, turn off the torch and use two pliers to pull apart the heated joint *(inset)*. Next, heat and separate the second joint. Remove a threaded valve from its union (if any), or cut the pipe on one side and unscrew the valve from the other pipe.

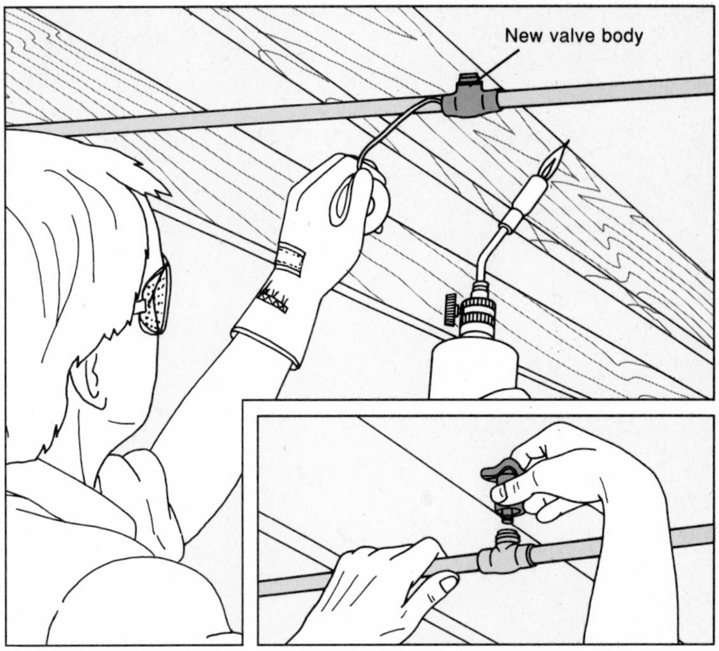

4 Installing the new valve. Solder the new valve body onto the pipe run *(page 140)*. When it has cooled, thread the stem into the valve opening by hand *(inset)*. Grip the valve body with a tape-covered wrench and tighten the bonnet with a second wrench. To install a threaded valve, screw it to the pipe by hand, then grip the pipe with one wrench and tighten the valve with a second wrench. Thread a new piece of pipe into the other side of the valve and add a union to complete the run *(page 101)*.

ADDING A FIXTURE SHUTOFF VALVE

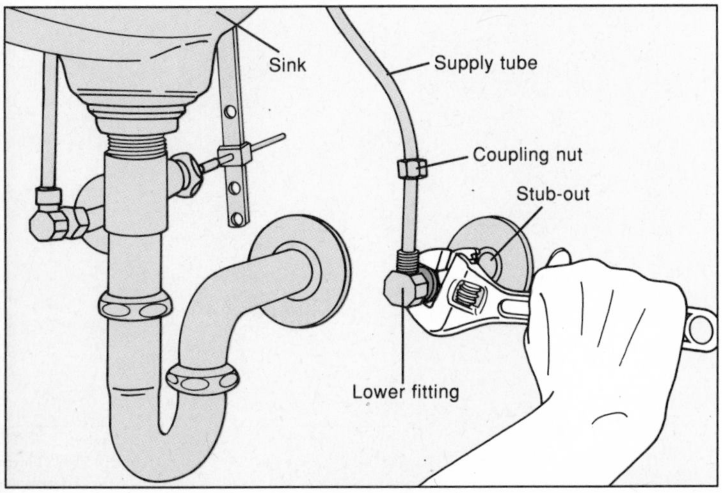

1 Removing the lower fitting. To add a shutoff valve to an existing fixture such as a sink faucet, you must also replace the supply tubes and their fittings. Close the main shutoff valve and drain the supply lines. With an adjustable wrench, unscrew the coupling nut on the lower fitting, then the nut on the stub-out *(above)*. Pull the fitting off the stub-out, then off the supply tube. Remove the compression ring by carefully cutting it off the the stub-out with a hacksaw.

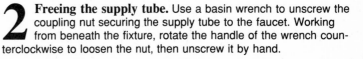

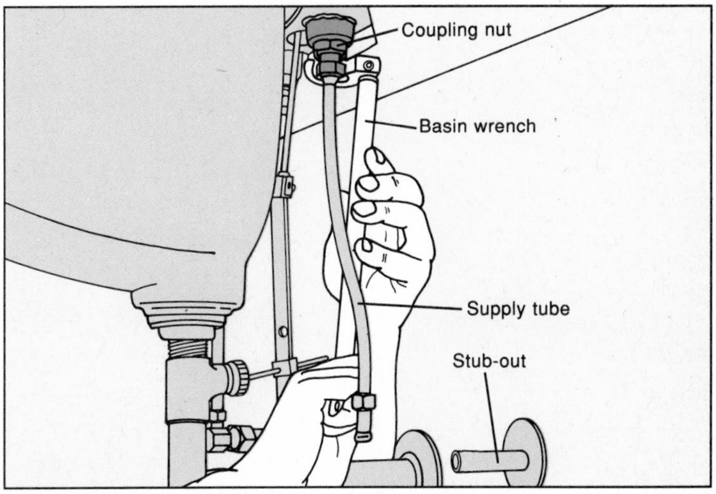

2 Freeing the supply tube. Use a basin wrench to unscrew the coupling nut securing the supply tube to the faucet. Working from beneath the fixture, rotate the handle of the wrench counterclockwise to loosen the nut, then unscrew it by hand.

ADDING A FIXTURE SHUTOFF VALVE (continued)

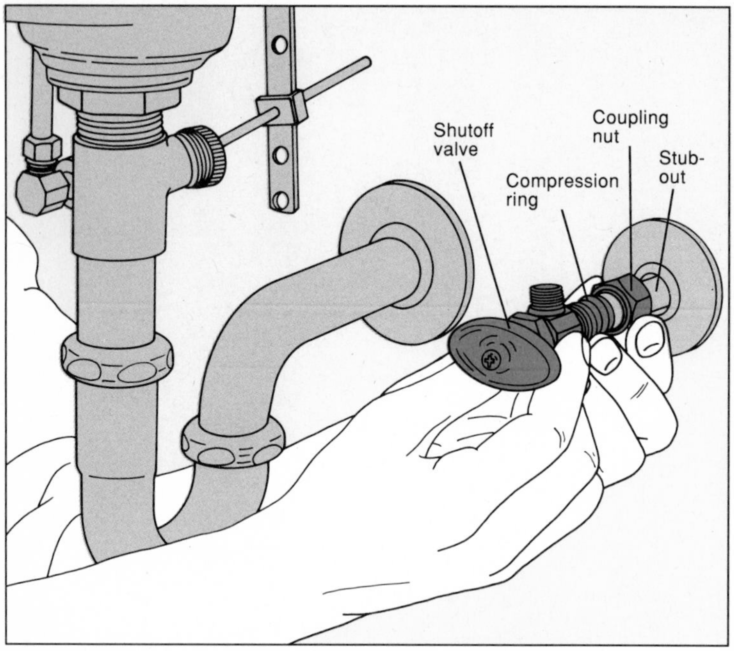

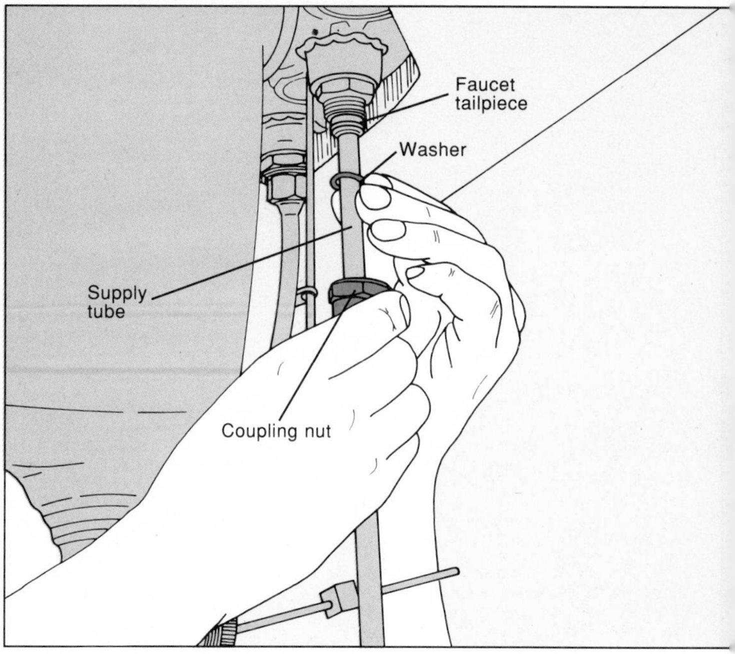

3 Adding the shutoff valve and flexible supply tube. Slip a new coupling nut and compression ring onto the stub-out, then fit the valve by hand *(above, left)* so that its outlet points up. Grip the valve with an adjustable wrench and tighten the coupling nut with a second wrench. (To solder a valve onto a copper stub-out, see page 140; to thread a valve onto steel pipe, see page 101.) Insert one end of the new supply tube into the faucet tailpiece. Slip the washer (if any) and plastic coupling nut up the tube *(above, right)* and tighten the connection by hand.

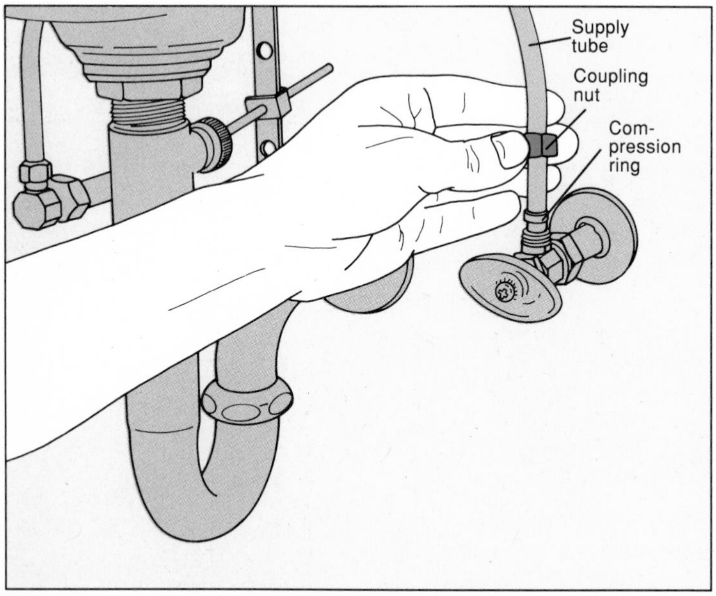

4 Connecting the valve. Slip the coupling nut and the compression ring onto the free end of the supply tube. Push the tube into the valve outlet as far as it will go. Slide the compression ring into the joint, making sure that it is squarely aligned, then slide the coupling nut over the fitting and screw it down by hand *(above)*.

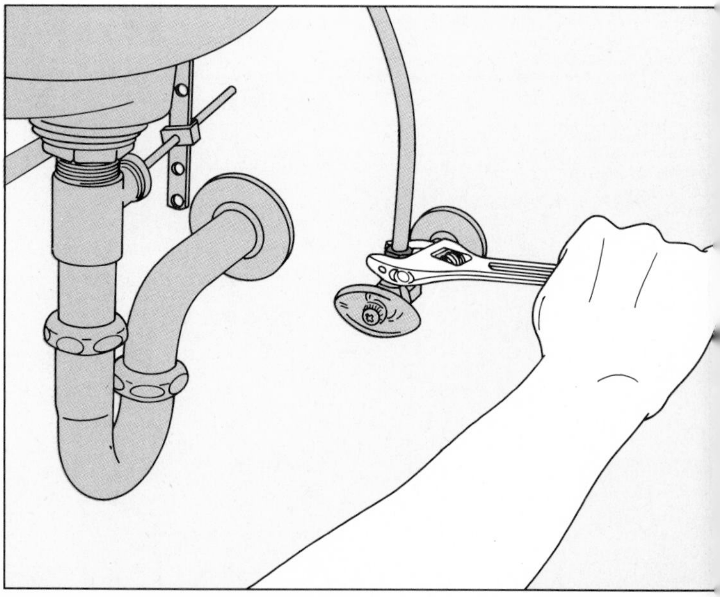

5 Testing the repair. Tighten the coupling nut with an adjustable wrench, as shown (some models may need only hand-tightening). Turn on the water. If a joint leaks, tighten the coupling nut one-half turn with a wrench and test again. Otherwise, you must disassemble the fitting and start again. If the faucet has an aerator, remove it and open the faucet to clear debris loosened by the repair.

ADDING A SHOCK ABSORBER

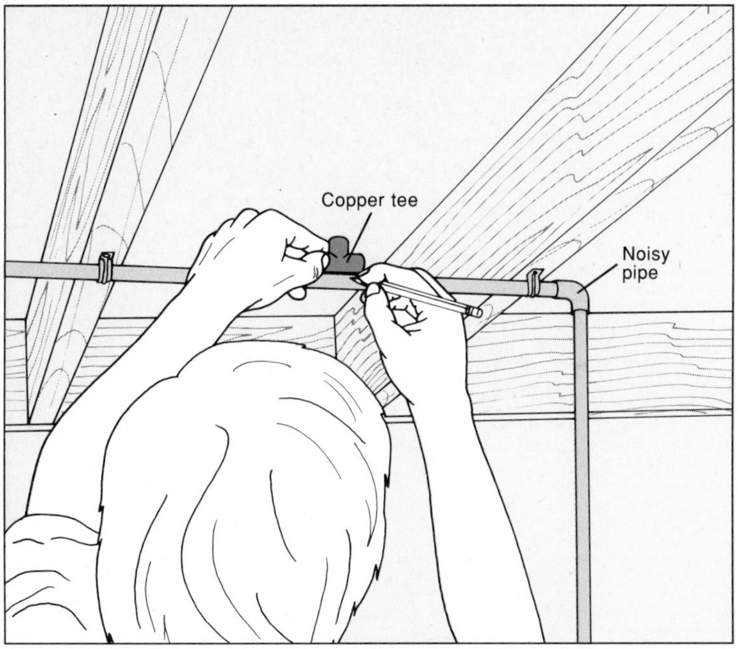

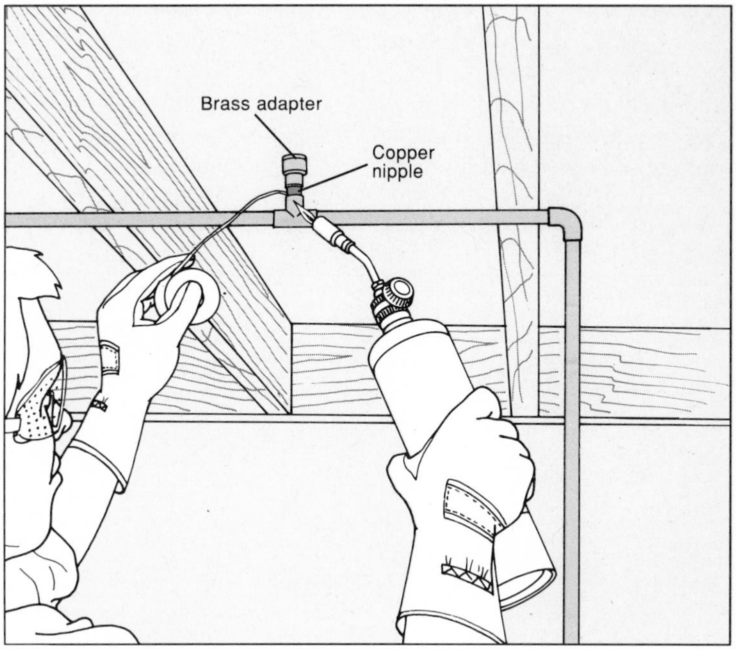

1 Installing the tee and adapter. Water hammer—a banging noise caused by vibrating pipes—often occurs when a valve abruptly stops the pressurized flow of incoming water. You can greatly reduce or eliminate this phenomenon by installing an air-filled shock absorber on a vibrating pipe. The shock absorber's correct air pressure is marked on its package; check and repressurize it if necessary at a service station. Close the main shutoff valve and drain the supply lines. From the noisy faucet, follow the supply pipe back to the nearest accessible location. Hold a copper tee above the pipe and mark its location *(above)*. Cut out the section with a tube cutter or hacksaw and solder the tee in place *(page 140)*. Next, solder a 2-inch nipple of copper pipe into the top of the tee. Finally, solder a female-threaded brass adapter into the copper pipe *(above)*. If there is not enough clearance above the pipe, the shock absorber can be installed upside-down.

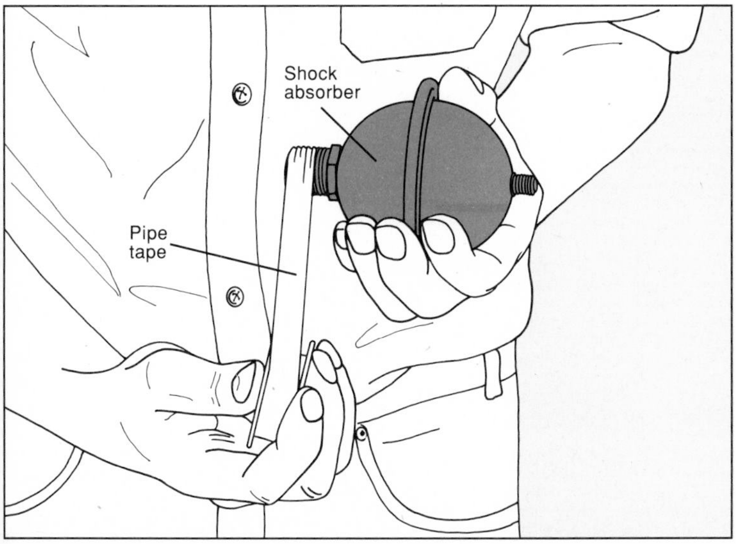

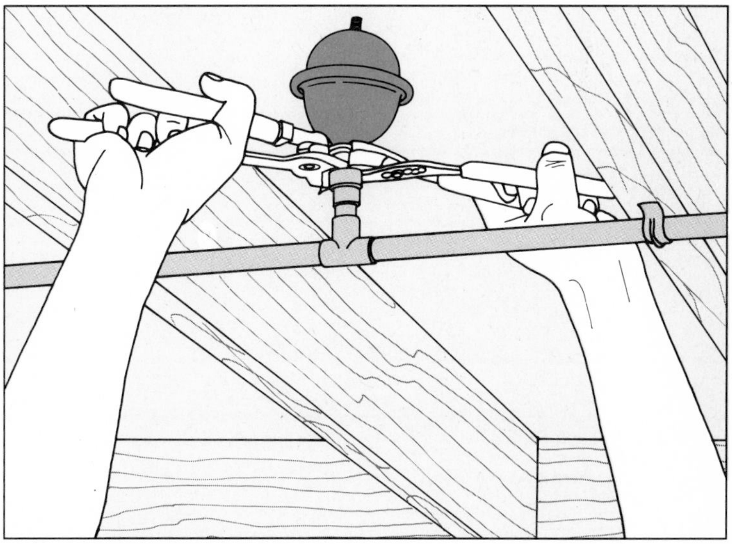

2 Attaching the shock absorber. Wrap pipe tape around the threads of the shock absorber *(above)* and screw it into the brass adapter by hand. Grip the adapter with a pair of channel-joint pliers and tighten the shock absorber with a second pair. Support the pipe on each side of the shock absorber with pipe hangers. Turn on the water, slowly at first. If there are leaks, tighten the shock absorber another half-turn with the two pliers.

GAS WATER HEATERS

A gas-fired water heater typically heats about 60 gallons of water to 140°F each day. When a hot water faucet is turned on, hot water flows from the tank through the hot water outlet as cold water enters through the dip tube. Sensing the drop in temperature, the thermostat opens a valve that sends gas to the burner, where it is ignited by the pilot or an electric spark. Heated air is vented from the burner chamber through the flu and its heat-retaining baffle and out the draft hood and vent

The number of gallons that can be heated from 50°F to 150°F in one hour is known as a heater's recovery capacity This figure, and the capacity of the tank, is stamped on a meta plate affixed to most heaters. The common complaint of insuf

TROUBLESHOOTING GUIDE

SYMPTOM	POSSIBLE CAUSE	PROCEDURE
No hot water	Pilot light out	Relight pilot *(p. 116)* □○
	Pilot light does not stay lit	Replace thermocouple *(p. 117)* ⬓○
	Repeated pilot outages	Check for floor drafts, check and clean flue and vent *(p. 118)* ⬓●
	Dial turned off	Reset temperature control dial *(p. 116)* □○
	Temperature control unit faulty	Have gas company check control unit and burner
Not enough hot water or water not hot enough	Heavy household demand for hot water	Stagger use of hot water or install a larger water heater
	Temperature control dial set too low	Raise temperature control setting *(p. 116)* □○
	Poor burner flame slows recovery time	Check and clean flue and vent *(p. 118)* ⬓●
	Sediment in tank slows heater recovery time	Drain and refill tank *(p. 119)* □◒ If rate of sedimentation is high, install a water softener
	Hot water faucet leaks	Repair faucet *(p. 18)*
	Loss of heat	Insulate tank and hot water pipes *(p. 119)* □○
	Incoming water too cold	Raise temperature control dial setting *(p. 116)* □○
	Dip tube broken	Call for service
Water too hot	Temperature control dial set too high or frequent short draws of hot water cause buildup of very hot water in top of tank	Lower temperature control dial setting *(p. 116)* □○
	Temperature control unit faulty	Call for service
Relief valve leaks continuously	Relief valve faulty	Test and replace valve *(p. 120)* ⬓◒
Relief valve leaks and hot water has not been used	Excessive pressure in supply water	Lower temperature control dial setting *(p. 116)* □○ Install pressure reducing valve *(call for service)*
Relief valve drips after heavy use of hot water	Water at top of tank too hot	Stagger hot water use or lower temperature *(p. 116)* □○ Test and replace valve *(p. 120)* ⬓◒
Drain valve leaks	Valve loose or faulty	Tighten valve handle; replace washer or valve *(p. 121)* ⬓◒
Cold water supply valve leaks	Valve faulty	Replace valve *(p. 110)* ⬓◒
Water pipes leak	Pipes broken or corroded	Replace pipes *(p. 93)*
Hot water rusty	Water heater tank corroded	Drain and refill tank *(p. 119)* ⬓◒ Check and replace anode rod *(p. 117)* ⬓◒ Lower temperature control dial setting *(p. 116)* □○ Replace water heater *(call for service)*
	Galvanized steel water pipes corroded	Replace pipes *(p. 93)*
Hot water smells bad	Magnesium anode rod reacting with sulfurous hot water	Replace with aluminum anode rod *(p. 117)* ⬓◒ Replace water heater *(call for service)*
Water heater rumbling	Sediment in tank traps pockets of very hot water	Drain and refill tank regularly *(p. 119)* □◒ Install water softener if sedimentation rate is high
Water heater sizzles	Water leaks into burner chamber	Replace water heater *(call for service)*
Water heater noisy	Burner pops when heater turned on or off	Have gas company check burner and gas pressure
	Noisy combustion from burner flame	Eliminate any strong floor drafts. If problem persists, have gas company check burner and gas pressure
Hot water or tank sooty	Improper ventilation	Check and clean flue and vent *(p. 118)* ⬓◒ If problem persists, have utility check burner and gas pressure

DEGREE OF DIFFICULTY: □ **Easy** ⬓ **Moderate** ■ **Complex**
ESTIMATED TIME: ○ **Less than 1 hour** ◒ **1 to 3 hours** ● **Over 3 hours**

ficient hot water is often caused by heavy demand on a too-small heater, rather than mechanical failure.

A few routine maintenance chores will increase the heater's efficiency. Vacuum under and around the heater to keep dust from clogging the pilot and the burner. (Be careful not to snuff the pilot, however). Drain the tank periodically to check for sediment. Once a year, disassemble and clean the vent and draft hood, and test the heater's relief valve.

Before working on a gas water heater, be sure to close the gas shutoff valve. For repairs to the control unit, pilot, burner and gas supply lines—or to replace a leaking or corroded tank—call for professional service.

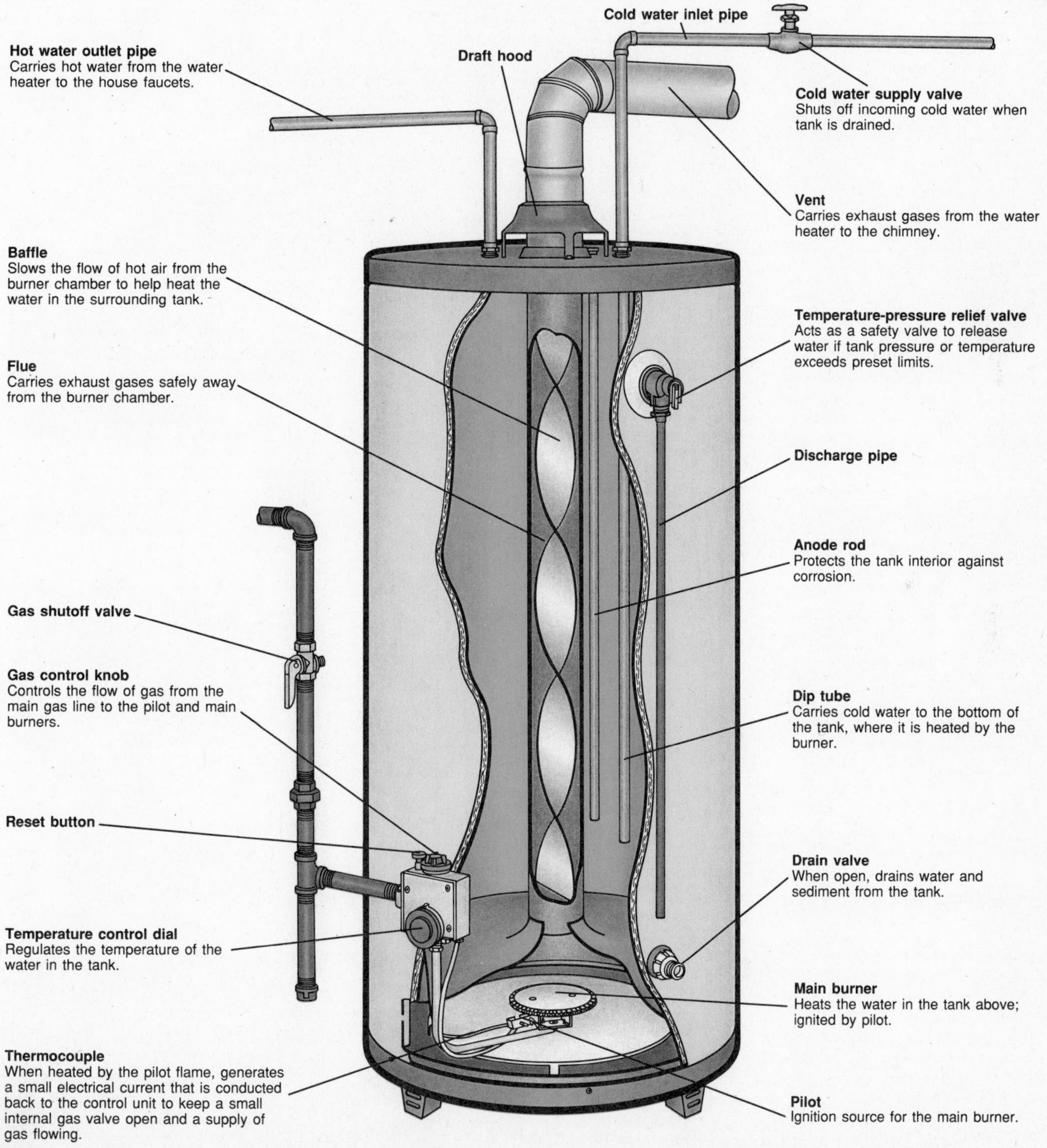

Cold water inlet pipe
Hot water outlet pipe
Carries hot water from the water heater to the house faucets.
Draft hood
Cold water supply valve
Shuts off incoming cold water when tank is drained.
Vent
Carries exhaust gases from the water heater to the chimney.
Baffle
Slows the flow of hot air from the burner chamber to help heat the water in the surrounding tank.
Temperature-pressure relief valve
Acts as a safety valve to release water if tank pressure or temperature exceeds preset limits.
Flue
Carries exhaust gases safely away from the burner chamber.
Discharge pipe
Anode rod
Protects the tank interior against corrosion.
Gas shutoff valve
Gas control knob
Controls the flow of gas from the main gas line to the pilot and main burners.
Dip tube
Carries cold water to the bottom of the tank, where it is heated by the burner.
Reset button
Drain valve
When open, drains water and sediment from the tank.
Temperature control dial
Regulates the temperature of the water in the tank.
Main burner
Heats the water in the tank above; ignited by pilot.
Thermocouple
When heated by the pilot flame, generates a small electrical current that is conducted back to the control unit to keep a small internal gas valve open and a supply of gas flowing.
Pilot
Ignition source for the main burner.

LIGHTING A GAS WATER HEATER

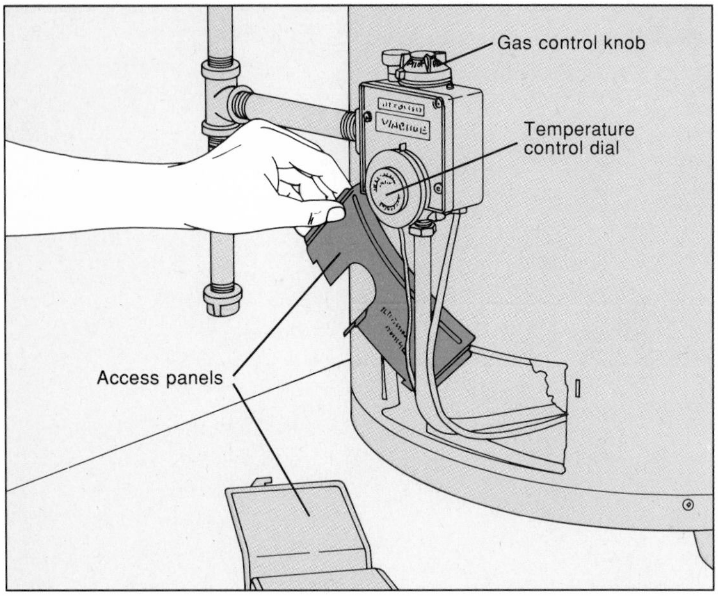

1 Gaining access to the pilot. Carefully remove the burner access panels by lifting them off the heater *(above)* or sliding them sideways. To light a pilot that has blown out (there will be no flame visible in front of the burner), turn the temperature control dial to its lowest setting and the gas control knob to OFF. Wait at least five minutes for the gas to clear. Then, if there is no gas odor, relight the pilot *(next step)*. If gas odor lingers, close the gas shutoff valve to the heater, ventilate the room and call the gas company.

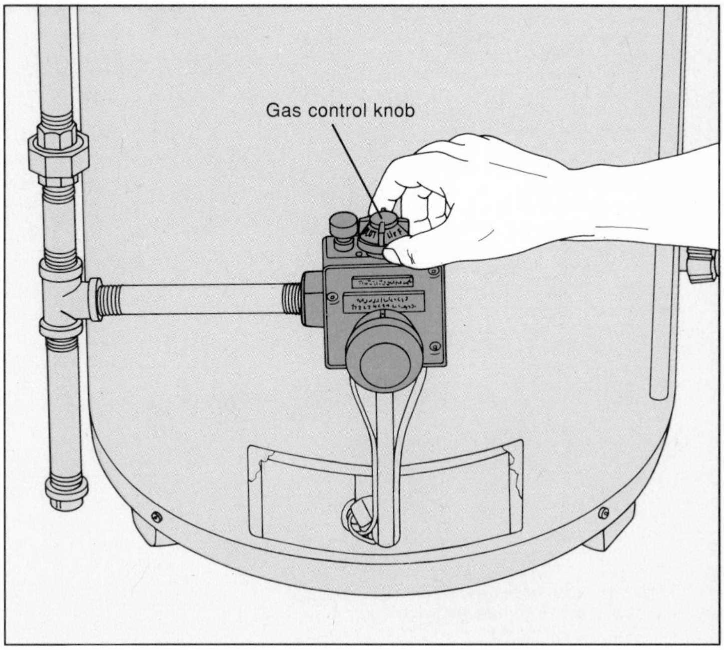

2 Checking the gas flow to the pilot. If the gas shutoff valve is already open, turn the gas control knob to PILOT *(above)*. But if the gas shutoff valve was closed, open it and wait about five minutes for the gas to reach the control unit before turning the control knob to PILOT.

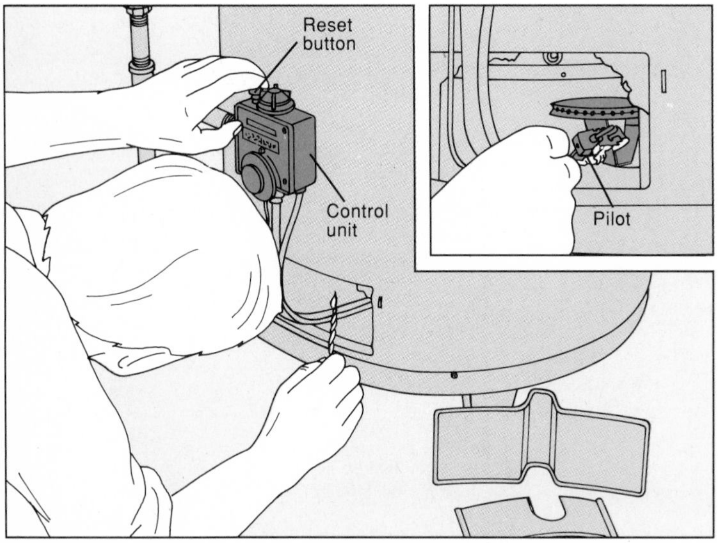

3 Lighting the pilot. Depress the reset button *(above)* while placing a lighted match near the tip of the pilot *(inset)*. If the pilot is hard to reach, make a taper from a tightly rolled piece of paper. When there is no reset button, hold down the gas control knob while lighting the pilot. Should the pilot fail to light immediately, close the gas shutoff valve and call the gas company. If it lights, continue to depress the reset button or gas control knob for one minute, then release it and set the temperature control *(step 4)*. If the pilot goes out, turn the gas control knob to OFF. Try tightening the hexagonal nut connecting the thermocouple to the base of the control unit, first by hand, then by giving it a quarter-turn with an open-end wrench. Relight the pilot. If it goes out again, service the thermocouple *(page 117)*.

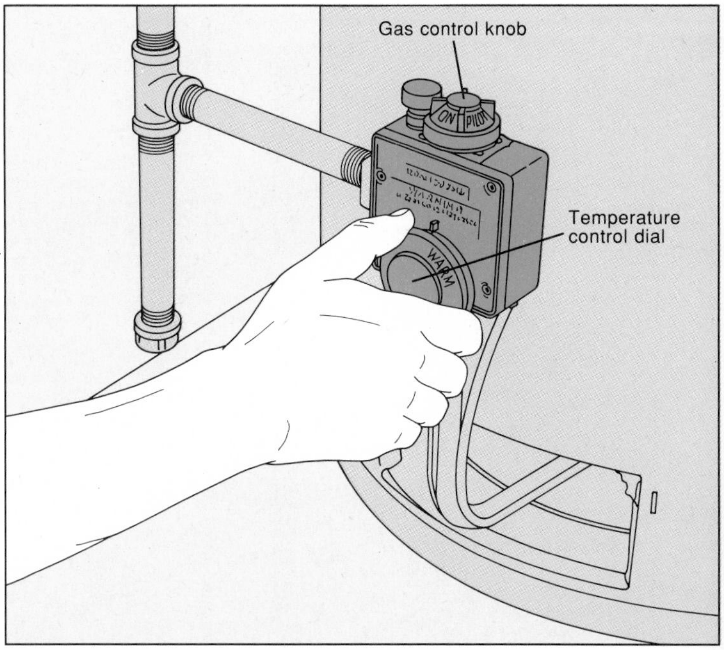

4 Setting the temperature. Turn the gas control knob to ON and the temperature control dial between 120°F and 130°F (or just above WARM), as shown. A moderate setting lowers heating costs, prolongs tank life, and reduces the risk of scalding. Replace the access panels.

REPLACING THE THERMOCOUPLE

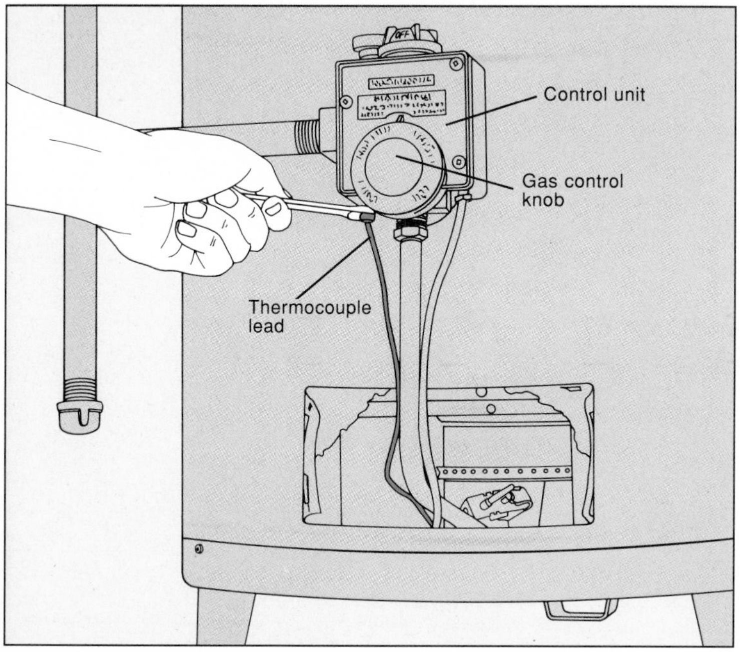

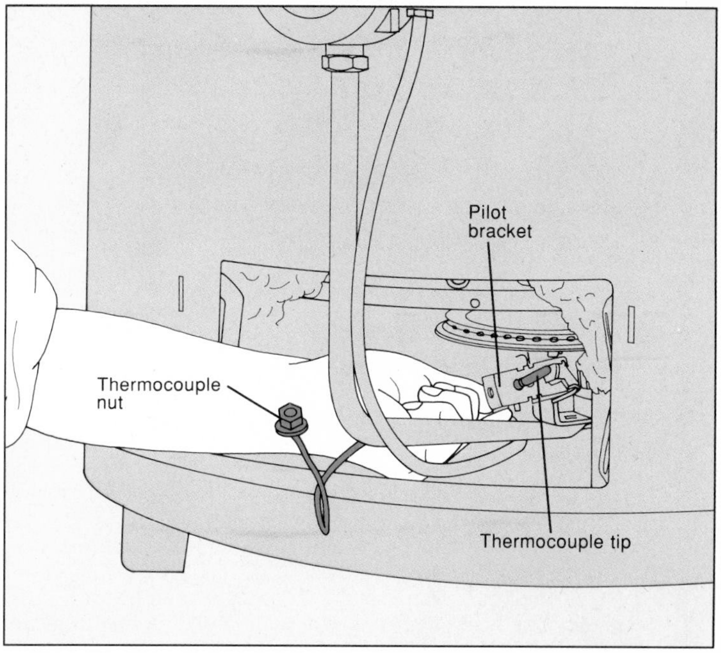

Disconnecting the thermocouple. Turn the gas control knob to OFF and close the gas shutoff valve. With an open-end wrench, loosen the nut that secures the thermocouple to the control unit *(above, left)*, then unscrew it by hand. Pull down on the copper lead to detach the end of the thermocouple from the control unit. There may be a second nut attaching the thermocouple tip to the pilot bracket; unscrew it and slide it back along the copper lead. Grip the base of the thermocouple *(above, right)* and pull firmly, sliding it out of the pilot bracket.

Buy an exact replacement for the old thermocouple at a plumbing or heating supplier. Push the tip of the new thermocouple into the pilot bracket clip as far as it will go. If there is a hexagonal nut at its tip, screw it to the bracket. Run the lead out and bend it up into a gentle curve. Screw the nut on the end of the thermocouple to the control unit by hand, then give it a quarter-turn with an open-end wrench. Relight the pilot *(page 116)*. If it does not stay lit, close the gas shutoff valve and call for service.

REPLACING THE ANODE ROD

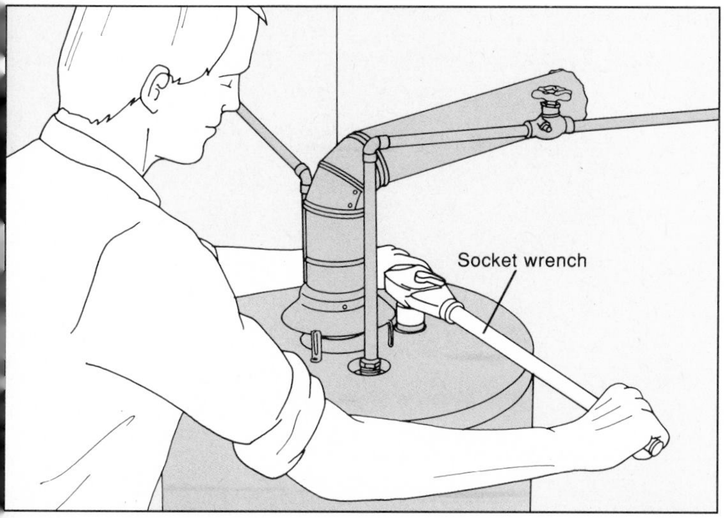

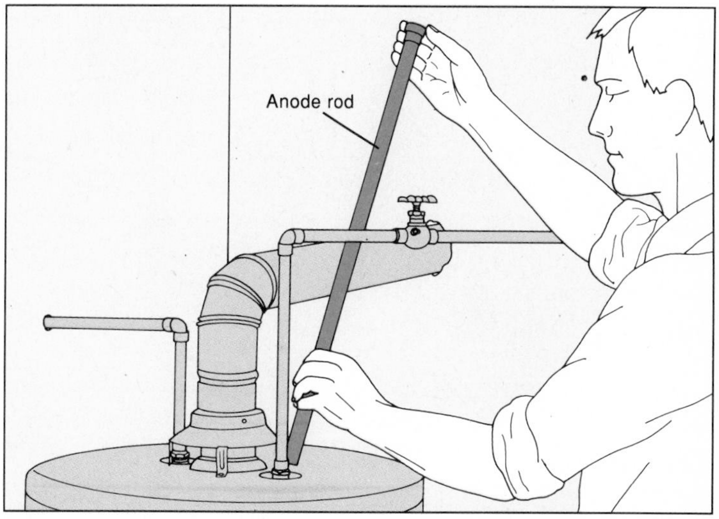

1 **Removing the anode plug.** Before removing the old anode rod, buy an exact replacement from the manufacturer or a heating supply company. Close the cold water supply valve and turn the gas control knob to OFF. Since the hexagonal plug that secures the anode may be rusted tight to the tank, borrow or rent a 24-inch socket wrench for better leverage. Drain two to three gallons of water from the tank *(page 119)*. Fit the socket over the anode plug and apply strong, even pressure to turn the ratchet counterclockwise *(above)*, while someone else braces the tank if necessary.

2 **Replacing the anode rod.** Raise the anode rod as far as possible with the socket wrench, then unscrew the last few inches by hand. Lift the rod straight up and out of the tank *(above)*. Apply only a single width of pipe tape to the threaded upper end of the new rod. Insert the rod into the tank, screw it in as far as possible by hand, then tighten it clockwise with the socket wrench. Open the cold water supply valve and relight the pilot *(page 116)*.

SERVICING THE FLUE AND VENT

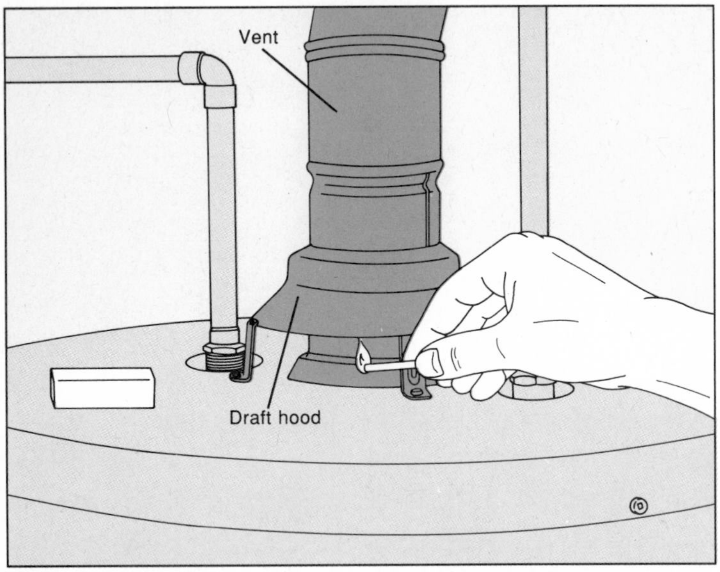

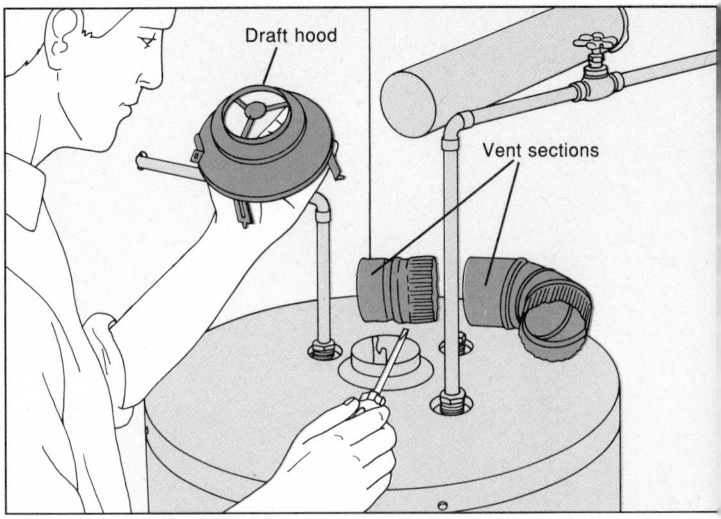

1 Testing the vent. Wait until the burner has been on for 5 to 10 minutes, then hold a lighted match under the draft hood *(above)*. If the vent is working properly, the match flame will be drawn in under the edge of the hood. If the flame is blown away from the draft hood or snuffed out, there may be a blockage within the vent. Clean the vent once a year *(step 2)* to ensure proper ventilation and prevent the backup of dangerous carbon monoxide fumes.

2 Disassembling the vent. Turn the gas control knob to OFF and close the gas shutoff valve. Remove the burner access panels *(page 116)* and cover the burner and floor with newspapers to catch soot and debris. Mark the vent sections for reassembly and tie up any overhead runs of ductwork with a cord or wire to prevent them from falling while you remove the vertical sections below. Finally, unscrew and remove the draft hood from the top of the tank *(above)*. Shake the hood and vent sections over the newspapers to release dirt, then scrub the insides with a wire brush. Replace any ductwork that is rusted or perforated.

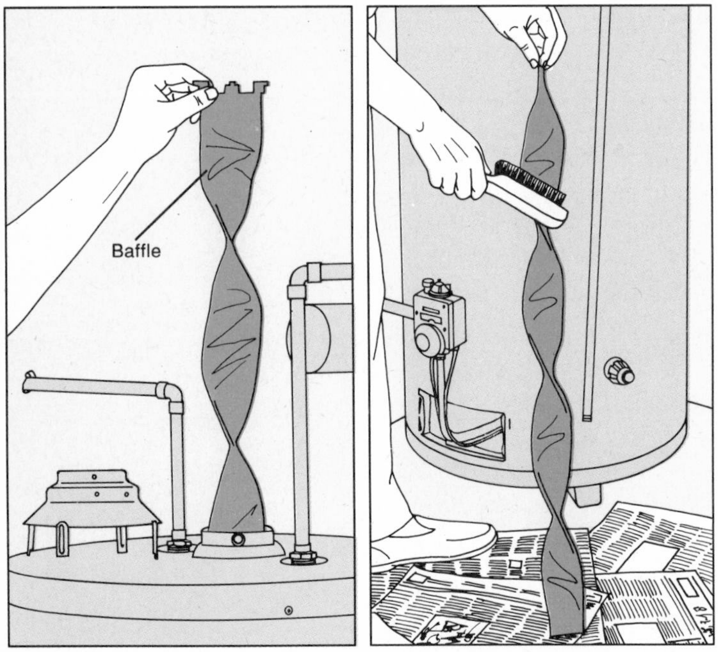

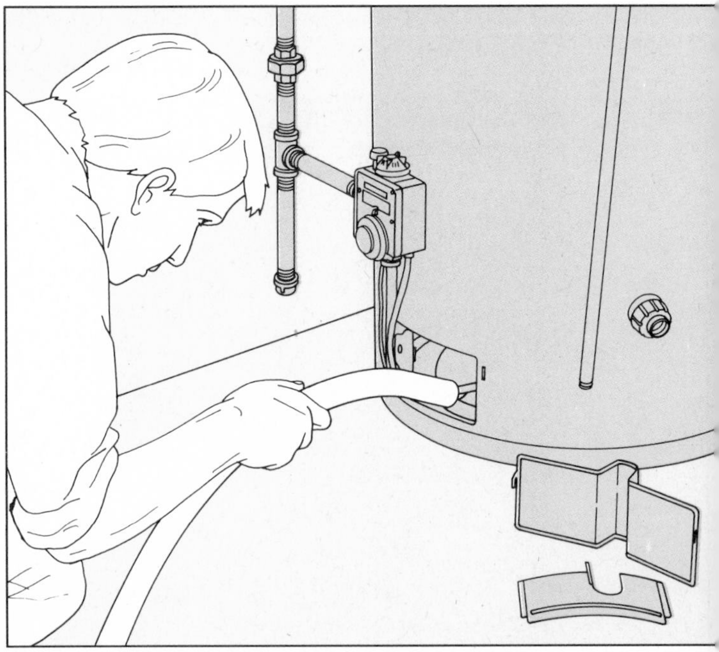

3 Cleaning the baffle. With the vent removed, you now have access to the heater flue and its removable baffle. Lift the baffle from the flue *(above, left)* and scrub it with a wire brush *(above, right)* to remove dust and soot. If there is not enough room to pull the baffle all the way out, lift it as high as possible, clean it, then rattle it to dislodge debris.

4 Cleaning the combustion chamber. Vacuum the inside of the combustion chamber *(above)*, then clean the burner and its ports with a soft brush. Use an old toothbrush to clean around the pilot. Replace the baffle, draft hood, and vent, then vacuum the combustion chamber again. Relight the pilot *(page 116)*, and test the vent with a lighted match as in step 1. If the flame is not drawn up the vent, turn the gas control knob to OFF, close the gas shutoff valve, then recheck the pilot, burner and vent. Turn on the gas, relight the pilot and test again. If the test fails, there may be a blockage in the main chimney—call for service.

INSULATING THE WATER HEATER

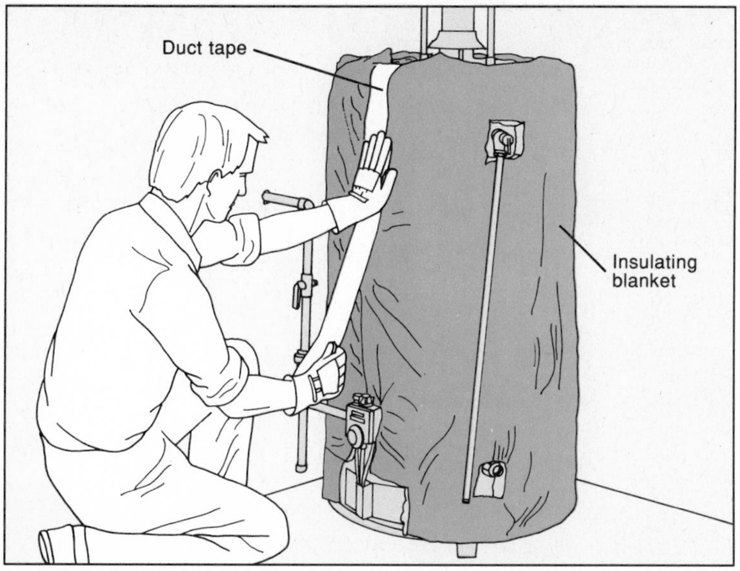

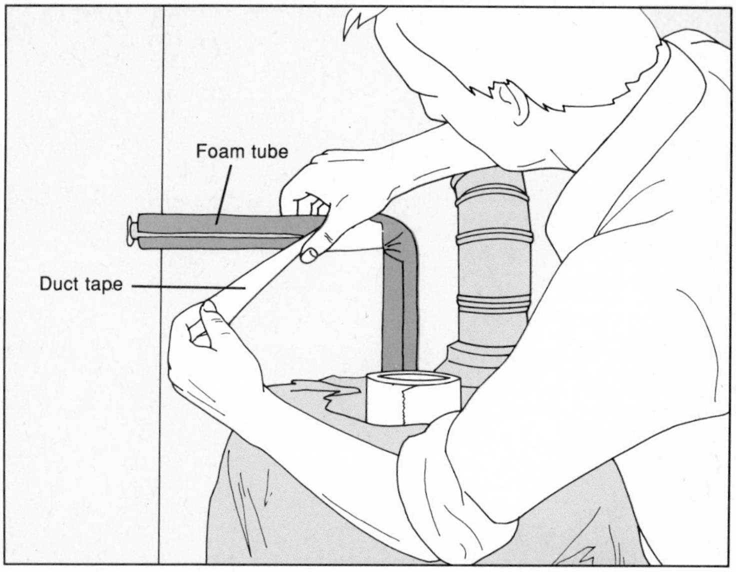

Insulating the tank and pipes. Energy-saving foam or fiberglass blankets are available for various sizes of water heaters. For a gas heater, first turn the gas control knob to OFF and close the gas shutoff valve. (On an electric heater, shut off power at the main service panel.) With waterproof duct tape, fasten the top of the insulating blanket to the top of the tank, then tape the edges of the blanket together to form a neat, vertical seam *(above, left)*. Do not cover the top of the tank. Use a utility knife to cut away all insulation from around the access panels, relief valve, drain valve, control panel and the space between the tank and the floor. Turn the gas or electricity back on and restart the heater. To prevent heat loss, you can also insulate runs of hot water pipe that pass through unheated areas with adhesive-backed foam, fiberglass tape or pre-slit foam tubes secured with duct tape *(above, right)*. It is not necessary to turn off the gas or electricity first.

DRAINING AND FILLING THE TANK

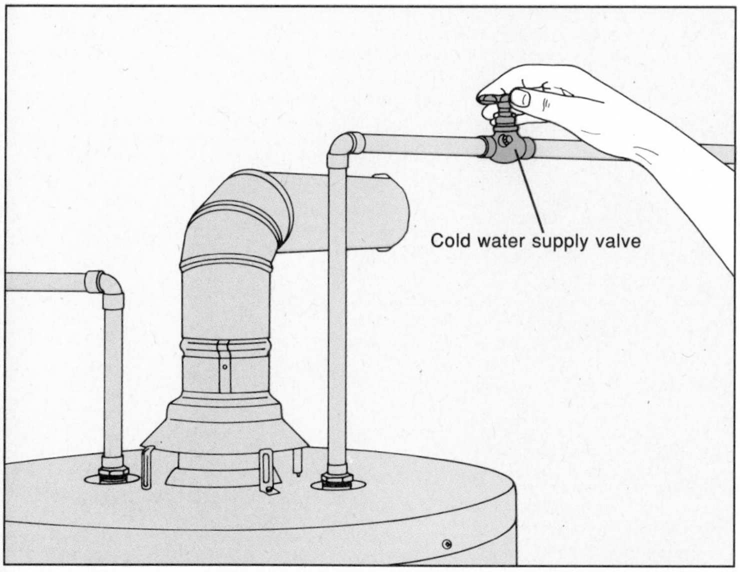

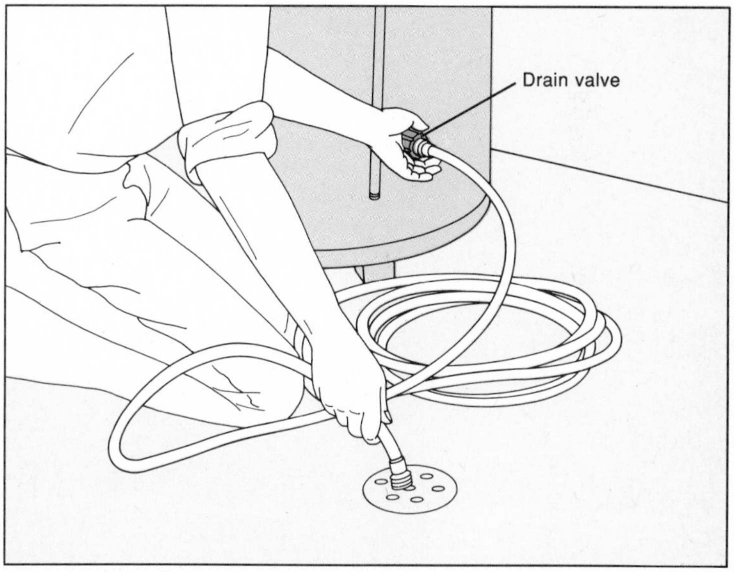

Draining the water heater. Turn the gas control knob to OFF and close the gas shutoff valve. (For an electric water heater, shut off power at the main service panel.) Close the cold water supply valve *(above, left)* and open a hot water faucet in the house to speed draining. Attach a hose to the drain valve and run it to a nearby floor drain *(above, right)* or into a bucket beneath the drain valve. Turn the drain valve clockwise to open it. As the tank empties, the valve may clog with sediment; open the cold water supply valve for a few minutes to allow the water pressure to clear the blockage. If you are using a bucket, watch carefully and turn the drain valve off before the bucket overflows; the process may take up to an hour. To refill the tank, close the drain valve tightly, open the cold water supply valve and open the hot water faucet farthest from the tank. When water flows from that faucet, the tank is full; close the faucet. Be sure the heater is full before turning on the gas or electricity, then relight the pilot *(page 116)* or turn the power back on. Two or three gallons can be drawn off in this manner every few months to check for sediment.

SERVICING THE TEMPERATURE-PRESSURE RELIEF VALVE

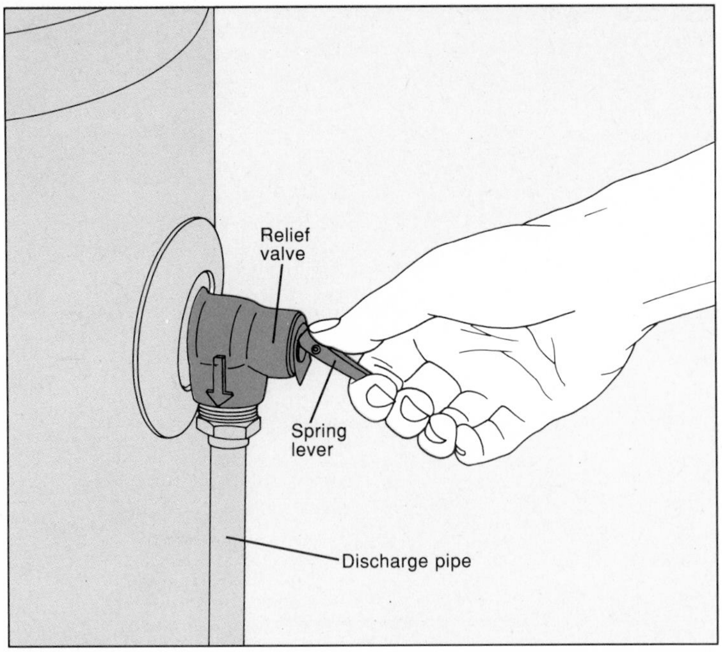

1 Testing the relief valve. In the unlikely event that the temperature or water pressure rises too high inside the water heater, the relief valve opens to prevent the tank from exploding. To test the valve, simply lift the spring lever *(above)*, keeping clear of the outlet or discharge pipe as hot water escapes. One-half to one cup of water should spurt out. Lift the lever several times to clear the valve of sediment. If no water spurts out, or if water continues to drip after the valve is released, it should be replaced. If there is a discharge pipe, remove it first *(step 2)*; if not, go to step 3.

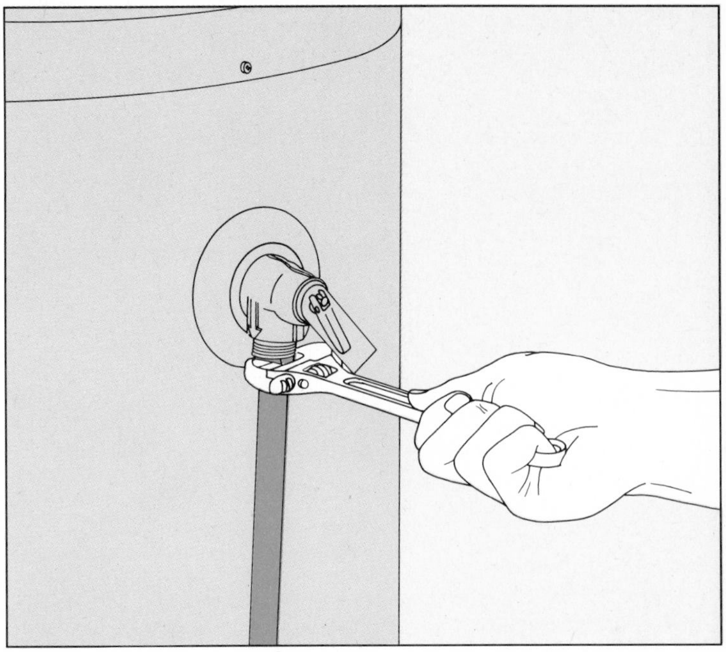

2 Removing the discharge pipe. Turn the gas control knob to OFF and close the gas shutoff valve. (For an electric water heater, shut off power at the main service panel.) Close the cold water supply valve. If the relief valve is located on top of the water heater, drain one gallon of water from the tank *(page 119)*; drain four to five gallons if the valve is on the side of the tank. Loosen and remove the discharge pipe, which may be threaded directly to the relief valve or to an adapter *(above)*.

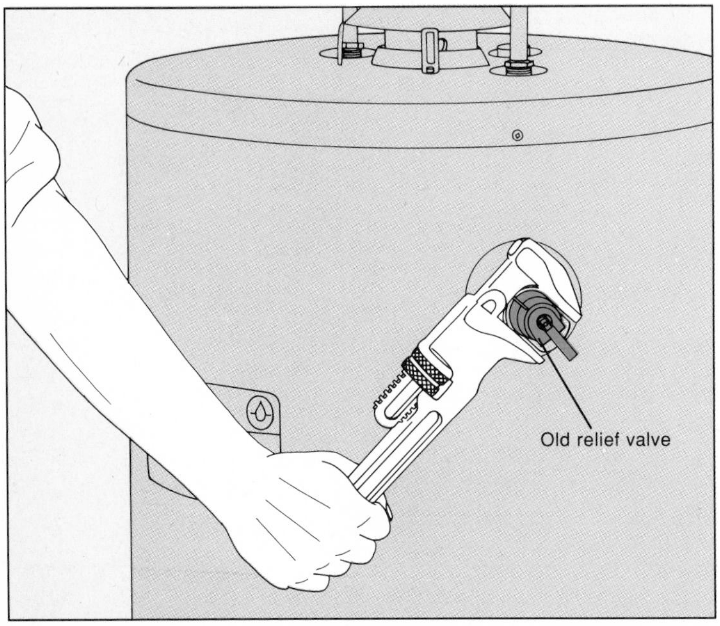

3 Removing the old relief valve. Fit a pipe wrench over the old relief valve and turn counterclockwise to unscrew the valve from the tank *(above)*. If the tank is old, the valve may be difficult to remove. Use firm, steady pressure (have a helper brace the tank if necessary), but do not jerk the valve—you might damage the tank. When the valve is loose, unscrew it by hand and pull it out.

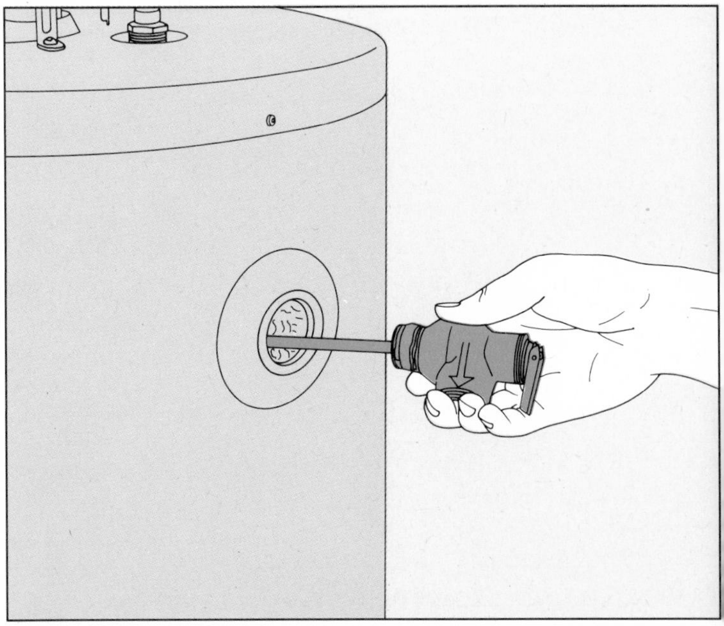

4 Installing a new relief valve. Take the old valve with you to buy an exact replacement. (The model type and size often appear on a metal tag hanging from the old valve.) Apply pipe tape to the threads of the new relief valve and screw it into the tank by hand, then tighten with a pipe wrench. Screw the discharge pipe (if any) back into the valve outlet. Refill the water heater and relight the pilot *(page 116)* or turn the electricity back on. If the valve continues to leak, have a plumber check the house water pressure.

SERVICING THE DRAIN VALVE

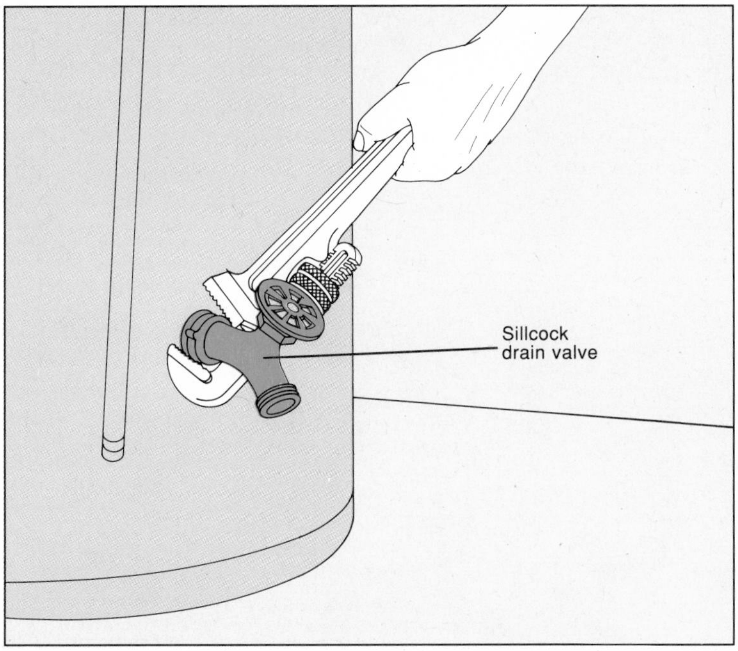

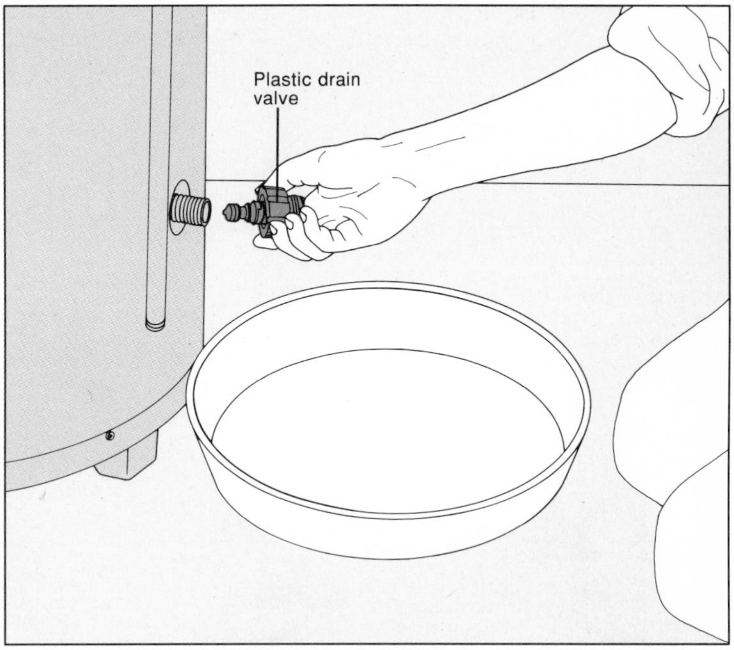

1 Repairing a leaking drain valve. Turn the gas control knob to OFF and close the gas shutoff valve. (For an electric water heater, shut off power at the main service panel.) Close the cold water supply valve and drain the heater completely *(page 119)*. If the drain valve has a removable handle, unscrew it and replace the washer behind it. Reassemble the handle, refill the tank and check the valve for leaks. If the handle cannot be removed, or the drain valve is made of plastic, replace the entire valve. Fit a pipe wrench over the base of the drain valve and turn it counterclockwise to unscrew the valve from the tank *(above, left)*, then go to step 2. If the valve is plastic, first turn the handle counterclockwise by hand, four complete revolutions. Then, while pulling firmly on the handle, turn the valve handle clockwise six complete turns to free it from the tank *(above, right)*. Replace it with an identical valve by pushing down on the handle and turning counterclockwise six times, then clockwise four times. You can also replace a plastic valve with a sillcock valve *(step 2)*.

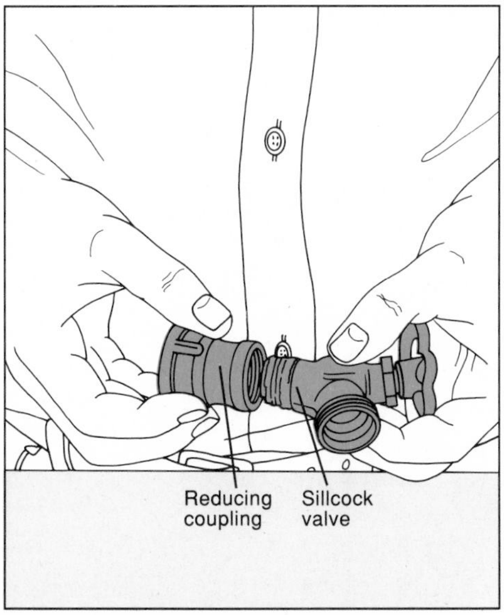

2 Assembling the new drain valve. Because it is both durable and contains a washer (and can therefore be serviced), a sillcock valve is the recommended replacement. When selecting the valve, also buy a 3/4-to-1/2-inch reducing coupling to match the valve to the tank. Wrap pipe tape around the threaded end of the sillcock valve and screw it into the 1/2-inch end of the reducing coupling *(above)*.

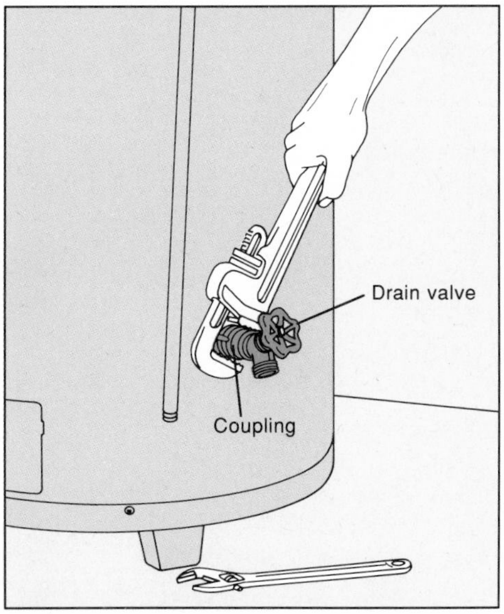

3 Installing the new valve. Apply pipe tape to the threaded end of the nipple on the water heater tank. Screw the coupling and valve onto the nipple and tighten as far as possible by hand. Finish tightenening the coupling with a pipe wrench *(above)*.

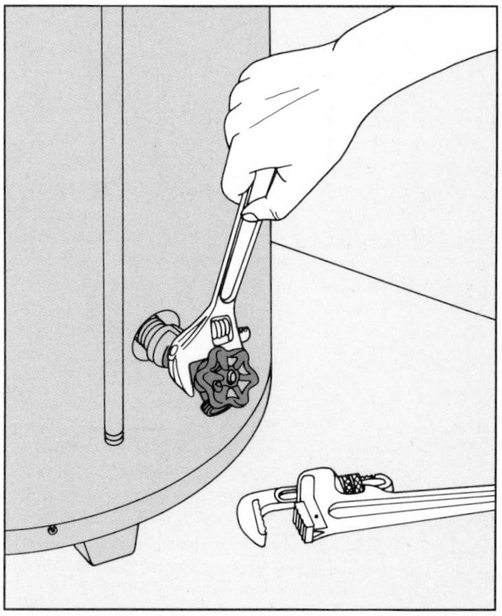

4 Aligning the valve. Fit an adjustable wrench over the body of the sillcock valve (but not over its outlet), and turn it clockwise to tighten the valve so that it faces down toward the floor *(above)*. Refill the tank and relight the pilot *(page 116)* or turn the electricity back on.

ELECTRIC WATER HEATERS

Electric water heaters usually have an upper and a lower heating element, each controlled by its own thermostat. The upper element has, in addition, a high-limit temperature cutoff to keep hot water from reaching the boiling point. Basic maintenance and repair—testing the relief valve, replacing the anode—are the same as for a gas water heater. The Troubleshooting Guide on page 123 will refer you to that chapter *(page 114)*. To test the electrical components, use a multitester as directed in the repair steps on pages 124-127.

First, adjust the multitester to zero. Set the selector switch to RX1, then touch the probes together. Using the ohms-adjust dial, align the needle exactly over 0. A reading of 0 ohms

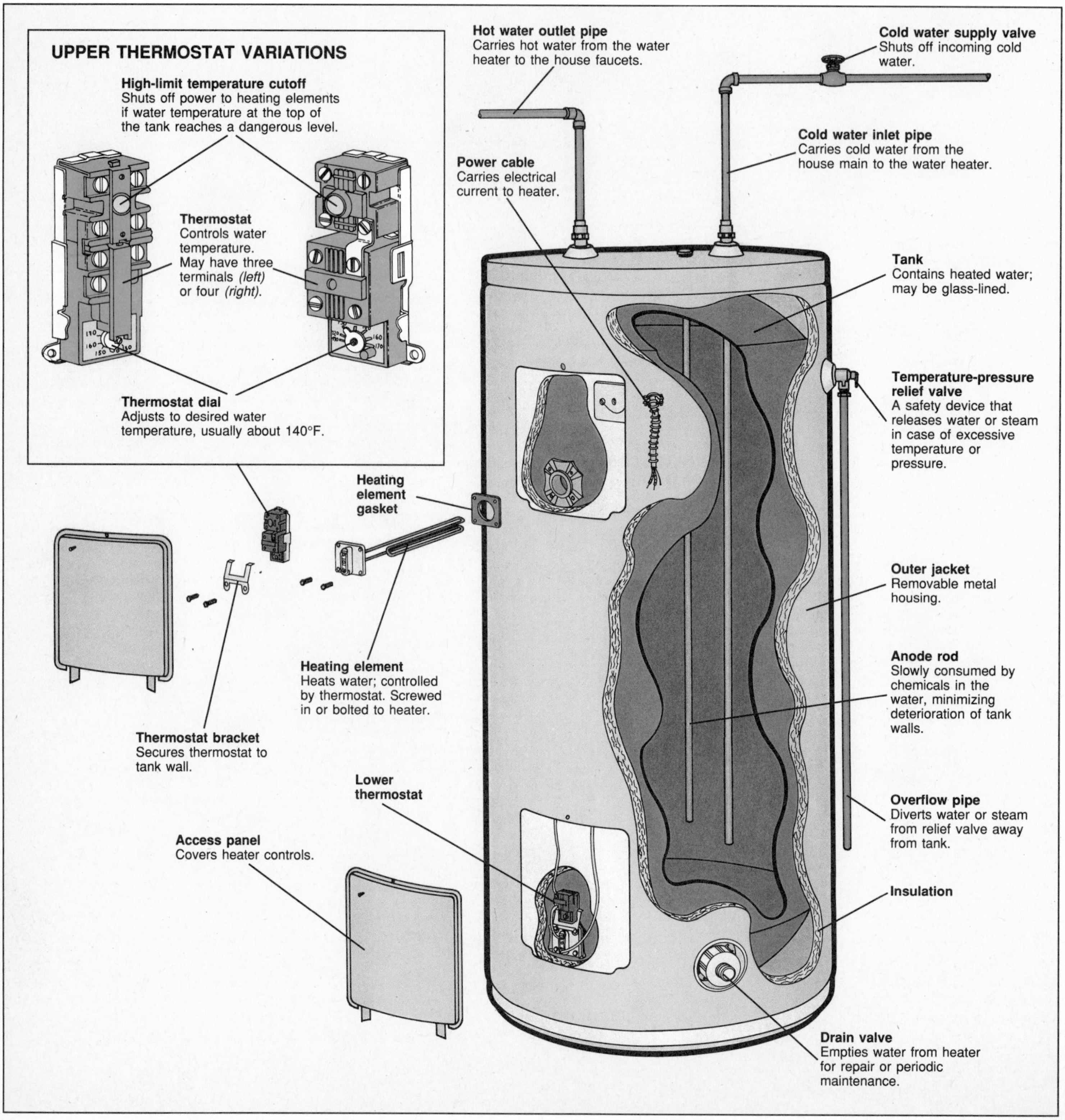

UPPER THERMOSTAT VARIATIONS
High-limit temperature cutoff
Shuts off power to heating elements if water temperature at the top of the tank reaches a dangerous level.
Thermostat
Controls water temperature. May have three terminals (left) or four (right).
Thermostat dial
Adjusts to desired water temperature, usually about 140°F.
Hot water outlet pipe
Carries hot water from the water heater to the house faucets.
Cold water supply valve
Shuts off incoming cold water.
Cold water inlet pipe
Carries cold water from the house main to the water heater.
Power cable
Carries electrical current to heater.
Tank
Contains heated water; may be glass-lined.
Temperature-pressure relief valve
A safety device that releases water or steam in case of excessive temperature or pressure.
Heating element gasket
Outer jacket
Removable metal housing.
Heating element
Heats water; controlled by thermostat. Screwed in or bolted to heater.
Anode rod
Slowly consumed by chemicals in the water, minimizing deterioration of tank walls.
Thermostat bracket
Secures thermostat to tank wall.
Lower thermostat
Overflow pipe
Diverts water or steam from relief valve away from tank.
Access panel
Covers heater controls.
Insulation
Drain valve
Empties water from heater for repair or periodic maintenance.

indicates continuity, or a completed circuit, while a reading of infinity indicates resistance, a total lack of current flow. The heating elements must have partial resistance; the needle should move to the middle range of the scale. To test for voltage, set the multitester at 250 volts AC, and read the results on the AC scale.

Observe strict safety precautions when working on an electric heater. Wear rubber soled shoes, and make sure the floor around the heater is dry. Do not skip steps in doublechecking that power to the heater is indeed turned off *(page 124)*. Post a sign on the disconnect switch box or service panel to prevent someone from turning on the power while you work.

TROUBLESHOOTING GUIDE

SYMPTOM	POSSIBLE CAUSE	PROCEDURE
No hot water	Heater disconnect switch off	Turn switch on *(p. 124)* □○
	Fuse blown or circuit breaker tripped	Replace fuse or reset breaker *(p. 124)* □○
	High-limit temperature cutoff faulty	Test cutoff *(p. 125)* ⬓◒▲
	Thermostat faulty	Test thermostats *(p. 126)* ⬓◒▲
	Element faulty	Test element *(p. 127)* ⬓◒▲
Not enough hot water or water not hot enough	Heavy household demand for hot water or cold incoming water slows recovery time	Stagger use of hot water or install a larger water heater
	Washing machine mixing valve or shower combination valve faulty	Check for hot water in the toilet tank refill tube Replace valve *(p. 110)* ⬓◒
	Thermostat control dial set too low	Raise dial setting to 140°F *(p. 126)* □○
	High-limit temperature cutoff faulty	Test cutoff *(p. 125)* ⬓◒▲
	Thermostat faulty	Test lower thermostat *(p. 126)* ⬓◒▲
	Sediment in tank slows heater recovery time	Drain and refill tank *(p. 119)* □◒
	Heating element faulty	Test heating element *(p. 127)* ⬓◒▲
	Loss of heat	Insulate tank and hot water pipes *(p. 119)* □○
Water too hot	Thermostat control dial set too high	Lower thermostat control setting to 140°F *(p. 126)* □○
	Thermostat improperly mounted	Reposition lower thermostat *(p. 126)* ⬓◒▲
	High-limit temperature cutoff faulty	Test cutoff *(p. 125)* ⬓◒▲
	Thermostat faulty	Test thermostat *(p. 126)* ⬓◒▲
	Heating element fautly	Test heating element *(p. 127)* ⬓◒▲
Fuse blows or breaker trips repeatedly	Wire in heater loose or broken	Check heater wiring *(p. 126)* □○
	House wiring faulty	Call an electrician
Temperature-pressure relief valve leaks continuously	Relief valve unseated	Pop spring lever to reseat valve *(p. 120)* □○
	Relief valve faulty	Replace valve *(p. 120)* ⬓◒
	Thermostat faulty	Test thermostat *(p. 126)* ⬓◒▲
Drain valve leaks	Valve loose or faulty	Tighten valve handle; replace washer or valve *(p. 121)* ⬓◒
Heater wet or dripping	Tank leaks	Replace water heater *(call for service)*
Insulation wet	Leak around element	Tighten loose element; change gasket *(p. 127)* ⬓◒▲
Hot water discolored	Sediment in tank	Drain and refill tank *(p. 119)* □◒
	Anode rod deteriorated	Replace anode rod *(p. 117)* ⬓◒
	Tank interior corroded	Replace water heater *(call for service)*
Hot water smells bad	Magnesium anode rod reacting with sulfurous hot water	Replace with aluminum anode rod *(p. 117)* ⬓◒ Replace water heater *(call for service)*
Water heater noisy	Sediment in tank	Drain and refill tank *(p. 119)* □◒
	Temperature-pressure relief valve faulty	Test valve *(p. 120)* ⬓◒
	Mineral scale on element	Clean or replace element *(p. 127)* ⬓◒▲

DEGREE OF DIFFICULTY: □ **Easy** ⬓ **Moderate** ■ **Complex**
ESTIMATED TIME: ○ **Less than 1 hour** ◒ **1 to 3 hours** ● **Over 3 hours** ▲ **Multitester required**

CHECKING THE VOLTAGE

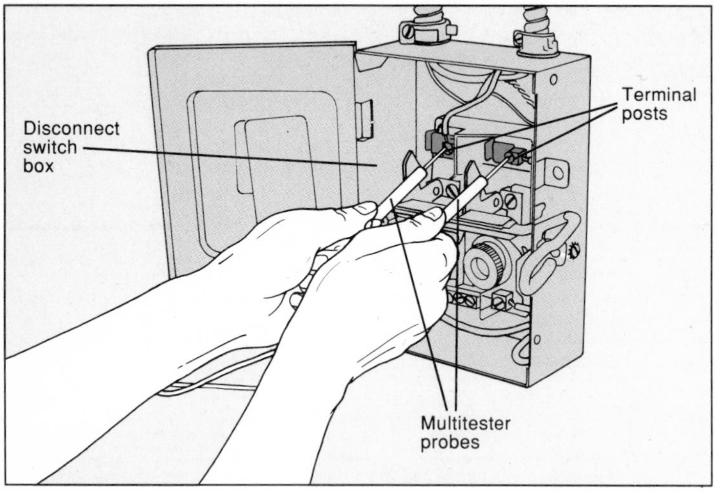

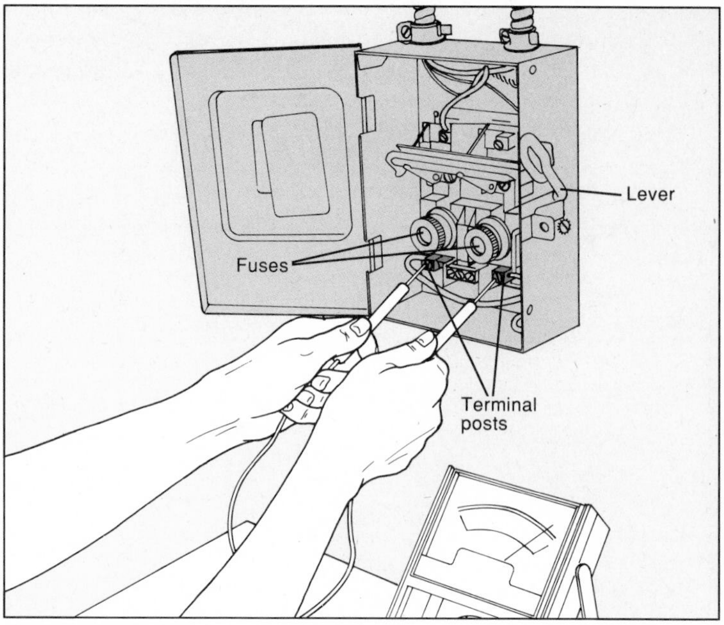

1 Checking the voltage at the disconnect switch box. If your water heater does not have its own switch box, check the main service panel for a blown fuse or tripped circuit breaker. (A 240-volt heater may have two fuses or breakers.) Replace or reset them, if necessary. If the heater still does not work, shut off the house's main power switch at the service panel, and label it so that no one will turn it on. Then access the controls *(step 1, below)*. If the water heater has a disconnect switch box, turn it off and replace any blown fuses. Then use a multitester set at 250 volts AC to verify incoming power to the switch box. **Caution**: Do not touch the switch box. Hold the tester probes by the insulated handles only and touch one probe to each of the upper terminals, as shown. The tester should read between 200 and 250 volts. Next touch one probe to the left terminal and the other to the grounding screw on the back of the box. The tester should read about 120 volts. Test the right terminal the same way. If all results are what they should be, go to step 2. If not, call an electrician.

2 Testing the lower terminals. With the power on, test the two lower terminals as you tested the upper terminals in step 1; the results should be the same. If all the readings in both steps are what they should be, enough power is getting to the heater. If not, call an electrician. Before working on any part of the heater, test that the power is off: Shift the lever arm of the disconnect switch box to the OFF position and test the lower terminals as you tested the upper terminals in step 1. This time the tester should show 0 volts in all cases. If the tester shows that the disconnect switch box conducts any power at all while in the OFF position, do not work on the switch box or the water heater; call an electrician.

ACCESS TO THE CONTROLS

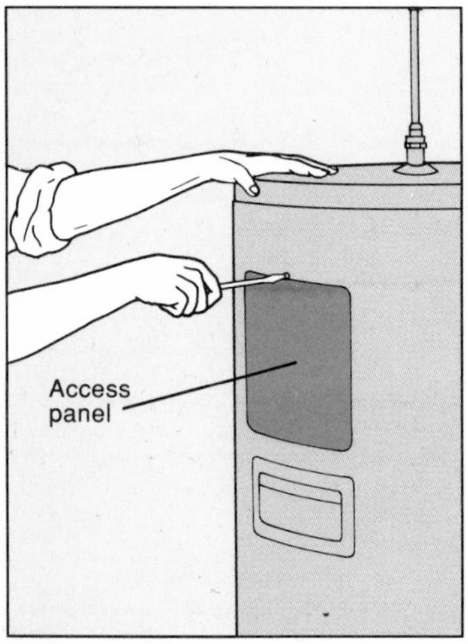

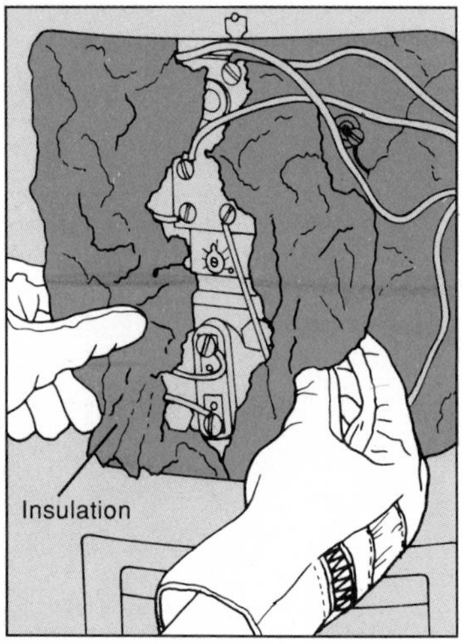

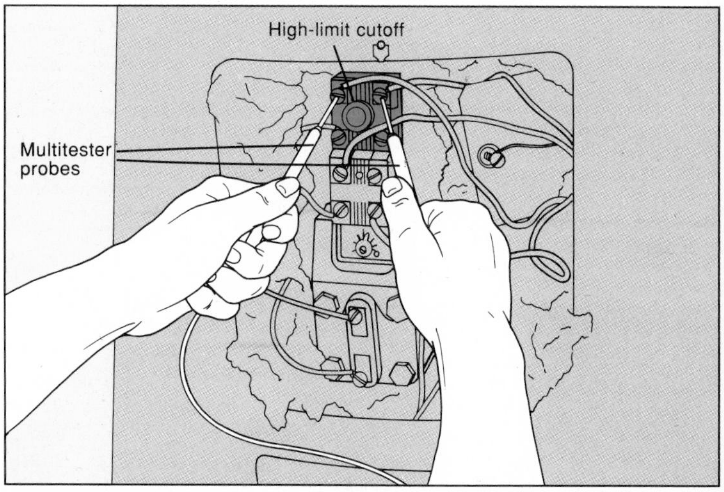

1 Accessing the controls. Shut off power to the heater at the main service panel or at the disconnect switch box, and test that it is off *(above)*. To reach the heater controls, unscrew and remove the upper and lower access panels *(above, left)*. Wearing gloves, turn aside insulation that is pre-slit *(above, right)*, or cut the insulation away with a serrated knife, being careful not to damage the controls behind it. Save any insulation you remove. You now have access to the high-limit cutoff, thermostats and heating elements.

2 Verifying power shutoff at the heater. Set a multitester at 250 volts AC and touch a probe to each of the upper terminals of the high-limit cutoff, as shown. Then touch one probe to the exposed interior tank wall and the other to each terminal, in turn. The tester should show 0 volts each time. If the power is off and the readings are not 0, do not work on the heater; call an electrician. Tighten any loose mounting bracket bolts and any loose electrical connections at the upper and lower controls.

TESTING THE HIGH-LIMIT TEMPERATURE CUTOFF

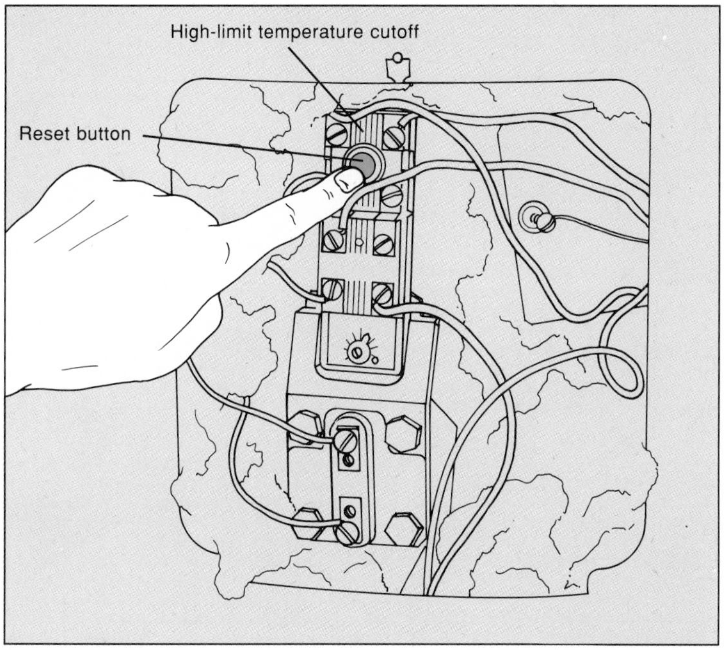

1 Checking the reset button. Disconnect power to the heater at the main service panel or switch box and verify that it is off, then remove the upper access panel and verify that power is off *(page 124)*. If the high-limit cutoff's reset button has popped out, push it in, as shown, and listen for a click. Turn on the power and wait three hours. If the interior tank wall feels warm near the bottom, turn off the power, replace the insulation and access panels and turn the power back on. If not, test the high-limit cutoff.

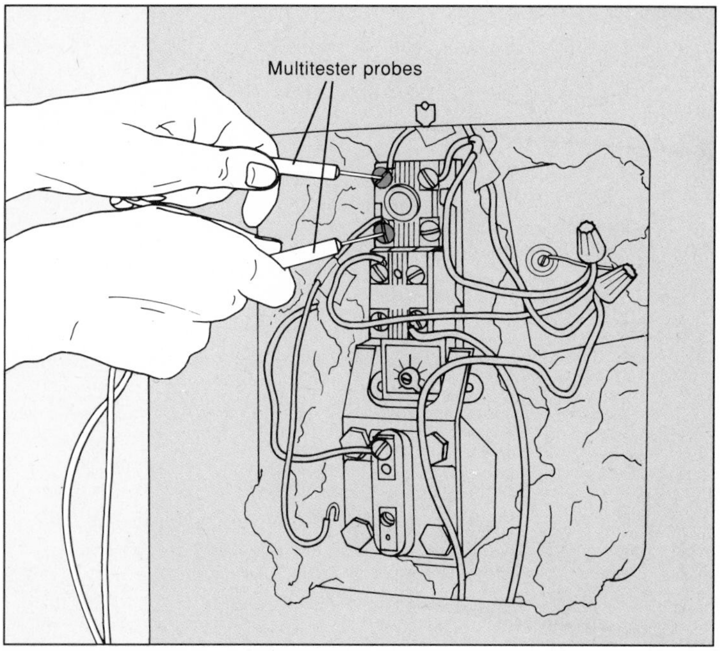

2 Testing the cutoff for continuity. With the power off, label the position of one of the element wires with masking tape, and disconnect it by removing its terminal screw. With a multitester set at RX1, touch a probe to each of the cutoff's two left terminals, as shown, and then to the two right terminals. The tester needle should sweep to 0 each time, indicating continuity. If the cutoff shows continuity, test the thermostat *(page 126)*; if not, replace the cutoff *(step 3)*.

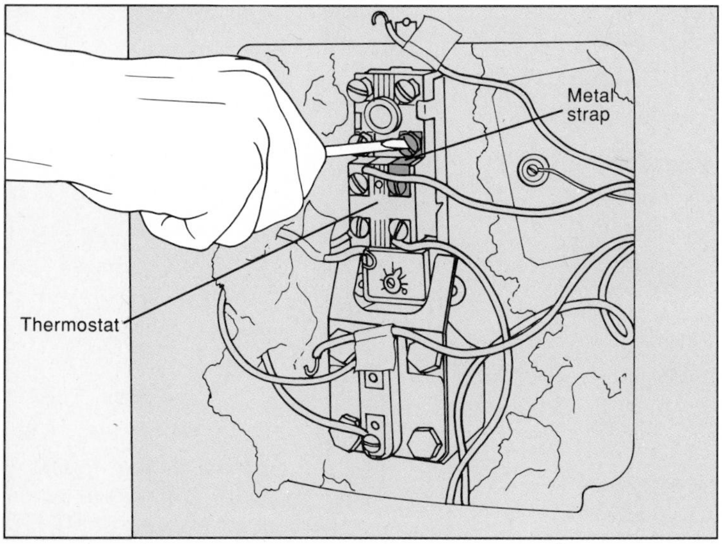

3 Removing the cutoff. Tag the wires to each of the cutoff terminals with masking tape to identify their positions for reassembly. To disconnect the cutoff's wires, loosen the terminal screws and gently unhook the wires. Remove the screws that hold the metal strap connecting the cutoff to the thermostat; the strap may be at the front *(above)* or the side. Take off the strap. Pull the cutoff up to release it from the spring clips that hold it to the heater, or pry it free with a screwdriver.

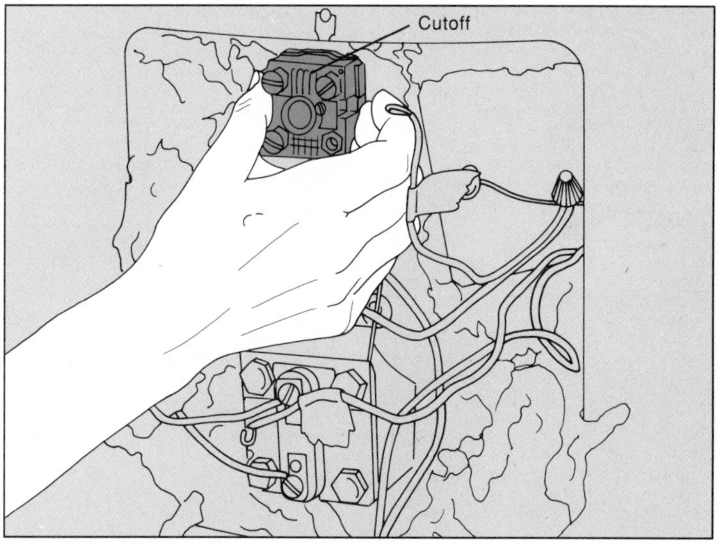

4 Replacing the cutoff. Buy a new cutoff of the same make and model from a heating or plumbing supplier. Before installing it, depress the reset button and test it for continuity *(step 2)*. Snap the new cutoff into place *(above)*. Reconnect the wires and metal strap in their proper positions, then turn on the power and wait three hours. If the interior tank feels warm near the lower element, the heater is working properly. Turn off the power, repack the insulation, install the access panels and turn the power on. If the tank does not become warm, test the thermostats *(page 126)* and elements *(page 127)*.

TESTING THE THERMOSTATS

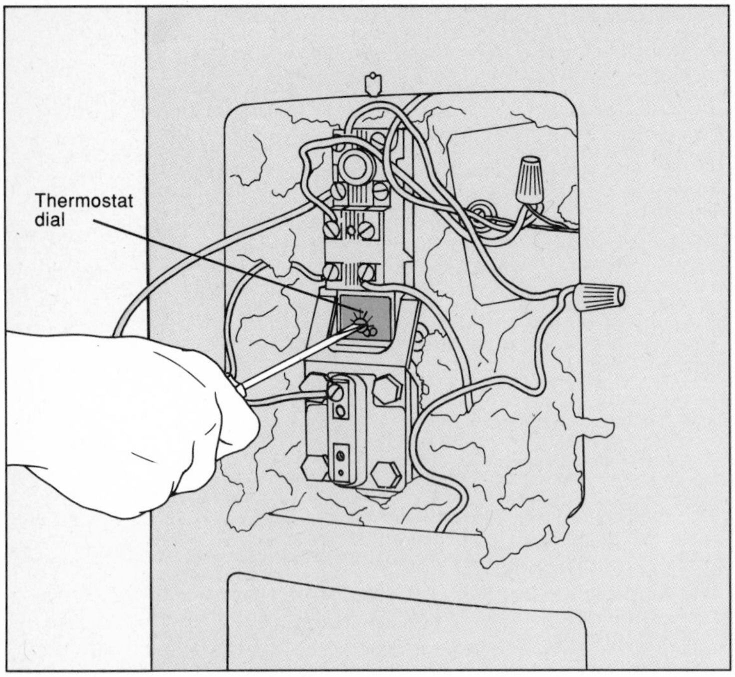

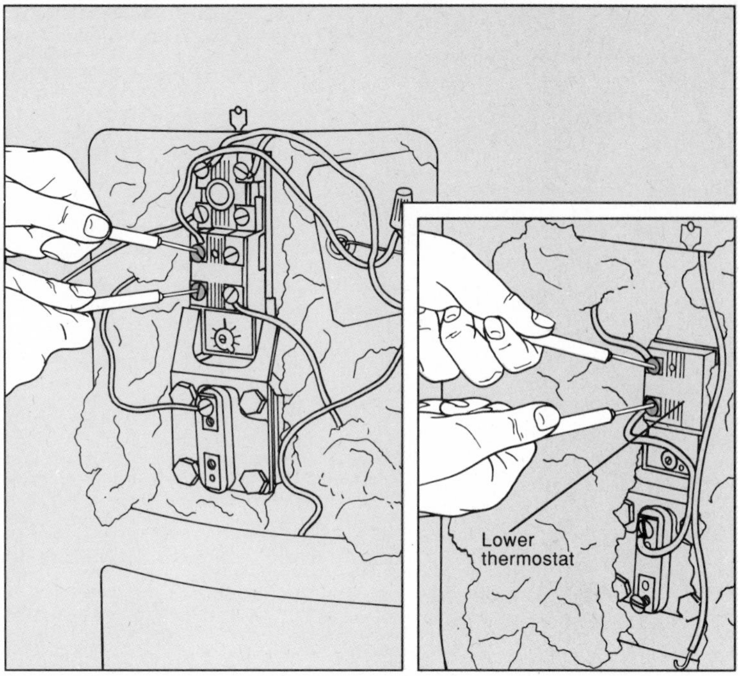

1 **Testing the thermostats.** Turn off power to the water heater at the disconnect switch box or the main service panel and test that it is off, then access the upper and lower controls *(page 124)*. Turn the thermostat dial counterclockwise to its lowest temperature setting *(above, left)* and listen for a click. If you hear no click, turn the dial clockwise to its highest point, run a hot water tap until the water runs lukewarm, and move the dial to its lowest setting; you should now hear a click. Label and disconnect the wire to the upper element. With a multitester set at RX1, touch a probe to each of the left terminals *(above, right)*. The tester needle should remain at infinity. Then touch a probe to each of the two right terminals on a four-screw thermostat, or to the upper left and upper right terminals on a 3-screw model. The tester needle should swing to 0. Adjust the thermostat to its highest setting; you should hear a click. Repeat the two tests; this time the results should be reversed. To test the lower thermostat *(inset)*, first adjust the upper thermostat to its lowest setting. Then turn the lower dial to its lowest setting; the tester needle should remain at infinity. Finally, turn the dial to the highest setting; the needle should swing to 0. If any of your results differ, replace the thermostat *(step 2)*. If the thermostats test OK, test the elements *(page 127)*.

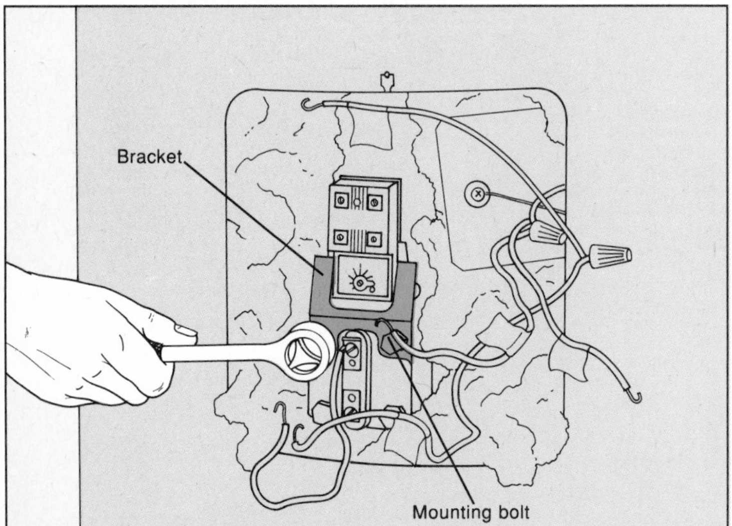

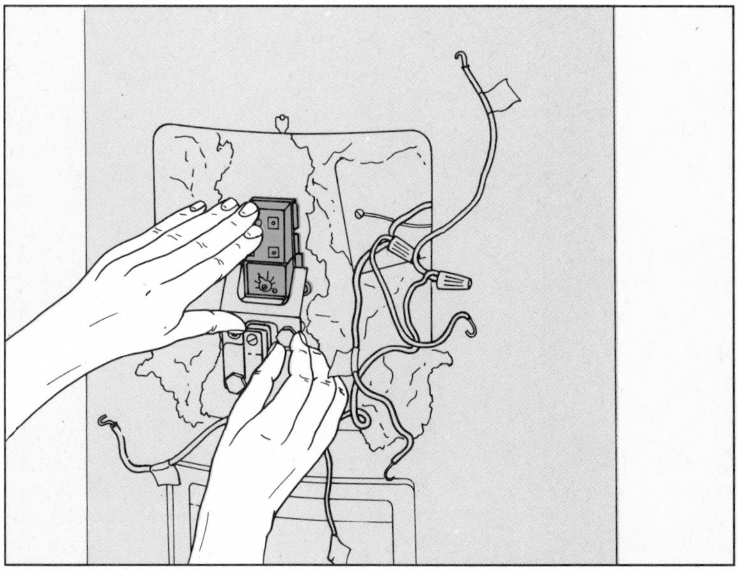

2 **Replacing the thermostat.** Remove the cutoff *(page 125)*, and label and disconnect the wires to the thermostat. Using a socket wrench, loosen the two bolts on the thermostat mounting bracket *(above, left)*. Slip the thermostat up and out of the bracket. Buy a new thermostat of the same make and model at a heating or plumbing supplier, and test it for continuity *(step 1)*. Insert the new thermostat behind the bracket *(above, right)* and tighten the bolts, making sure the back of the thermostat fits flush with the heater wall. Adjust the thermostat to the medium setting, or 140°F. If the heater has two thermostats, adjust both to the same setting. Reinstall the cutoff *(page 125)*, reconnect all the wires and turn on the power. Wait three hours and, if the exposed tank wall feels warm near the lower element, the heater is working properly. Turn off the power, repack the insulation uniformly, replace the access panels and turn the power back on. If the tank is not warm, test the heating elements *(page 127)*.

TESTING THE HEATING ELEMENTS

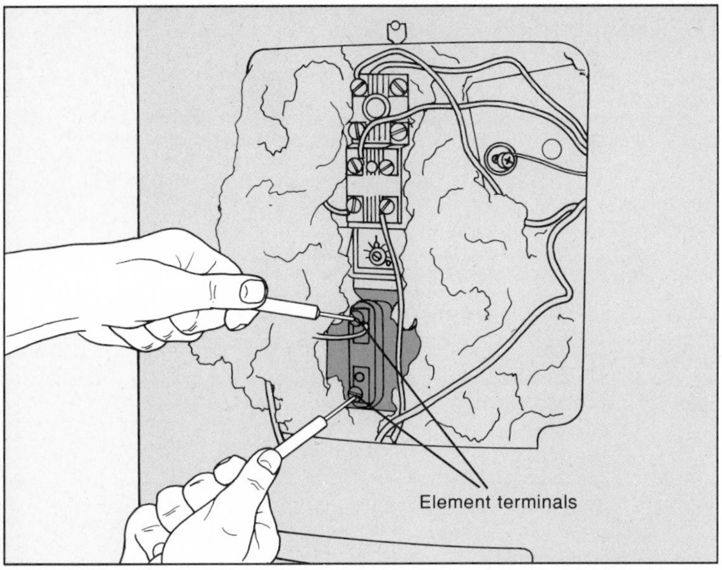

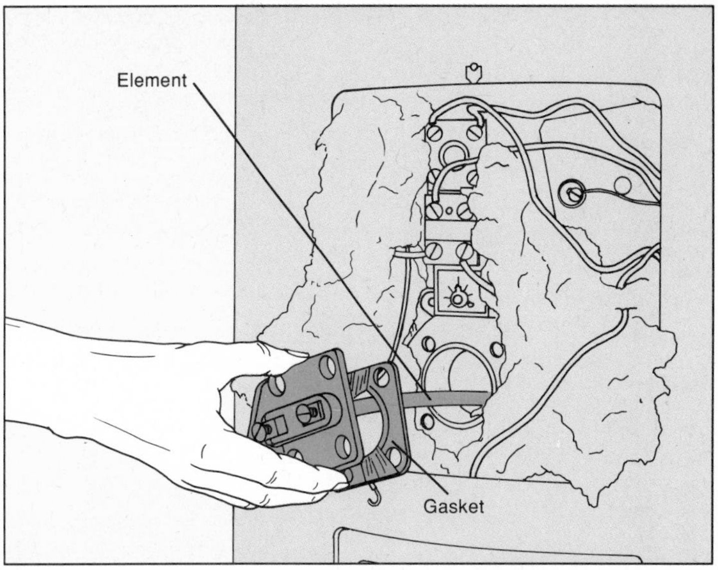

1 Testing the element. Turn off power to the water heater, and test that it is off *(page 124)*. To test the upper or lower element, disconnect one of its wires. Using a multitester set at RX1000, touch one probe to an element mounting bolt or the thermostat bracket, and the other to each element terminal screw in turn. If the tester needle moves at all, the element is grounded and should be replaced *(step 2)*. To test whether the element works, set the multitester at RX1 and touch a probe to each of the two terminal screws *(above)*. The tester should indicate resistance in the medium range of the ohms scale; if not, replace it *(step 2)*.

2 Removing the element. Drain the heater *(page 119)*. Disconnect the remaining element wire and label the position of each wire with masking tape. If the element is held by mounting bolts, use a socket wrench to remove them. Carefully lift off the thermostat bracket—the thermostat will hang by its wires. If you do not see mounting bolts, the element itself is screwed in; use a socket wrench to unscrew it. To remove either type of element, pull it straight out, as shown, working it loose gently if the shape has become distorted. Clean mineral scale from a working element by soaking it in vinegar for several hours, then use an old knife to chip off the scale.

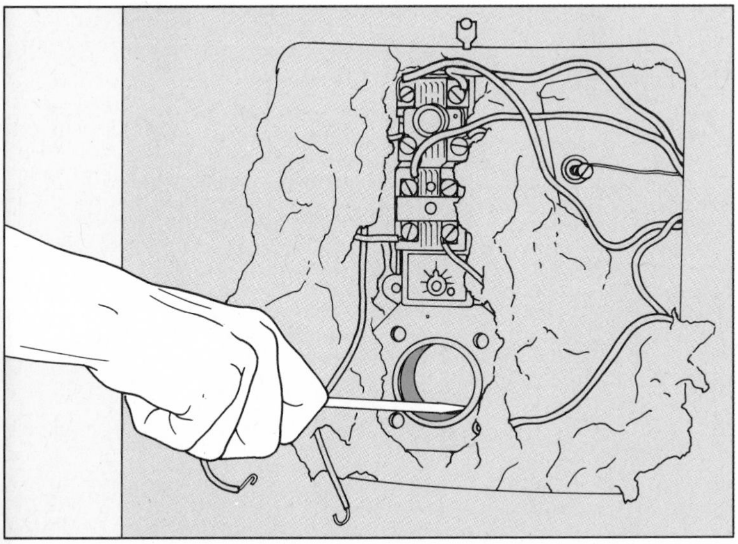

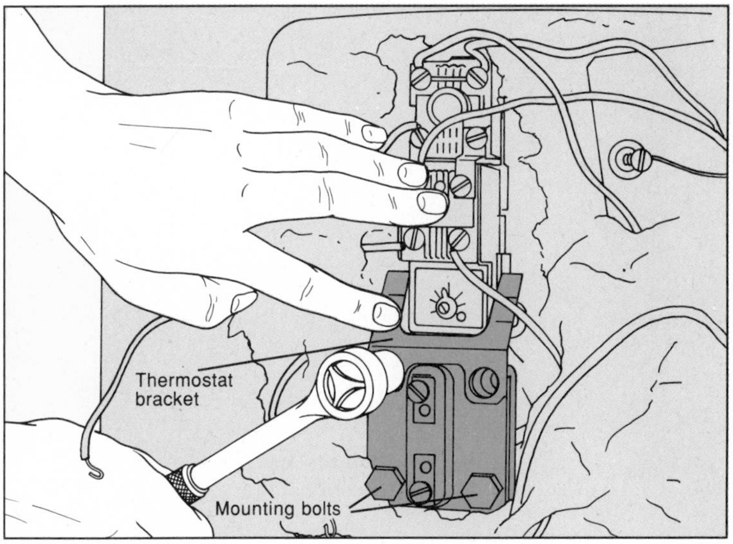

3 Replacing the element. Test a new element *(step 1, above)* before you install it. Whether replacing an element or reinstalling a cleaned one, buy a new gasket. Using an old screwdriver, scrape scale and rust from the inside surface of the element fitting *(above, left)* so the gasket can form a tight seal. Remove the old gasket and thread the new gasket onto the element, then ease the element into the heater. If the element was bolted in place, install the two lower mounting bolts, then reposition the thermostat and thermostat bracket *(above, right)* and tighten the two upper bolts. If the element is a screw-in type, install it using a socket wrench. Install a lower element the same way, and reconnect all the wires. Refill the heater *(page 119)*, turn it on and wait three hours; if the exposed tank wall feels warm near the lower element, the heater is working properly. Turn off the power, repack the insulation uniformly, replace the access panels and turn the power back on.

WATER SOFTENERS

Hard water combines with soap to leave a dull film on fixtures, tubs and sinks. Mineral buildup can also block pipes and corrode water heaters, washing machines and faucets. A water softener effectively reduces hardness by exchanging magnesium and calcium ions, which react with soap, for sodium ions, which do not react.

Although its various cycles may seem complex, the water softener is a simple plumbing fixture to maintain. On an automatic softener, check the brine line and injector for blockages every six months, and clean or replace the injector screen. If a softener is installed in a warm place, such as near a furnace, condensation may form in the brine tank. This results in compacted salt—known as salt bridging—which blocks the flow of salt water from the brine tank to the resin tank. Break up or remove a salt bridge when this occurs. Iron, if heavily concentrated in supply water, breaks down the resin beads, which may then block hoses or valves. If your water contains up to 3 ppm of iron, add a bag of resin cleaner to the resin tank; install a charcoal-activated iron filter on the supply pipe if the water contains more than 3 ppm of iron *(page 16)*.

TROUBLESHOOTING GUIDE

PROBLEM	PROCEDURE
Water remains hard	Check and refill salt in brine tank
	Set timer to regenerate more often
	Have water checked for iron content and install iron filter if necessary
Salt compacts to form a bridge	Poke salt bridge with broom handle to loosen salt pellets. If salt is hard, disconnect softener, empty tank, and lift out the salt bridge. Replenish with fresh salt
Softener not drawing brine	Flush brine line *(p. 129)* □○
	Clean or replace injector and filter screen *(p. 129)* □○
Softener doesn't work at all	Remove control unit and take to dealer to service motor and timer *(p. 129)* ⬓◒

DEGREE OF DIFFICULTY: □ Easy ⬓ Moderate ■ Complex
ESTIMATED TIME: ○ Less than 1 hour ◒ 1 to 3 hours

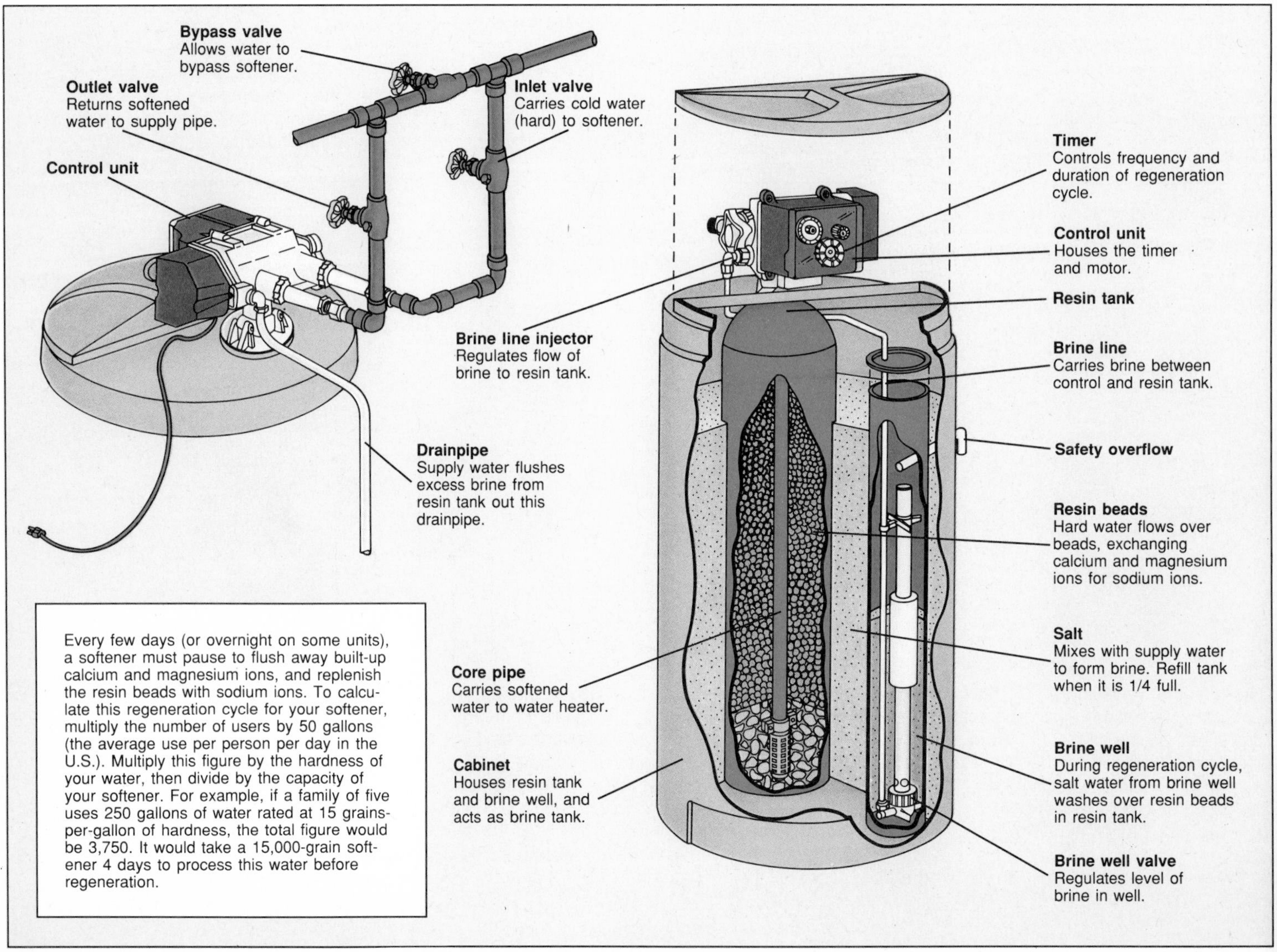
Bypass valve
Allows water to bypass softener.
Outlet valve
Returns softened water to supply pipe.
Inlet valve
Carries cold water (hard) to softener.
Control unit
Brine line injector
Regulates flow of brine to resin tank.
Drainpipe
Supply water flushes excess brine from resin tank out this drainpipe.
Timer
Controls frequency and duration of regeneration cycle.
Control unit
Houses the timer and motor.
Resin tank
Brine line
Carries brine between control and resin tank.
Safety overflow
Resin beads
Hard water flows over beads, exchanging calcium and magnesium ions for sodium ions.
Salt
Mixes with supply water to form brine. Refill tank when it is 1/4 full.
Core pipe
Carries softened water to water heater.
Cabinet
Houses resin tank and brine well, and acts as brine tank.
Brine well
During regeneration cycle, salt water from brine well washes over resin beads in resin tank.
Brine well valve
Regulates level of brine in well.

Every few days (or overnight on some units), a softener must pause to flush away built-up calcium and magnesium ions, and replenish the resin beads with sodium ions. To calculate this regeneration cycle for your softener, multiply the number of users by 50 gallons (the average use per person per day in the U.S.). Multiply this figure by the hardness of your water, then divide by the capacity of your softener. For example, if a family of five uses 250 gallons of water rated at 15 grains-per-gallon of hardness, the total figure would be 3,750. It would take a 15,000-grain softener 4 days to process this water before regeneration.

SERVICING THE BRINE LINE AND INJECTOR

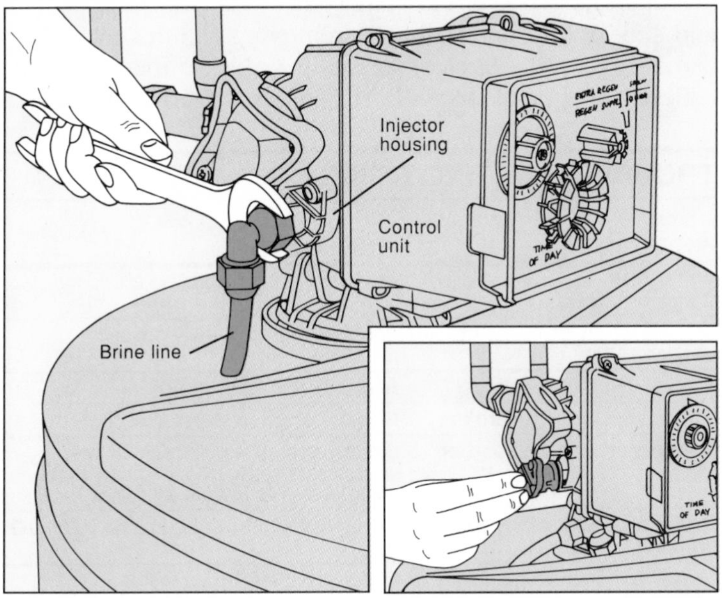

1 Inspecting the brine line. Unplug the softener. Close the main shutoff valve and open the nearest hot water faucet to lower the water level in the softener below the inlet valve. With a wrench, loosen the nut connecting the brine line to the injector housing *(above)* and gently pull the brine line free. If there is an obstruction in the line, use an oven baster to flush it out with warm water. Next, unscrew the injector housing from the control unit by hand *(inset).*

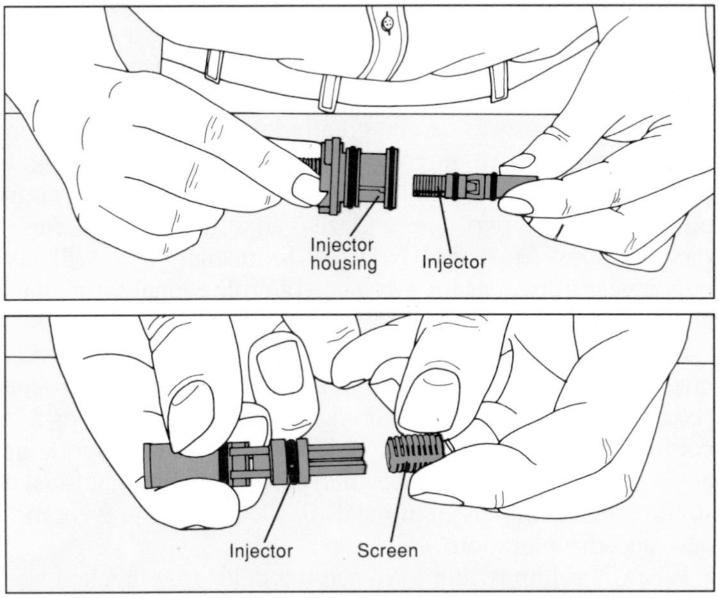

2 Cleaning the injector. Gently pull the injector from its housing *(above, top)*, then remove the small filter screen from the injector *(above, bottom).* Clean a clogged screen in warm, soapy water; replace a broken one. Next, remove any obstruction in the injector by blowing gently. (Do not use a sharp object to remove a blockage.) Replace the screen, then fit the injector back in its housing. Screw the injector housing to the control unit and reattach the brine line. Turn the water back on and plug in the softener.

SERVICING THE CONTROL UNIT

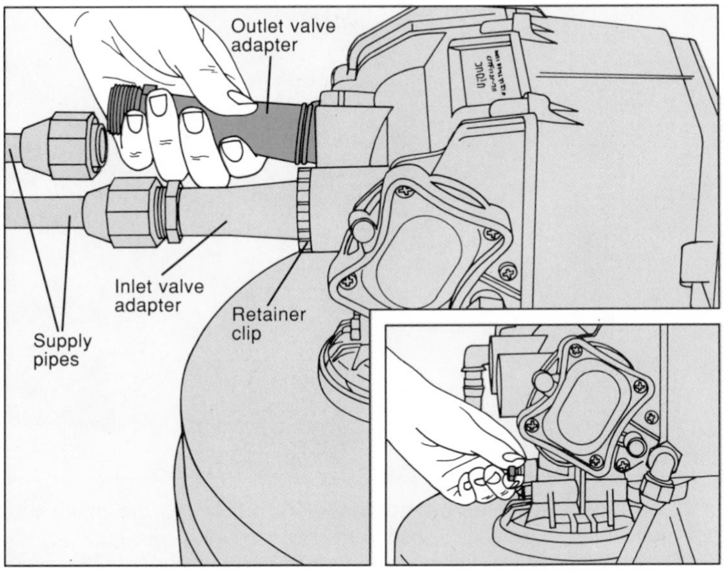

1 Disconnecting the control unit. Unplug the softener. Close the main shutoff valve and open the nearest hot water faucet to lower the water level in the softener below the inlet valve. Unscrew the fittings joining the supply pipes to the inlet and outlet valve adapters, then pry off the retainer clips that secure the adapters to the control unit *(above).* Push the pipes back to provide enough space to remove the control unit. Next, unscrew the nuts securing the locking bar rods beneath the inlet and outlet valves *(inset)*, then remove the brine line from the injector housing *(step 1, above).*

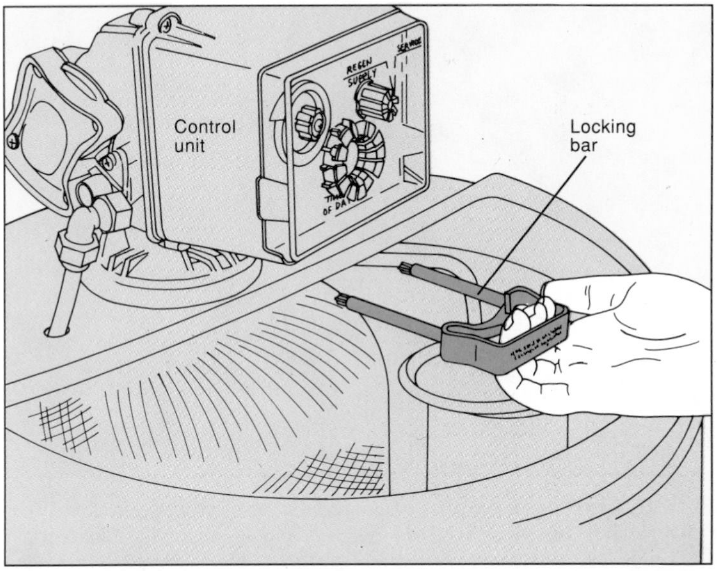

2 Replacing the control unit. Lift off the cabinet lid and pull out the locking bar from beneath the control unit *(above).* Grasp the control unit firmly and pull it up and off the cabinet. Take the unit to the dealer to repair a faulty timer or motor. To reinstall the unit, reposition it on top of the cabinet, insert the locking bar rod and replace the nuts. Push the inlet and outlet valve adapters back into the control unit and reattach their retainer clips. Push the pipes back in place and screw the fittings onto the adapters. Reconnect the brine line, turn on the water and plug in the softener.

TILE AND ACCESSORIES

While the sight of a chipped sink or bathtub, mildewed caulking, cracked tile or broken towel rack may make you wish you could renovate the entire bathroom, the expense and effort may seem formidable. Yet with minimal expense and little effort, these eyesores can be greatly improved or eliminated.

Chips or stains in an enamel or porcelain sink or bathtub can be repaired with epoxy touch-up paint, or refinished completely with two-part epoxy glaze—an attractive but temporary measure. Most do-it-yourself refinishing jobs will last only a year or two, as long as a costly professional refinish by a "bathtub doctor."

Repair damaged caulking to cure dampness or unsightly cracking at the joints between walls and bathtubs, and around fixtures. Replace grout to eliminate cracking, peeling or discoloration between ceramic wall and floor tiles. Sponging down tiles and grout joints regularly with a vinegar-and-water solution or a commercial tub-and-tile cleaner will prevent mildew and discoloration.

When it is impossible to match the color of a cracked tile, consider installing an accent tile of a contrasting color. A broken accessory, whether set flush against the wall, recessed or mounted on a plate, can be quickly replaced once the wall behind it has been cleaned of old adhesive and grout.

Prepare to work in the bathroom by covering the bathtub and sink drains, and taping or removing fixtures to prevent damage. Keep the area as dry as possible so that grout, tile adhesive and caulking will set and dry effectively.

TROUBLESHOOTING GUIDE

PROBLEM	PROCEDURE
Water leaks between bathtub and wall	Apply new caulking *(p. 131)* □○
Discoloration or moisture between tiles	Apply new grout between tiles *(p. 132)* □○
Tiles loose or broken	Replace tiles *(p. 133)* ⬓◒
Accessory loose or broken	Replace accessory *(pp. 134, 135)* □○
Bathtub or sink chipped or stained	Clean with alcohol, then cover with several thin coats of epoxy touch-up paint, available in various colors. □○
Bathtub or sink badly chipped or worn	Resurface with epoxy glaze *(p. 130)* ⬓●
Porcelain, enamel or tile stained or dirty	Clean surface with appropriate cleaning agent *(p. 138)*
DEGREE OF DIFFICULTY:	□ **Easy** ⬓ **Moderate** ■ **Complex**
ESTIMATED TIME:	○ **Less than 1 hour** ◒ **1 to 3 hours** ● **Over 3 hours**

REFINISHING A BATHTUB OR SINK

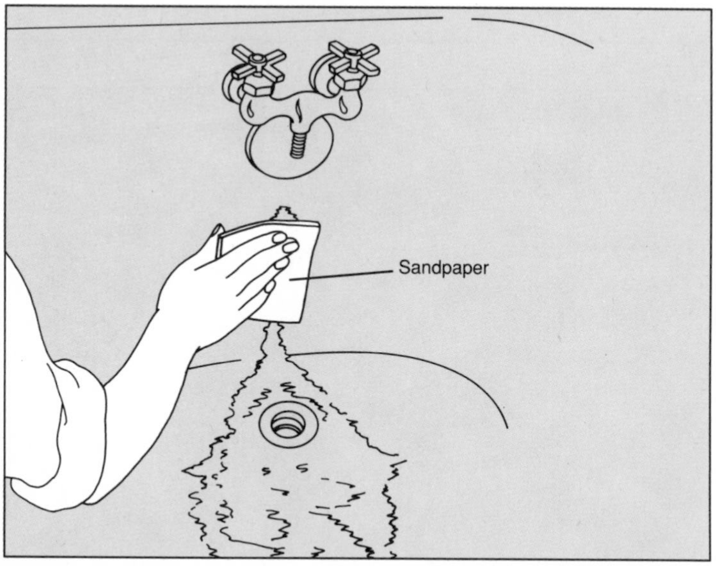

1 Preparing to paint. With medium-grit sandpaper, sand the area to be resurfaced—a bathtub in this example. Clean the tub thoroughly with a sponge and all-purpose cleaner, removing any peeling paint. Rinse thoroughly and allow the tub to dry. **(Caution:** Ventilate the bathroom while you work with the epoxy glaze since the fumes may make you dizzy.) Have two brushes on hand, a small one for touching up chips and painting around the fixtures, and a larger one for the tub itself. Put masking tape around the fixtures to protect them from the glaze. Following instructions on the cans of two-part epoxy glaze, mix just enough glaze in a plastic container to use as a first coat, then reseal the cans tightly. Allow the mixed glaze to stand for one hour before applying it.

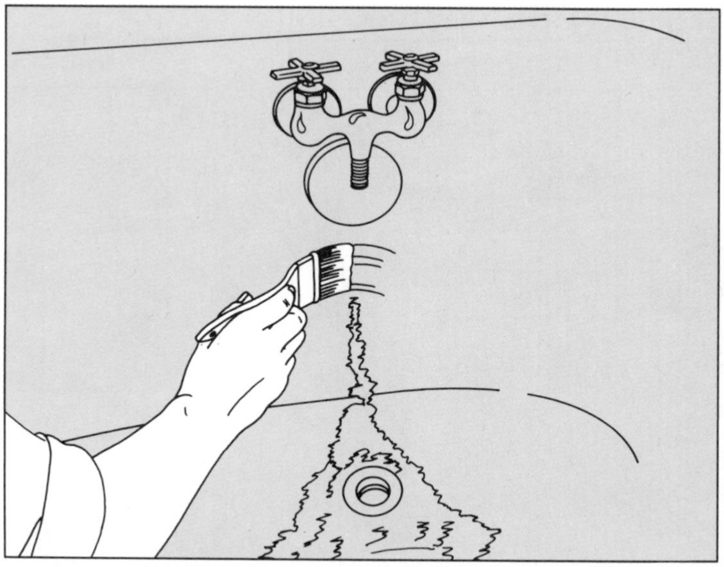

2 Refinishing the bathtub. Before refinishing the entire tub, cover any badly damaged areas with several thin coats of glaze *(above)*, allowing each coat to dry thoroughly. (The glaze takes 16 to 36 hours to dry between coats, depending on humidity.) Then apply the first complete coat with long strokes across the width, then length of the bathtub. Let the tub dry, then sand the surface lightly with medium-grit sandpaper and wipe with a clean sponge. Mix and apply a second light coat of glaze, allowing it to flow rather than brushing too vigorously. If a third coat is necessary, allow the second to dry, lightly sand and wipe the tub, then apply the glaze. Do not use the bathtub for five days after the last coat.

REPLACING WORN CAULKING

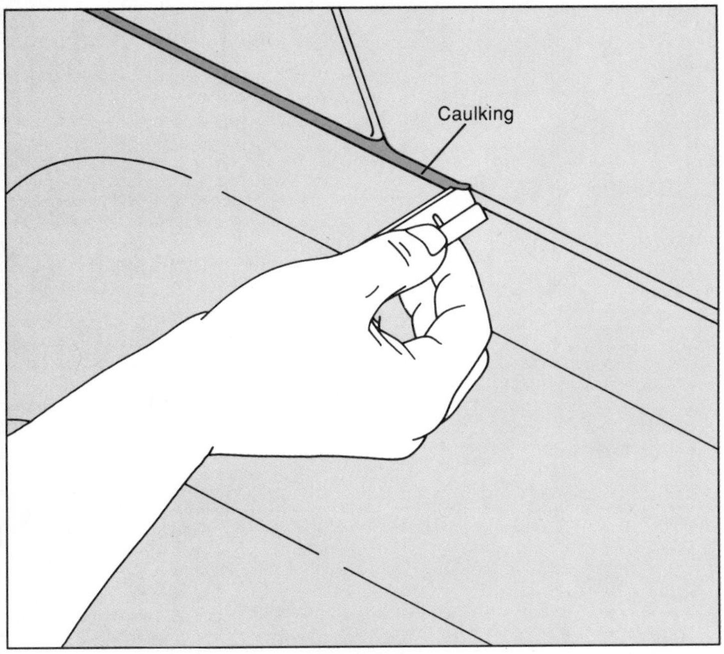

1 Removing the old caulking. Cut away cracked, peeling or discolored caulking with a single-edge razor blade, as shown, or with a putty knife. Be careful not to scratch the tile or enamel above or below the joint.

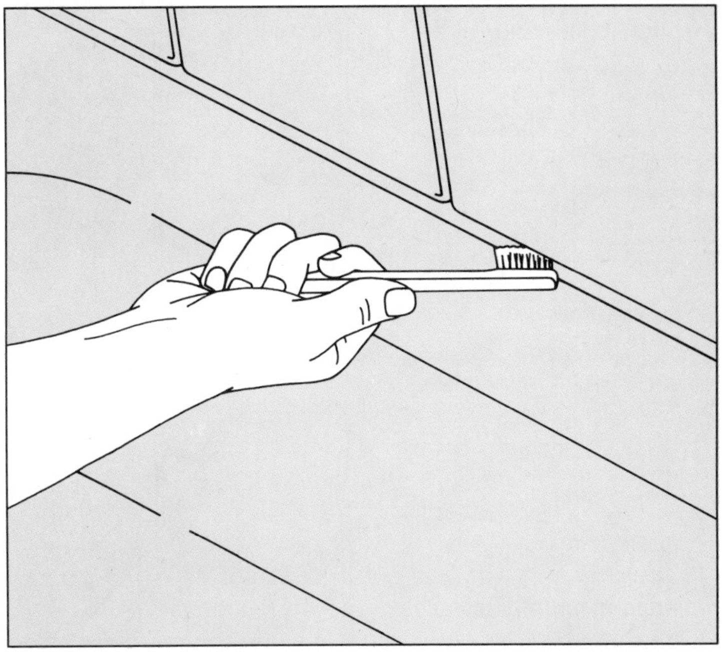

2 Preparing the joint. Remove any remaining caulking or soap residue by scrubbing with an old toothbrush and all-purpose cleaner, as shown. Wrap a clean rag dampened with alcohol around the blade of a putty knife and draw it along the joint to remove any grit and soap residue. Fill the bathtub so that the weight of the water will open the joint as wide as possible.

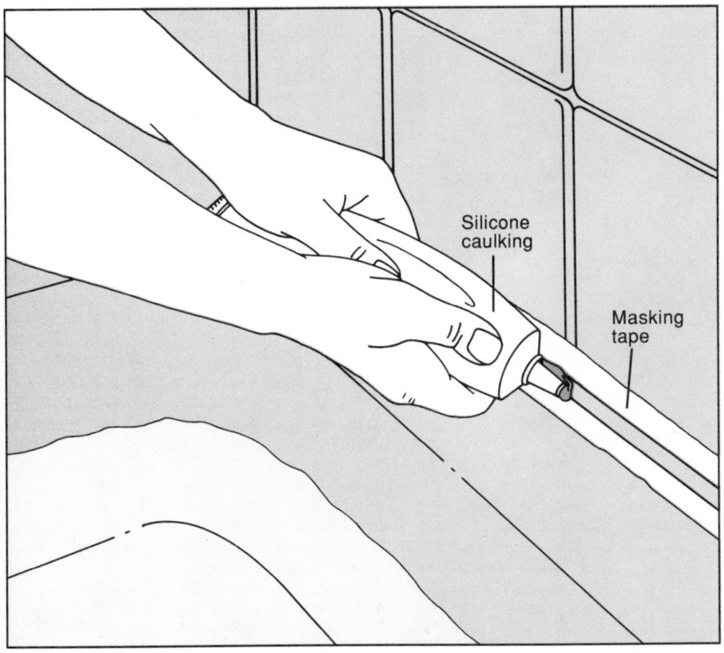

3 Applying silicone caulking. To help form an even line, tape the edges of the tub and tile with masking tape before caulking. Cut the nozzle of the caulking tube at an angle so that the opening is slightly larger than the width of the joint. Position the tube at the end of the exposed joint and push the tube ahead of you, while slowly squeezing it *(above)*. Remove the tape when the caulking is tacky.

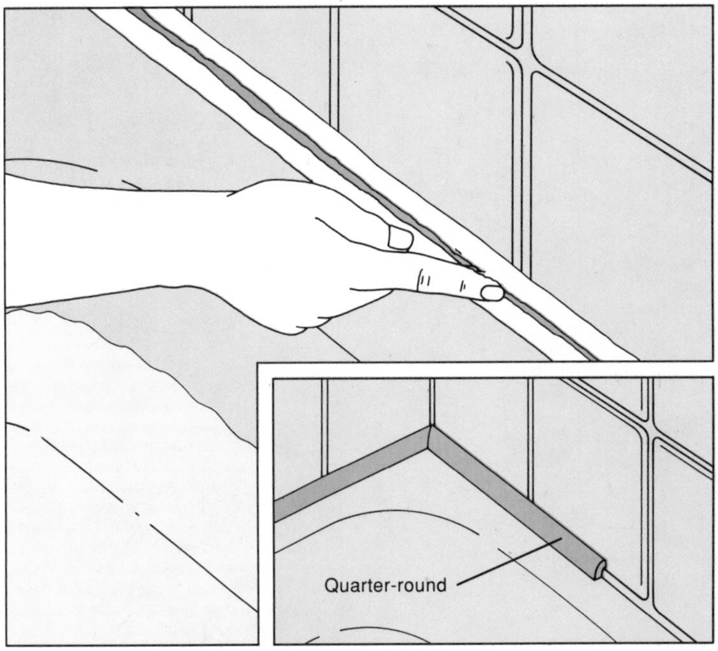

4 Smoothing the joint. Draw a wet finger lightly and evenly along the bead of caulking *(above)*. Let the caulking dry for several hours before using the bathtub. If the caulking does not hold, or pops out of the joint, you can install decorative quarter-round ceramic tile, available in plumbing supply stores, over the joint *(inset)*.

REPLACING WORN OR DISCOLORED GROUT

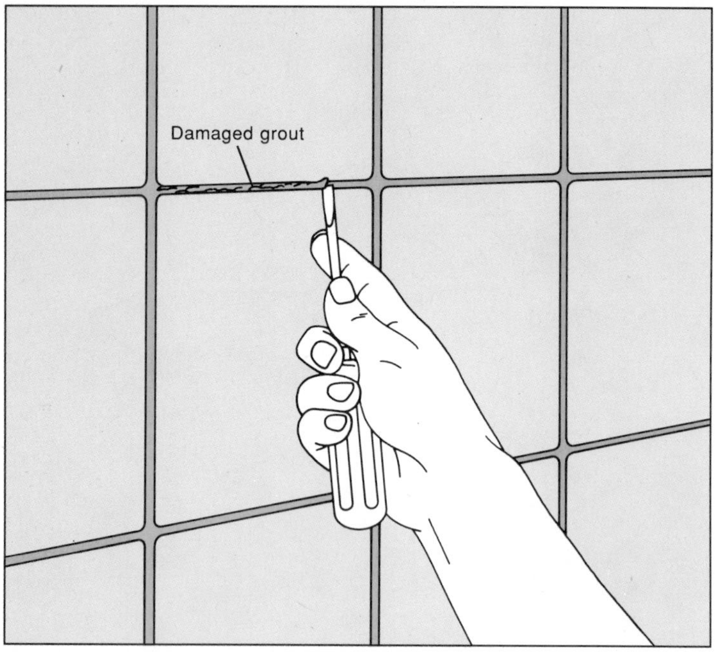

1 Digging out damaged grout. Replace grout where there is water discoloration or seepage between tiles. Pull the edge of an old screwdriver along the joint *(above)*, or push a carpet knife along the joint away from you. Be careful not to scratch the adjacent tiles. Scrub the joint and surrounding tiles with a damp sponge and alcohol to remove any remaining grout and to speed drying.

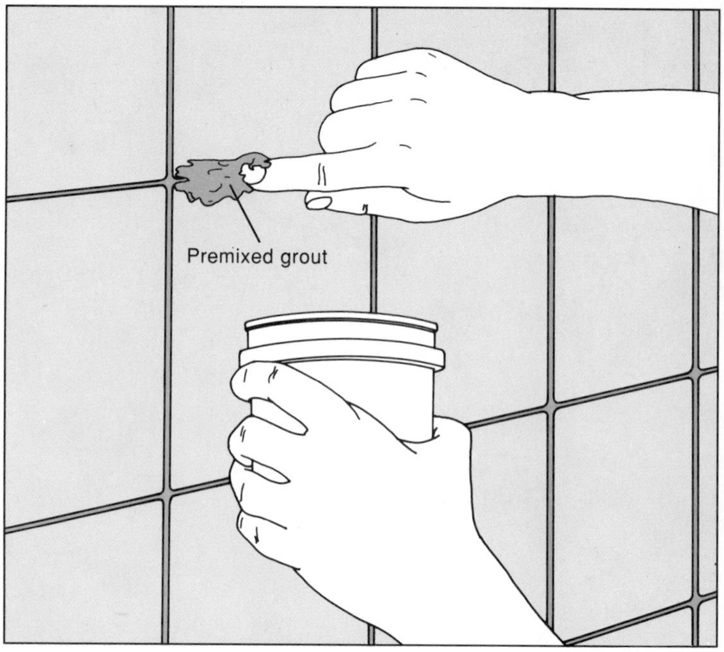

2 Applying the new grout. For small areas, press premixed latex tile grout into the joint with your finger, as shown. Work it firmly into the joint, leaving some on the tile surface. For larger areas, it is less expensive to buy grout in powdered form and mix as directed with water or a special latex solution. A pencil eraser or the end of an old toothbrush is also helpful in forcing grout into the joint.

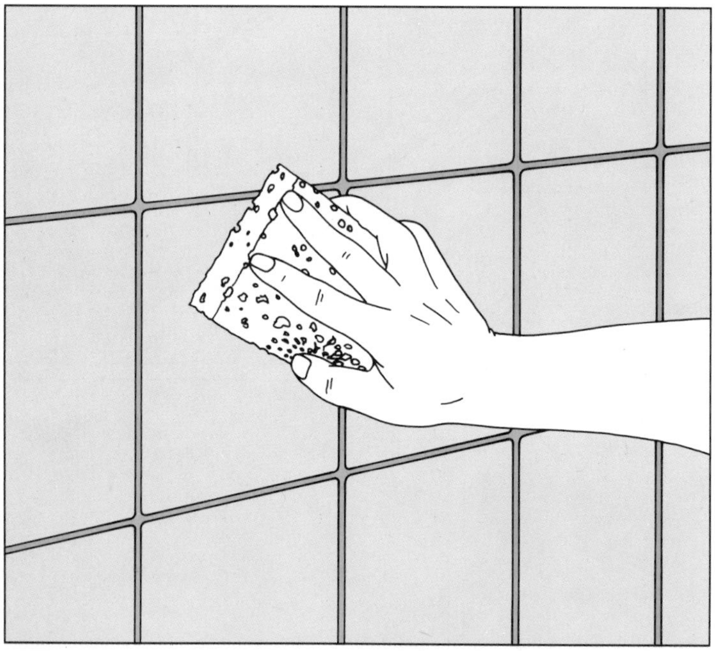

3 Sponging the tile. With a clean, damp sponge, wipe the excess grout diagonally across the adjacent tiles *(above)* to help make the joint more water-resistant. Rinse out the sponge and continue wiping until a thin, gritty film remains on the tiles. Wait 15 minutes for the grout to set, then wipe off any remaining film from the tile.

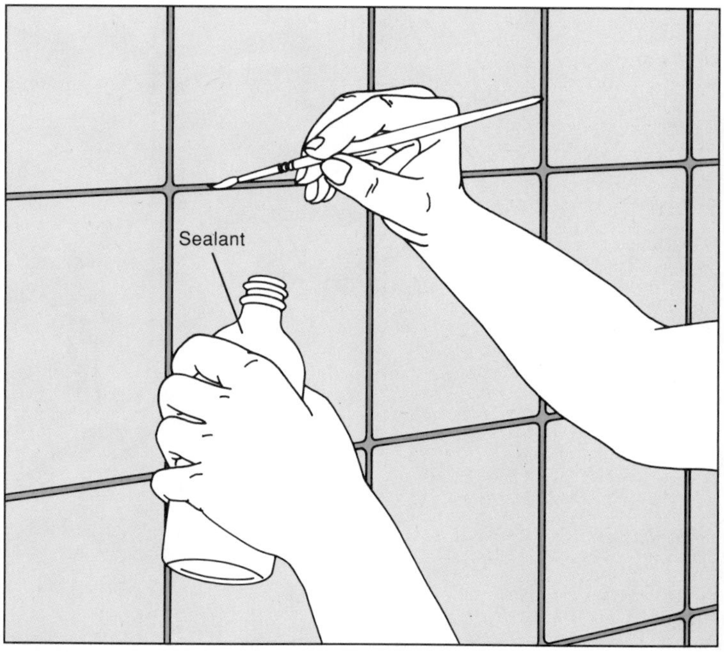

4 Sealing the grout. Wait 24 hours for the grout to harden before sealing it to protect against cracking, mildew and water spotting. Buff the tiles with a soft cloth. Brush a clear silicone tile sealant over the grout, as shown, using a brush with a tip that fits within the joints. Wipe the sealant across the tiles with a soft cloth to protect them also. If the sealant alters the sheen of the tile, seal all the tiles around the bathtub at the same time.

REPLACING LOOSE OR BROKEN CERAMIC TILES

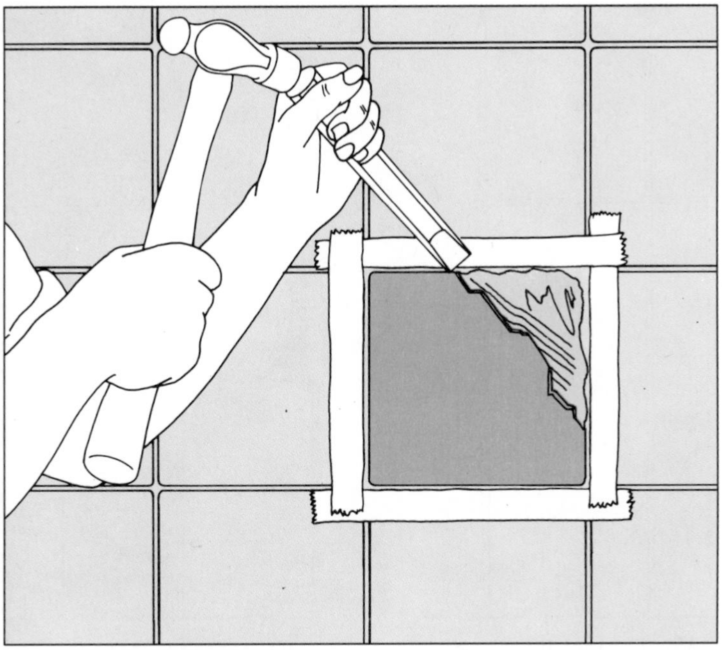

1 Removing the broken tile. Protect the edges of the adjacent tiles with masking tape. Scrape away loose tile and old grout with a putty knife. Chisel away the remaining pieces of tile with a ball-peen hammer and cold chisel *(above)*. If the tile does not easily break away, drill it out *(next step)*.

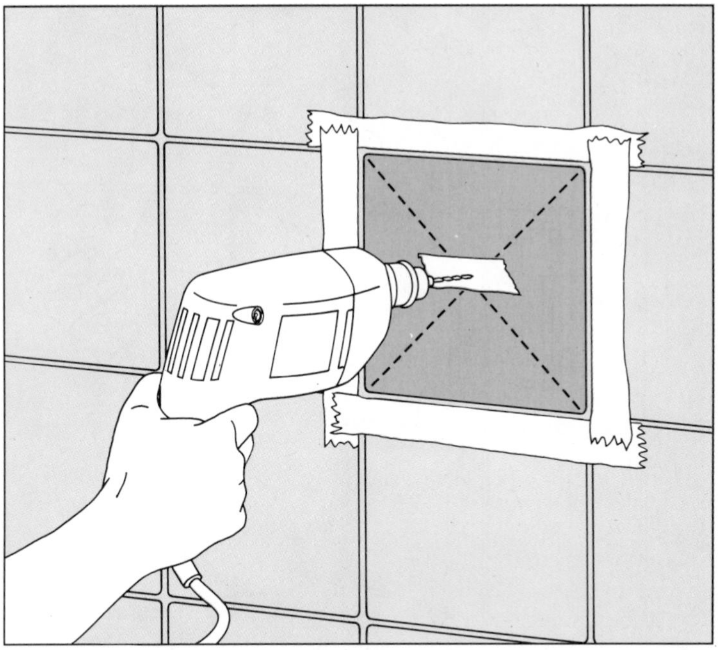

2 Removing a stubborn tile. Place a piece of masking tape across the center of the damaged tile. Wearing safety goggles, drill a hole through the tape, as shown, using a masonry bit on an electric drill set at slow speed. (**Caution**: Do not let the drill or cord come in contact with water.) Score an X across the hole with a single-edge razor blade, position the cold chisel in the center of the X and tap gently with the hammer to break apart the tile. Use the chisel to pry out the remaining pieces.

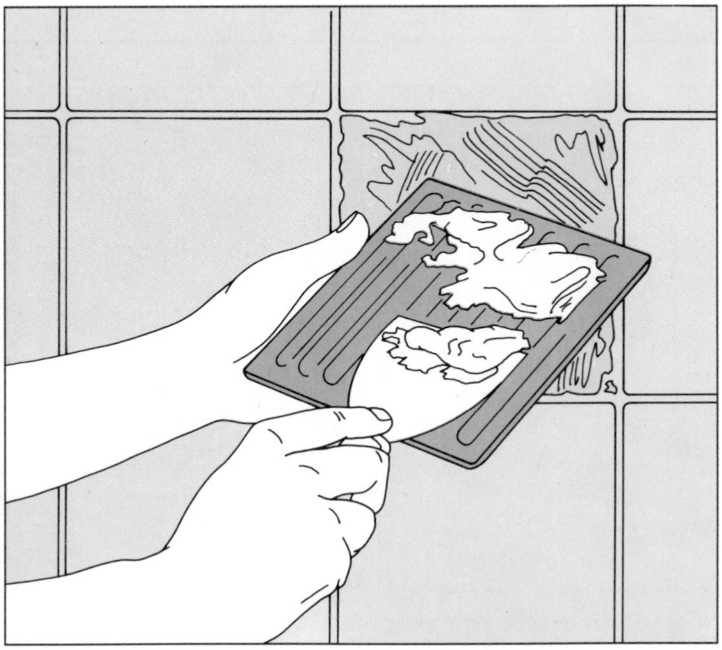

3 Cementing the new tile. Use a putty knife to scrape away any remaining adhesive and grout from the wall behind the damaged tile. Apply silicone tile adhesive to the back of the new tile, using the putty knife if the back of the tile is ridged *(above)*, or a serrated trowel if the back is flat.

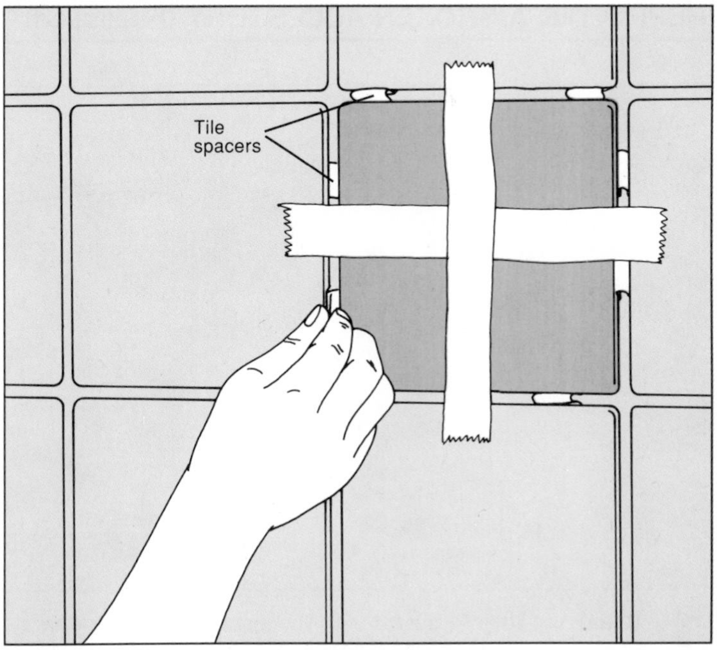

4 Setting the new tile in place. Press the tile firmly in place and set plastic tile spacers, available where tiles are sold, or finishing nails into each joint to evenly separate the tile from adjacent tiles. Secure the tile with masking tape while it sets overnight. Remove the spacers *(above)* and tape, and regrout around the tile *(page 132)*.

REPLACING A BROKEN ACCESSORY (Flush-set)

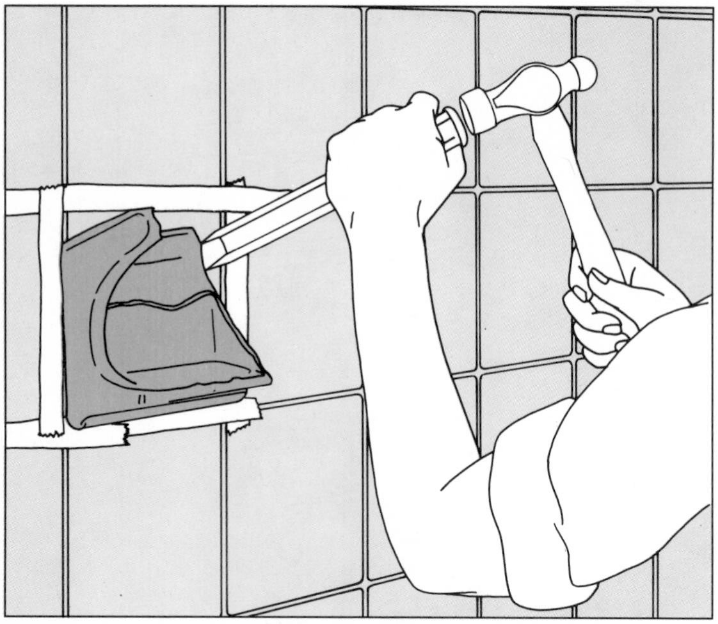

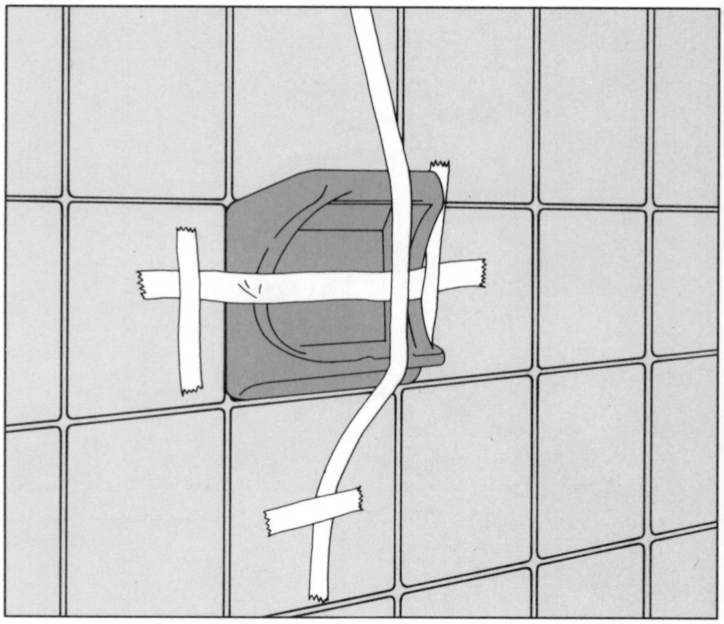

1 Removing the old accessory. Wear safety goggles and protect the edges of the adjacent tiles with masking tape. Gouge a groove in the joint around the accessory with a utility knife or old screwdriver. Set a cold chisel *(above)* into the groove and tap the accessory with a ball-peen hammer to force it away from the wall. With a putty knife, scrape out the remaining grout and adhesive.

2 Setting the new accessory in place. For the best result, choose a replacement accessory as large as or slightly larger than the original. Apply an even coat of silicone tile adhesive to the back of the new accessory. Press it firmly into place and secure it with masking tape *(above)* for 24 hours. Remove the tape and grout the surrounding joints *(page 132)*.

REPLACING A BROKEN ACCESSORY (Recessed)

1 Removing the old accessory. With a putty knife, scrape away any caulking from around the old accessory. If the accessory is attached with a mounting bracket in the wall, unscrew and remove the accessory, as shown. To avoid breaking adjacent tiles, leave the mounting bracket inside the wall. A new screw-mounted accessory will probably attach to it. If the old accesssory was not screwed in place, pry it away with a putty knife. If necessary, use a cold chisel and ball-peen hammer to knock the accessory out, wearing goggles and taping the adjacent tile first.

2 Attaching the new accessory. To attach a new accessory to the old mounting plate, replace one screw, then the other screw and tighten both. If you are replacing an accessory that did not have a mounting bracket, tuck the ends of the bracket between the tile and wall, its spring pressing against the wall *(above, left)*. Apply silicone caulking around the back of the accessory, then screw it to the bracket. To install an accessory with no bracket, fill the recess with crumpled newspaper, then caulk around the back of the accessory *(above, right)*. Press the accessory in place and secure it with masking tape for 24 hours.

REPLACING A BROKEN ACCESSORY (Surface-mounted)

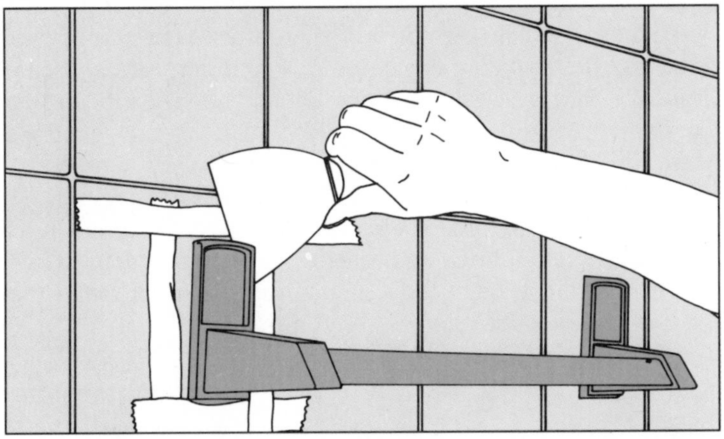

1 Removing the old accessory. Select a new accessory that is the same size as or larger than the old. Protect the adjacent tiles with masking tape. Pry away an adhesive-backed mounting plate with a putty knife *(above)* or unscrew a screw-mounted plate. Sand the tile beneath so that it is flush with the adjacent tiles.

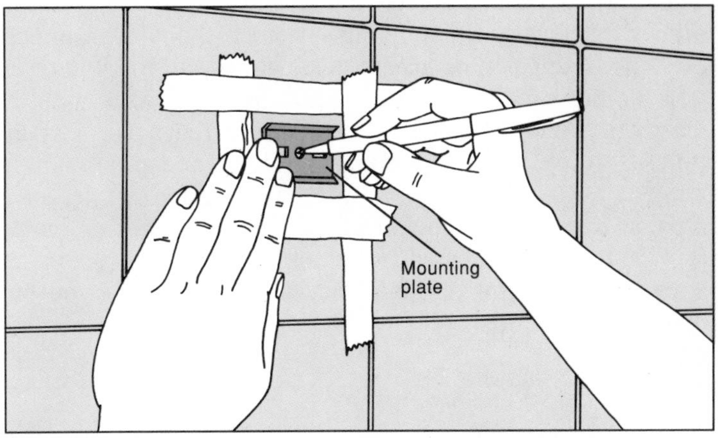

2 Positioning the new accessory. To install an adhesive-backed accessory, remove the protective strip and press it firmly onto the sanded tile. To install a screw-mounted accessory, position the mounting plate on the tile. If its holes align with those in the tile, simply screw on the mounting plate *(step 5)*. If not, check the position with a level, then mark the new screw holes, as shown.

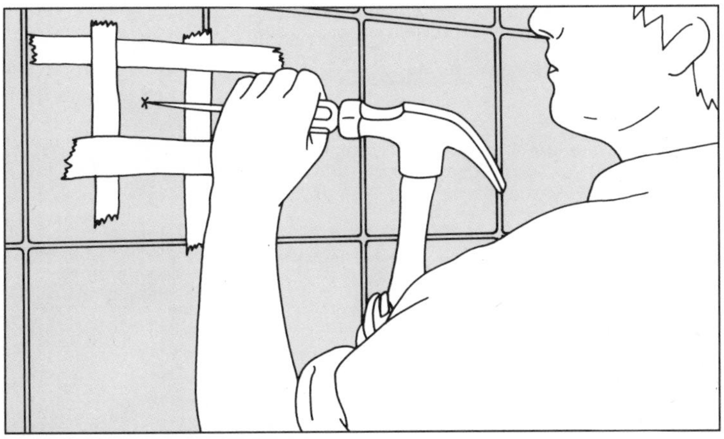

3 Preparing the wall or tile for drilling. With a punch and hammer *(above)*, make a slight indentation on each mark to keep the drill bit from wandering.

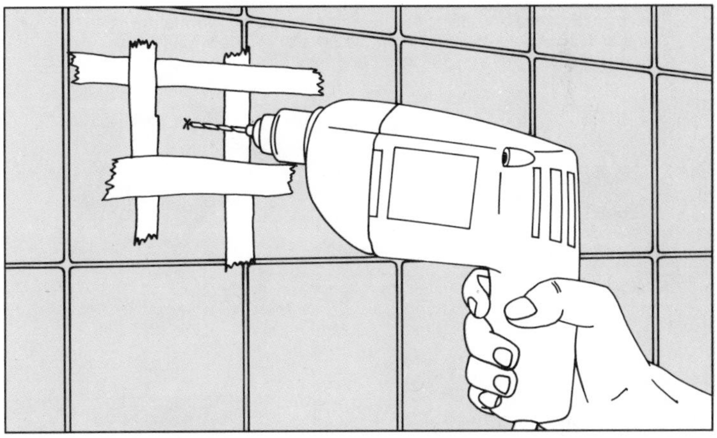

4 Drilling holes for the new mounting plate. Use an electric drill with a masonry bit to drill at slow speed through the mounting plate hole marks *(above)*.

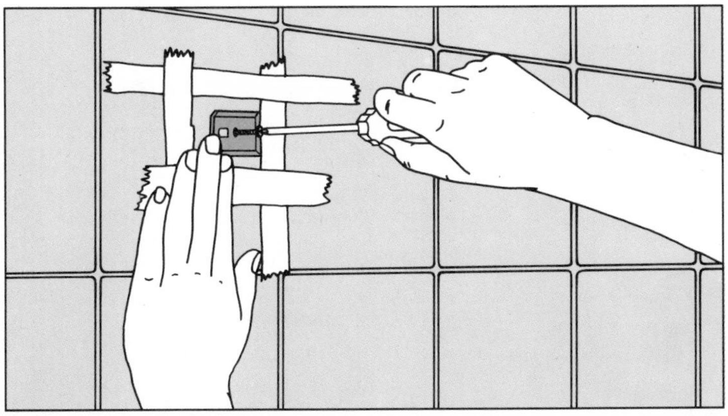

5 Securing the mounting plate. If you are attaching the mounting plate to a tile set in cement, use plastic or lead anchors to secure the screws. On drywall, use toggle bolts or hollow wall anchors. Position the mounting plate *(above)*. If you are installing a metal accessory, the angled edges of the plate should be at the top and bottom (not the sides). Screw on the mounting plate.

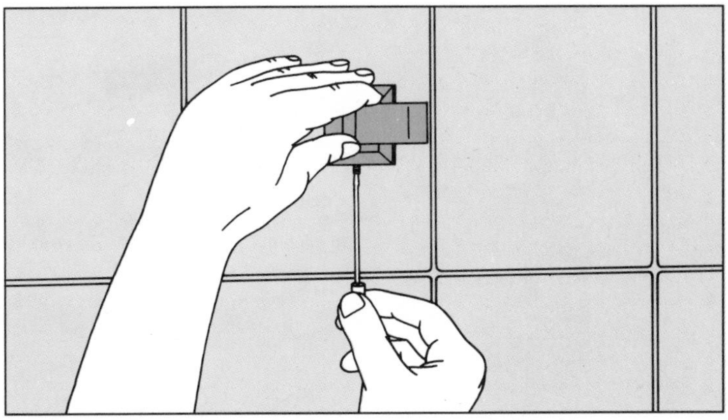

6 Installing the new accessory. If the accessory is metal, slip it over the top and bottom of the mounting plate and tighten the setscrew with a hex wrench or small screwdriver *(above)*. If the accessory is ceramic, slide it over the mounting plate and caulk around the edge *(page 131)*.

TOOLS & TECHNIQUES

This section introduces basic maintenance and repair techniques for household plumbing systems, from soldering copper pipes and patching access holes in drywall to winterizing your home. You can handle most plumbing repairs with the basic tool kit shown below. Apart from wrenches, pliers and pipefitting tools, keep a supply of hose and pipe clamps, washers, O-rings and slip fittings on hand for emergency repairs. Specialized tools, such as a power auger or ratchet pipe cutter, can be rented from tool rental shops or hardware stores. Some plumbing suppliers will lend tools if you buy materials from them.

It is usually more economical to buy a complete set of wrenches, screwdrivers or drill bits rather than a single tool here and there. Avoid cheap tools, which may slip and cause injury or damage work. Use the right tools for the job, and care for them properly. Clean metal tools with a rag moistened with a few drops of light oil. (Don't oil file or wrench handles.) To remove rust, rub with fine steel wool or emery cloth. The adjusting mechanisms of wrenches should be kept clean and free of grit, and lubricated occasionally with powdered graphite or light oil. Protect tools in a sturdy plastic or metal toolbox, with a secure lock if stored around children.

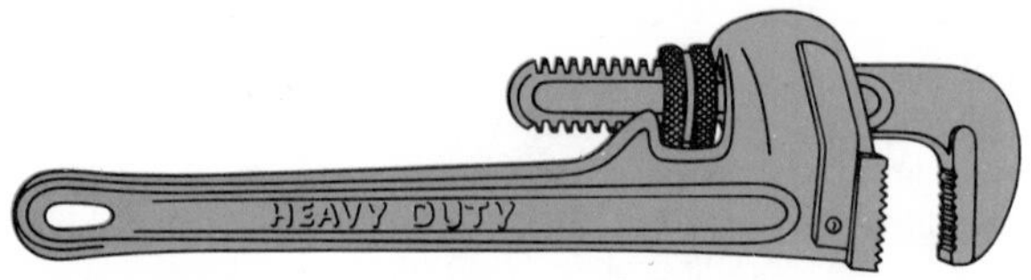

Pipe wrench
Adjustable tooth-jawed wrench designed to grip pipe; often used in pairs, one on the threaded pipe and the other on the fitting. Never use on chrome fittings without first protecting the metal with rags or tape. Avoid overtightening the jaws of a pipe wrench—the wrench grips tighter as it is turned and may crush the pipe.

Monkey wrench
Similar to pipe wrench but has no teeth. Prevents chrome or other visible fittings from being marred by jaws.

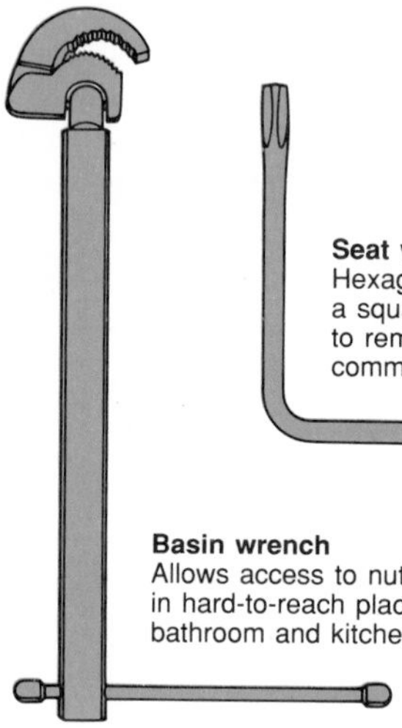

Basin wrench
Allows access to nuts and bolts in hard-to-reach places under bathroom and kitchen sinks.

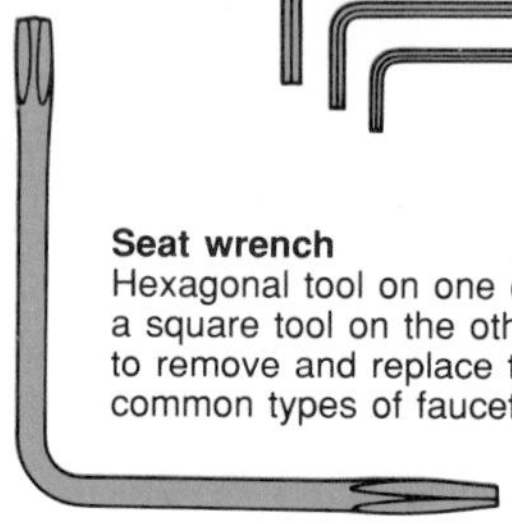

Seat wrench
Hexagonal tool on one end and a square tool on the other used to remove and replace the two common types of faucet seats.

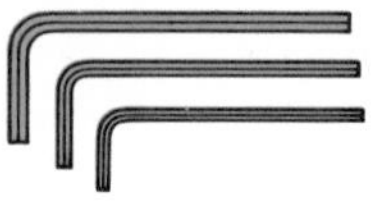

Hex wrenches
End of hexagonal wrench fits setscrews used to secure faucet handles, tub spouts and escutcheons.

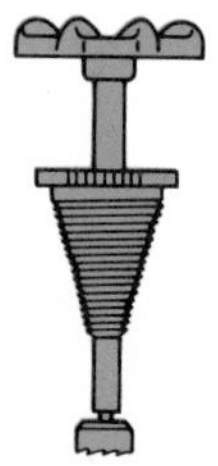

Valve-seat dresser
Cutting tip regrinds worn or damaged faucet seats that cannot be removed with a seat wrench.

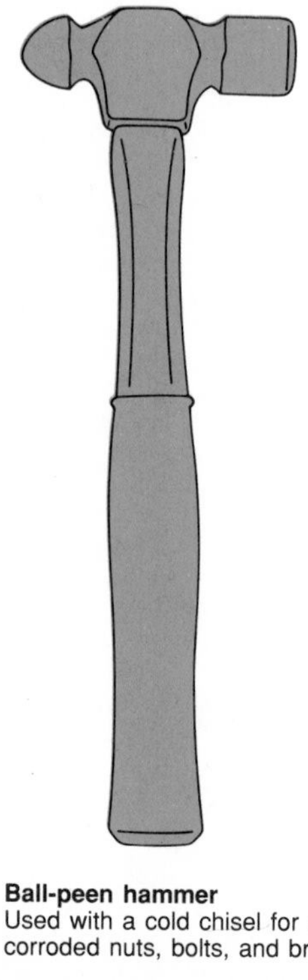

Ball-peen hammer
Used with a cold chisel for removing corroded nuts, bolts, and broken tiles.

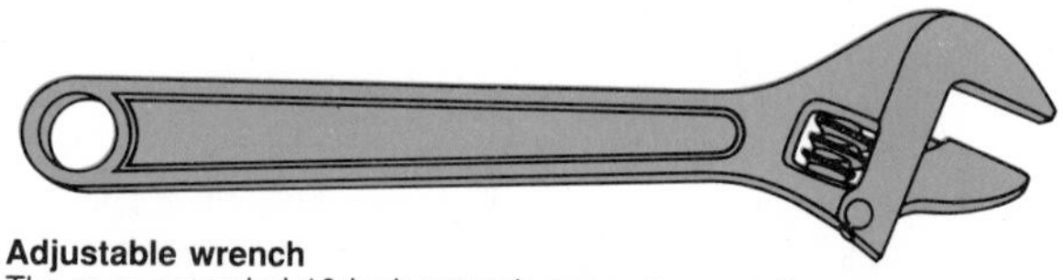

Adjustable wrench
The recommended 10-inch wrench opens to accept square or hexagonal nuts up to 1 inch across. With any wrench, always place the strain on the stronger, fixed jaw.

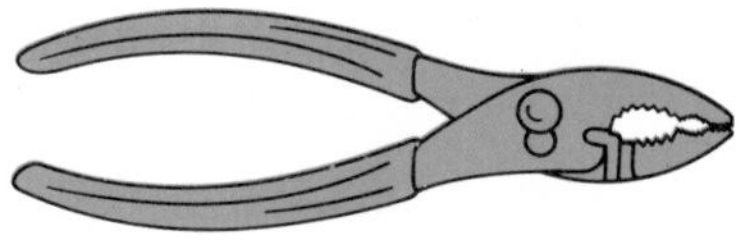

Slip-joint (utility) pliers
General-purpose gripping and bending tool; jaws can be set at two widths for gripping narrow or wide objects.

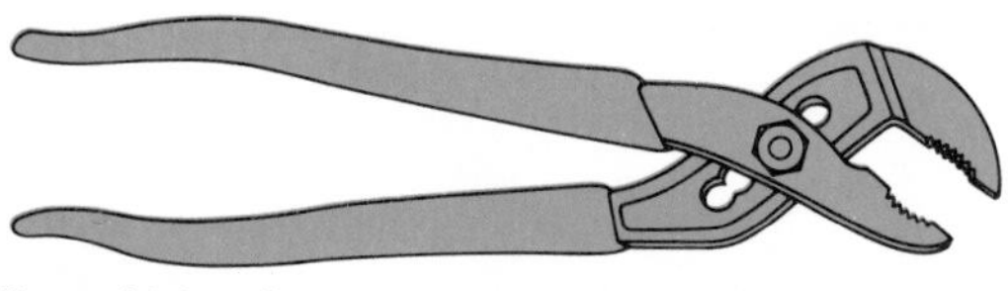

Channel-joint pliers
Perhaps the most versatile plumbing tool; pliers can be adjusted to keep the jaws parallel over a wide range of settings.

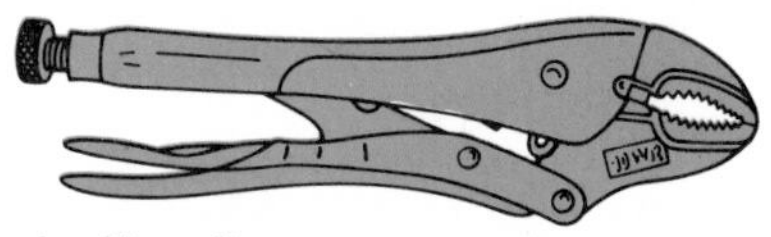

Locking pliers
A clamping action keeps the tool from slipping and frees both hands. Use with care—they have enough leverage to deform copper pipes and fittings.

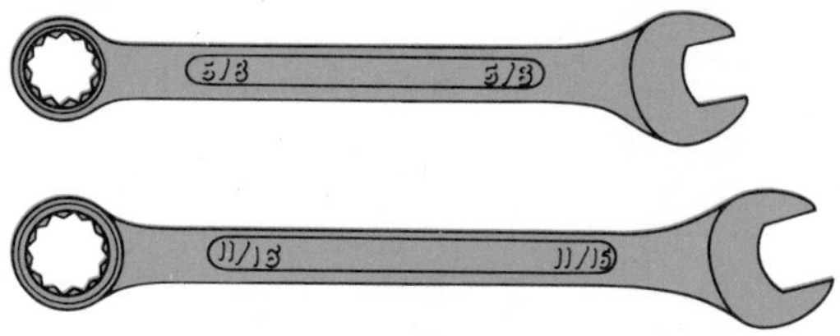

Combination wrenches
For turning nuts and bolts; a typical set fits nuts from 1/4 inch to 1 inch across.

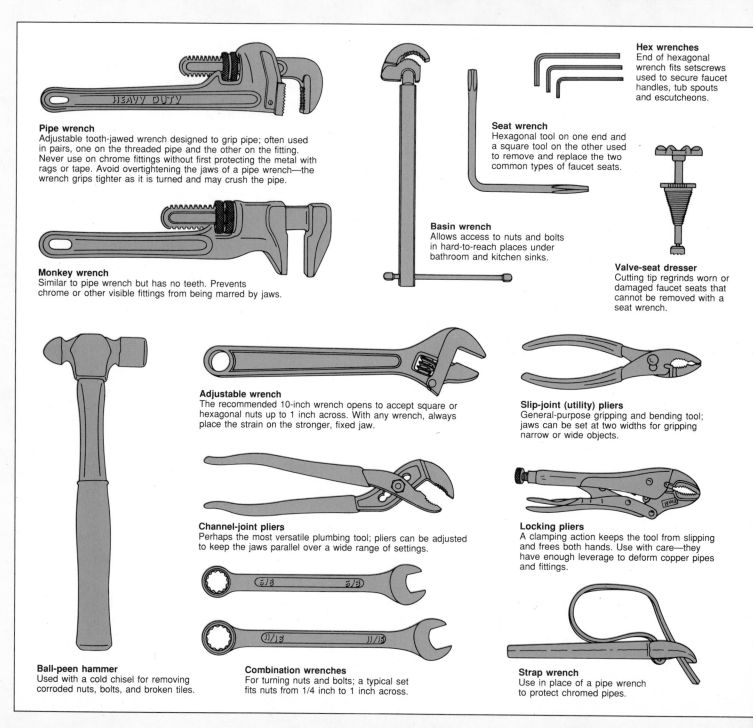

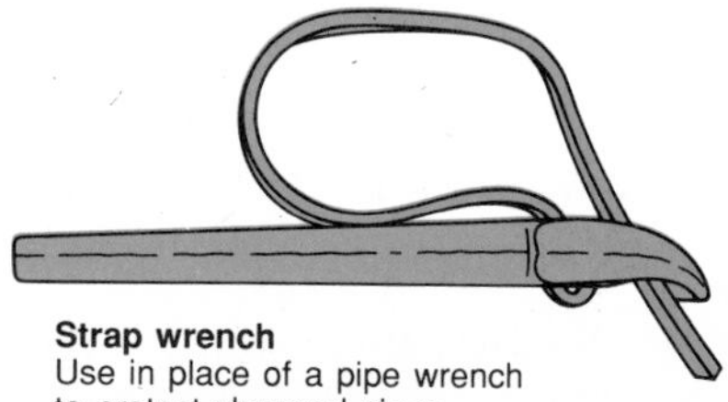

Strap wrench
Use in place of a pipe wrench to protect chromed pipes.

To prevent accidents, keep your work area clean and free of clutter. Wear safety goggles when soldering, chiseling or working with caustic liquids, and heavy work gloves to protect against cuts and burns. Wear rubber boots when working in damp areas and do not use electrical tools in those conditions. Avoid electrical shock by using pliers and screwdrivers with insulated handles, or handles wrapped in electrical tape.

To free a rusted or corroded fitting, first try tightening it before loosening it, or apply penetrating oil. The oil works its way between parts to dissolve rust and corrosion. Apply it liberally to the stubborn fitting, allow it to work for at least 15 minutes, then loosen with a wrench. If necessary, apply more oil and wait overnight before trying again. A cold chisel can be used for severing cast-iron pipes and chewing through rusted nuts and bolts. Pound on a chisel only with a ball-peen hammer, and always wear gloves and goggles.

When fitting new pipes, take time to leak-proof the joints with pipe tape or plumber's putty. Wind 1 1/2 turns of pipe tape clockwise around the male threads tightly enough for them to show through. Knead a small rope of putty between your hands, apply it to slip nuts or drain fittings, then press it flat before joining the pieces.

Hacksaw
For cutting steel or plastic pipe; use a mini-hacksaw in tight spaces where a regular hacksaw won't reach.

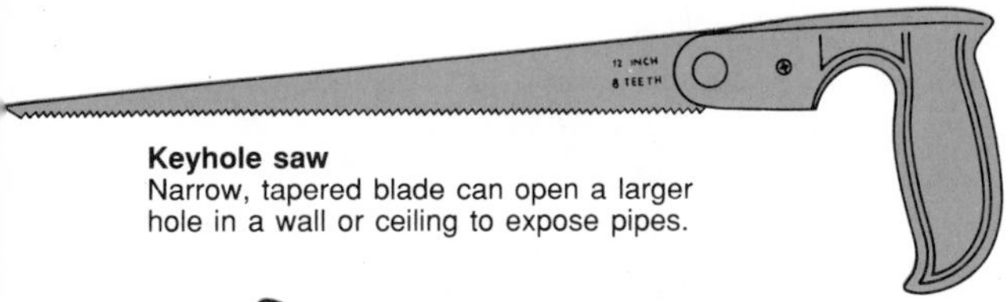

Keyhole saw
Narrow, tapered blade can open a larger hole in a wall or ceiling to expose pipes.

Penetrating oil
A chemical "tool" to help loosen corroded parts for disassembly.

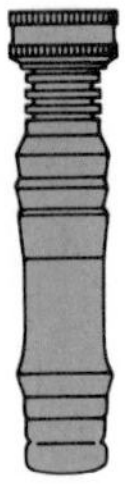

Expansion nozzle
Connected to garden hose and inserted in drain opening, nozzle expands to seal the pipe and water pulses to flush out blockages.

Propane torch
Used for sweat soldering copper pipe and removing rusted metal fittings; use with caution to thaw frozen pipes.

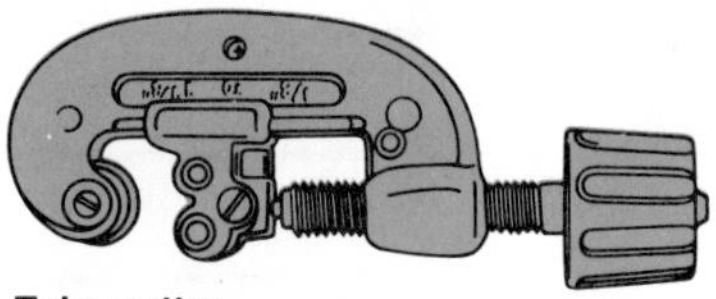

Tube cutter
Cuts copper and plastic pipe smoothly and squarely; blade attachment is used to deburr the ends of the cut pipe.

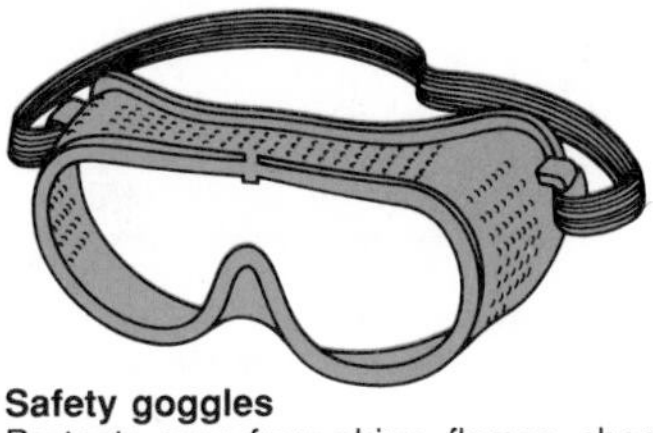

Safety goggles
Protects eyes from chips, flames, chemicals.

Pipe bender
Tightly wound steel coil that slips over flexible pipe to bend it without kinks.

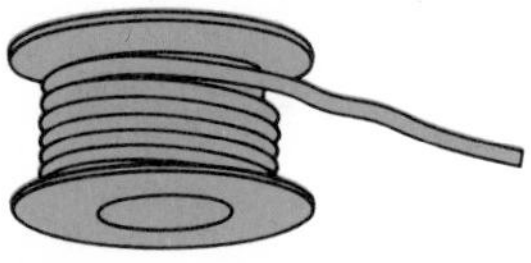

Solder
Joins copper pipe to fittings by sweat soldering; used with a propane torch and non-corrosive flux. Codes may require 90/10 tin/lead solder (called "low-lead").

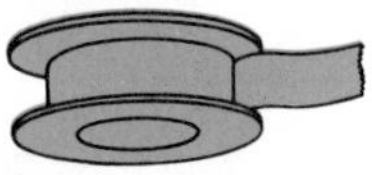

Pipe tape
Wrapped around male threads to ease threading and unthreading, and to make a watertight seal.

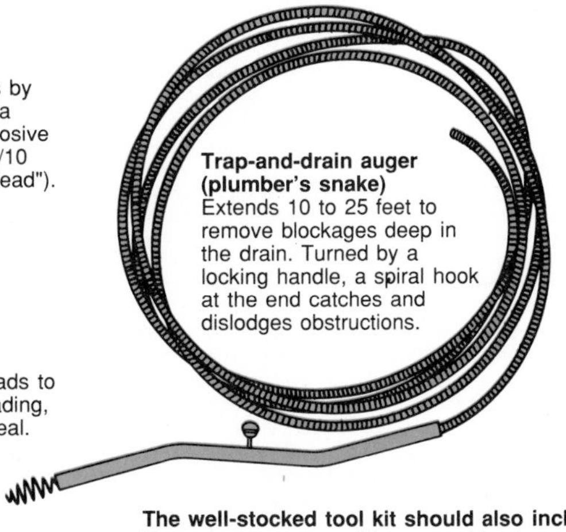

Trap-and-drain auger (plumber's snake)
Extends 10 to 25 feet to remove blockages deep in the drain. Turned by a locking handle, a spiral hook at the end catches and dislodges obstructions.

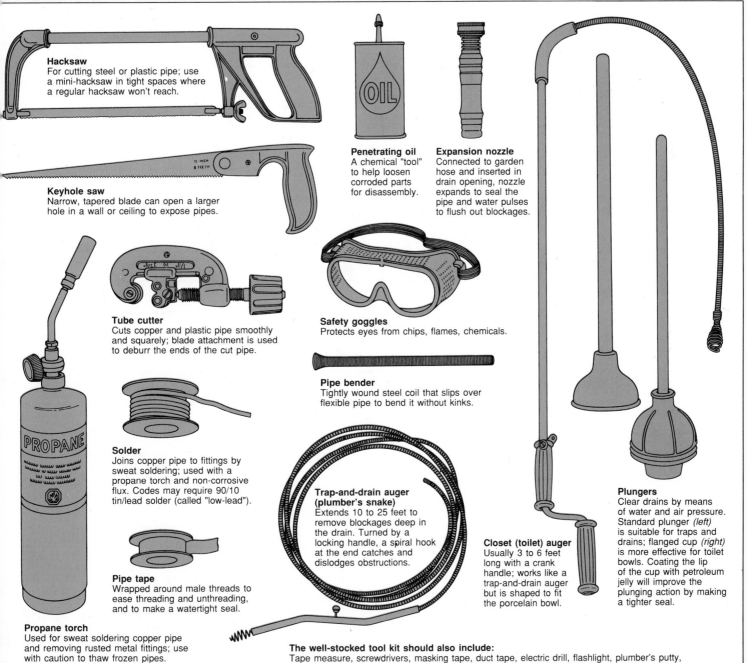

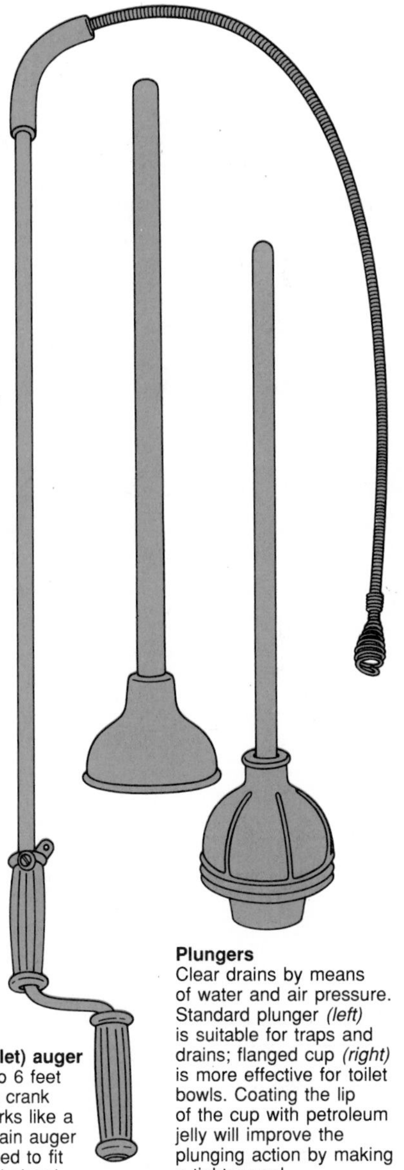

Closet (toilet) auger
Usually 3 to 6 feet long with a crank handle; works like a trap-and-drain auger but is shaped to fit the porcelain bowl.

Plungers
Clear drains by means of water and air pressure. Standard plunger *(left)* is suitable for traps and drains; flanged cup *(right)* is more effective for toilet bowls. Coating the lip of the cup with petroleum jelly will improve the plunging action by making a tighter seal.

The well-stocked tool kit should also include:
Tape measure, screwdrivers, masking tape, duct tape, electric drill, flashlight, plumber's putty, and the following safety gear: heavy work gloves, rubber gloves and a dry-chemical fire extinguisher.

CLEANING AND CARE

SYMPTOM	PROCEDURE
Tile adhesive on tile surface	If the adhesive is still damp, wash the tile immediately with a damp cloth. If the adhesive has set, scrape it off carefully with a single-edge razor blade or ice scraper. Rub the tile with a soft cloth or a brush and paint thinner. Rinse with water.
Tiles dirty or filmy	Scrub the seams with a toothbrush and mixture of 2 teaspoons of vinegar or heavy-duty laundry detergent in a pan of water. Rub heavily soiled tiles with a soft cloth dampened with kerosene, then rinse with detergent and water.
Mildew on tiles or grout	Wearing rubber gloves, scrub with a small, stiff brush and a mixture of 1/2 cup laundry bleach and 1 quart of water. Rinse with water, then rub the tile with a soft cloth and let dry overnight. Scrub any remaining mildew with a stiff brush and a commercial mildew remover containing sodium hypochlorite and sodium carbonate.
Tile discolored	Rub with a soft cloth and kerosene or a commercial tile cleaner containing phosphoric acid. For heavy discoloration, use a soft cloth to coat the tile with undiluted liquid neutral soap. Let dry overnight, rinse and rub again with a mixture of neutral soap and warm water. While the tile is still wet, scrub it with a stiff brush and scouring powder. Rinse with a sponge and water, then polish with a soft cloth.
Enamel bathtub surface rough, grainy	Scrub thoroughly with a stiff brush and vinegar to dissolve lime deposits.
Plastic laminate stained	Sponge away soil stains with a mild detergent. For tougher stains, mix baking soda and water into a paste and gently rub with a soft cloth, being careful not to scratch the surface.
Plastic laminate scratched	Polish minor scratches with an automotive body polish. For deeper scratches, rub with sandpaper before polishing with the wax. Or fill the scratch with colored plastic laminate filler, available from countertop makers.
Linoleum or vinyl marked with crayon	Rub with a soft cloth and silver polish until the mark is removed.
Wallpaper spotted with grease	For washable wallpaper, rub with a cloth dampened with isopropyl alcohol or a 1-to-10 solution of enzyme detergent and water. For nonwashable wallpaper, gently rub with cheesecloth sprinkled with turpentine or naphtha.
Enamel bathtub discolored	Vigorously scrub a paste of cream of tartar and hydrogen peroxide onto the tub with a small, stiff brush. For heavy discoloration, refinish the tub surface with an epoxy glaze *(page 130).*
Enamel bathtub or sink stained with iron rust	Wearing rubber gloves, scrub with a stiff brush and a cleanser containing phosphoric and oxalic acids. Let stand for 10 minutes before rinsing off. For tough stains, place paper towels across the bottom of the sink or bathtub and saturate with household bleach. Let set 30 minutes before rinsing.
Bathtub or sink stained green from copper	Wash with a mixture of soapsuds and ammonia. For tough stains, scrub with a mixture of 1 part oxalic acid to 20 parts water. Rinse immediately with water.
Stainless-steel sink stained with rust	Rub with lighter fluid until the rust marks disappear, then wipe sink with all-purpose kitchen surface cleanser.
Stainless-steel sink or fixtures spotted from water	Wipe with a cloth dampened in rubbing alcohol or white vinegar.
Fiberglass tub or shower surround stained	Gently rub a paste of non-abrasive cleansing powder and water over stain. Restore dull surfaces with an automotive body polish.
Ring around porcelain-on-steel bathtubs	Scrub with a sudsy cloth and a non-abrasive scouring powder. For tough stains, scour the tub with a stiff brush and a mixture of 1 part oxalic acid and 10 parts water.
Adhesive bathtub decals worn	Soak with mineral spirits, nail polish remover or lighter fluid. Gently scrape off the decals with a razor blade or knife, being careful not to scratch the tub's surface. When completely removed, spray the tub surface with an all-purpose bathroom cleanser and rub with an abrasive sponge. Finish by waxing with an automotive body polish.
Toilet bowl discolored	Flush to wet the sides of the bowl. Apply a paste of toilet bowl cleaner or borax and lemon juice. Let set for two hours, then scrub thoroughly with a toilet brush. For stubborn stains, rub with wet sandpaper until the stain disappears.
Shower curtains mildewed	Rub mildewed areas with a mixture of baking soda and water. For heavy mildew, remove the curtains, soak in salt water, rub with lemon juice and rinse before rehanging.
Glass shower door filmy	Rub with a sponge dampened in white vinegar or household ammonia.
Brass bathroom fixtures tarnished	Remove light tarnish with a commercial brass cleaner. For heavy tarnish, remove the fixtures and soak overnight in a plastic container filled with 1 quart water, 1/2 pint vinegar, and 4 tablespoons table salt. When the tarnish is removed, buff the brass with a paste polish and a soft cloth. Coat with a commercial acrylic spray to prevent further tarnishing.
Chromed faucets tarnished	Wash with a mild soap or detergent. Do not use metal polishes or abrasive powders; they will damage the plating. Polish with a clean cloth. Use vinegar to remove any mineral deposits.

PATCHING AN ACCESS HOLE

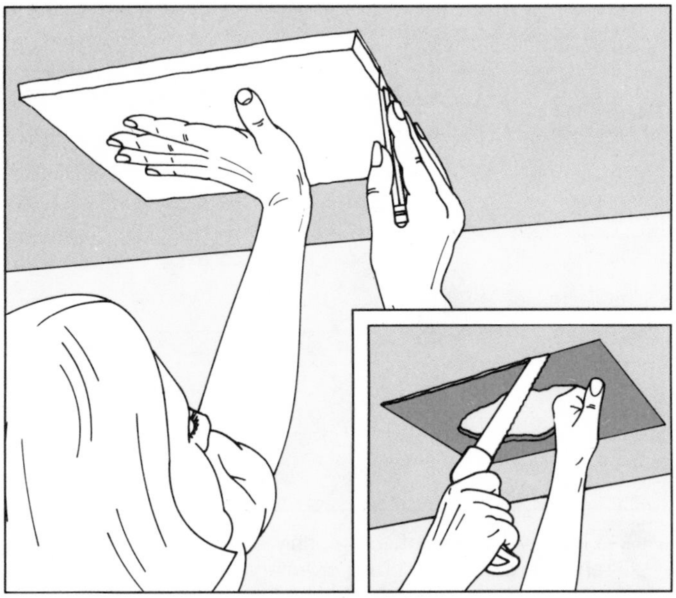

1 Cutting the patch. Patching a hole made to get at a leaky pipe is often the last step of a plumbing repair. All it requires is some scrap drywall, panel adhesive and joint compound. Using a pencil and straightedge, rule off a rectangular section around the hole. Transfer these measurements to a piece of drywall, cut the patch with a keyhole saw and hold it up to the wall or ceiling to trace a more precise hole *(above)*. Cut out the traced area *(inset)*.

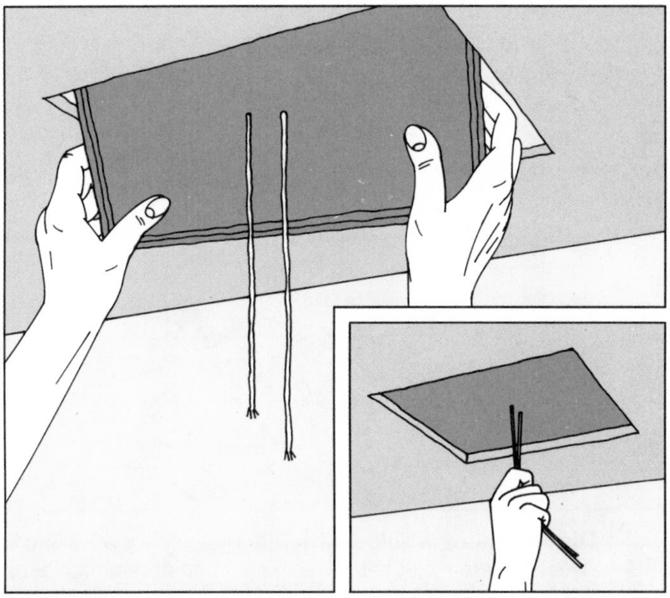

2 Fitting the backing. Cut a second piece of scrap drywall slightly larger than the hole to be used as a backing for the patch. With a bent coat hanger, poke two small holes in the backing, then thread a string through the holes. Use a caulking gun to apply panel adhesive around the edges of the backing. Place the glued backing into the wall or ceiling behind the hole *(above)* and pull on the string to set the backing in place *(inset)*.

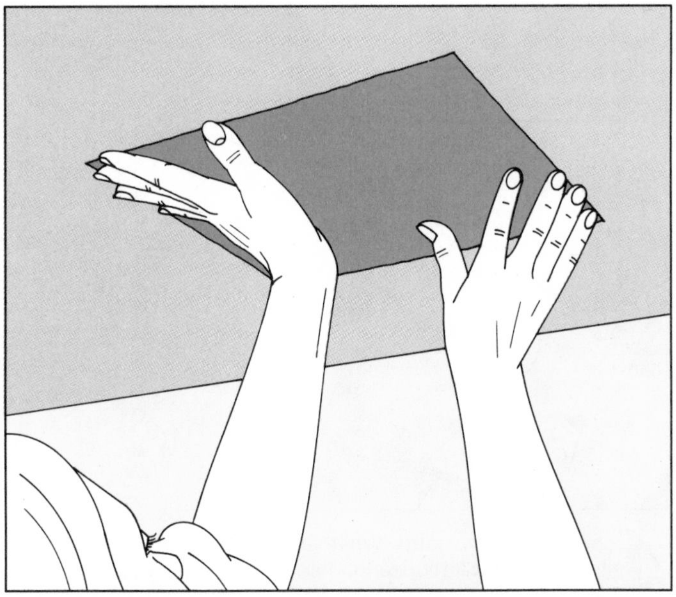

3 Fitting the patch. Remove the string from the backing. Apply panel adhesive to the edges of the patch, and several thin beads of adhesive along the back. Fit the patch into the hole and press it firmly into place *(above)*.

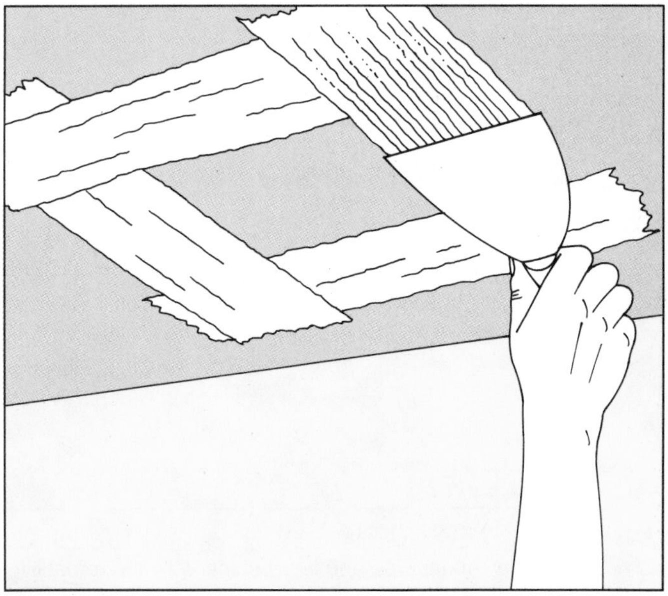

4 Filling the joints. Using a flexible putty knife, apply premixed joint compound to fill the crack around the patch. With even pressure and lateral strokes, smooth out the joint compound to conceal the seam. Allow the joint to dry overnight. If necessary, apply a second coat of compound with a wider putty knife to feather the joint. When completely dry, use sandpaper for a smooth final finish.

SWEAT SOLDERING COPPER PIPE

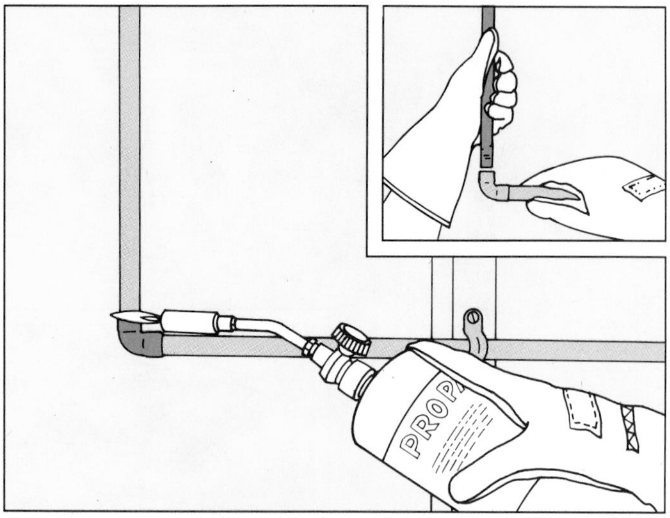

1 Disassembling a soldered joint. The cure for a leaking sweat-soldered joint is to take the joint apart and resolder it. Wear safety goggles and heavy work gloves, protect flammable materials with a fireproof shield, and have a dry-chemical fire extinguisher at hand. Turn off the water and drain the pipe. Ignite the propane torch and play the flame over the fitting until the old solder melts *(above)*, then quickly pull the pipe from the fitting *(inset)*. Do this again to separate the other fitting.

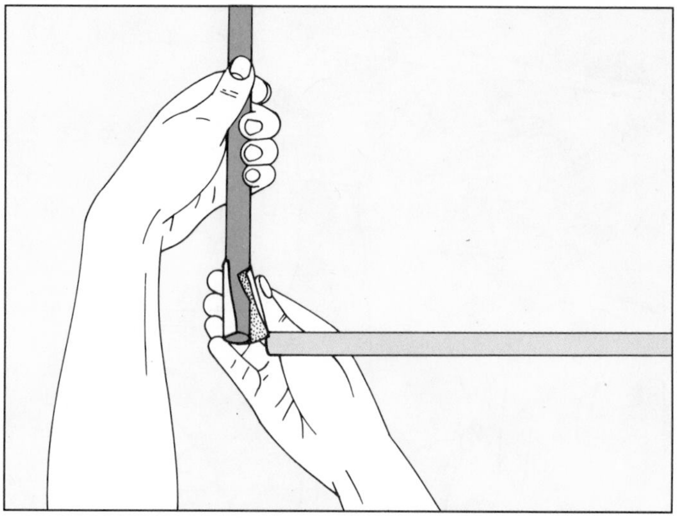

2 Cleaning the joint. Thoroughly burnish the fitting inside and out with a wire brush or emery cloth, as shown, then wipe with a clean cloth. Be careful not to touch the polished surfaces since dirt or grease, even a fingerprint, may interfere with the capillary action of the solder. Cleaning the pipe and fitting helps the solder flow evenly to form a watertight joint.

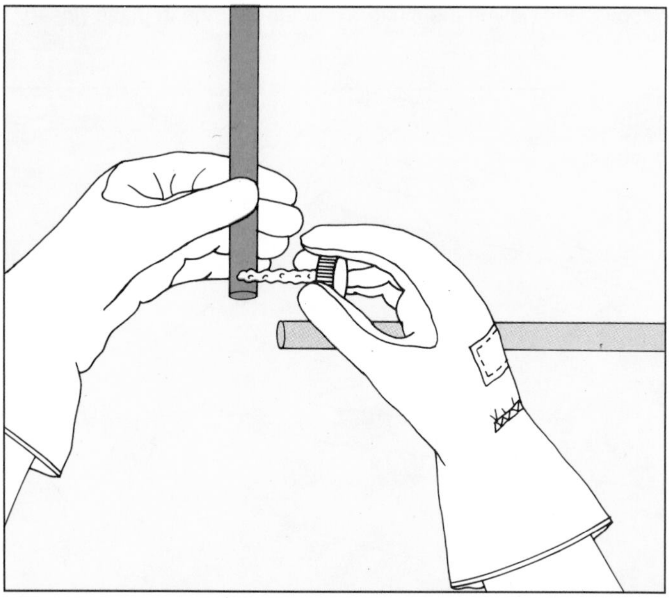

3 Applying flux. Available as a paste or liquid, flux both cleans the copper pipe and helps solder flow more readily. With an applicator or small brush, apply a light coat of flux to the inside of the fitting and a heavier coat to the outside ends of the pipes *(above)*. If water drips from either pipe, make a temporary dam using a ball of white bread. Later, it will dissolve and drain away. Push the pipes into the fitting.

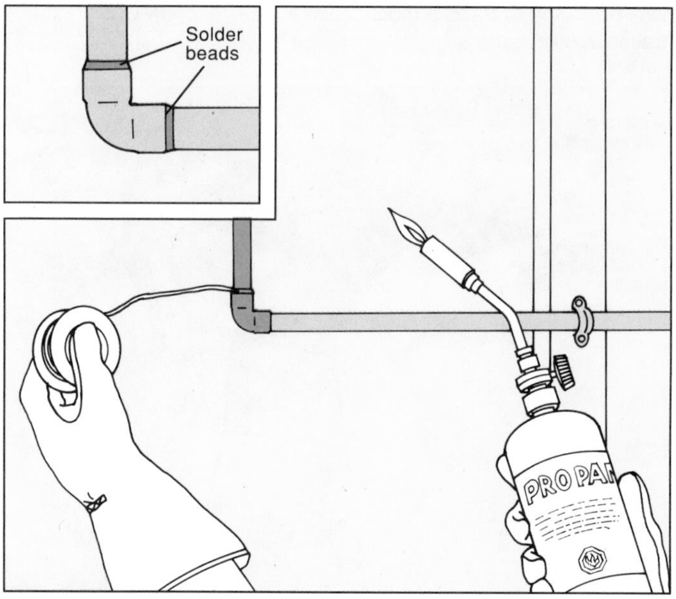

4 Soldering the joint. Whether required by code or not, low-lead solder will reduce the risk of lead entering the plumbing system. Ignite the propane torch and play the inner core of the flame over the fitting, but do not touch the flame to the solder. To test if the fitting is hot enough, touch the tip of the solder to the joint *(above)*. If it melts, apply more solder. The capillary action will pull the solder into the joint to seal the connection. Apply solder until a bead of metal appears completely around the rim and starts to drip. (Any gaps will probably leak.) Repeat to join the other end of the pipe and fitting. Wait a few minutes for the connection to cool, then turn on the water, slowly at first, to test for leaks.

BACK-SIPHONAGE

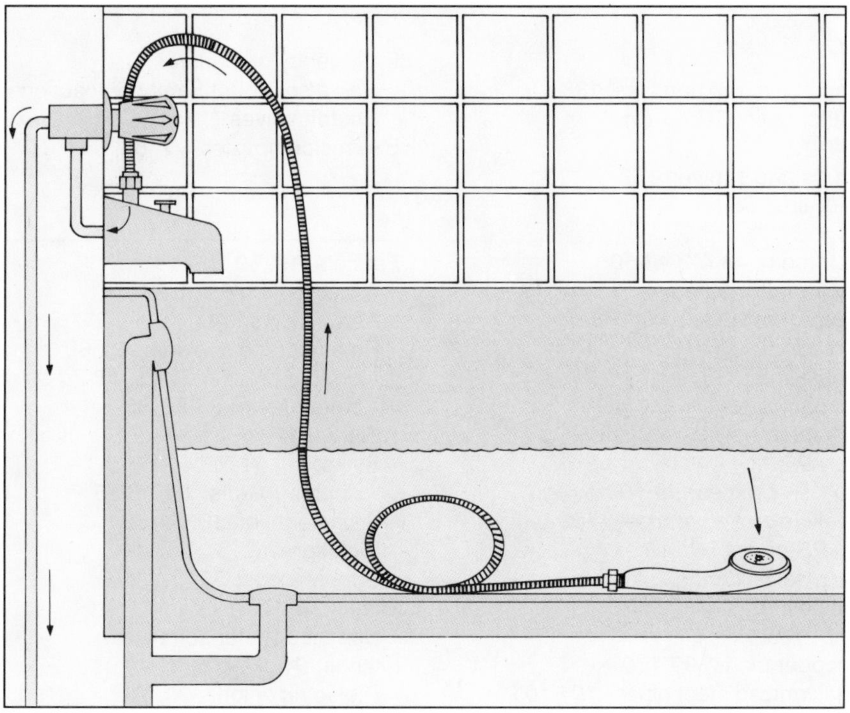

Preventing hazardous cross-connections. Back-siphonage is the reverse flow of contaminated water into supply pipes. It takes place through what is called a cross-connection—something as simple as a garden hose left in a swimming pool or a hand-held shower attachment lying in a filled bathtub *(left)*. If pressure in the supply lines is low, a vacuum could be created that would actually draw contaminated water back into the supply lines. Most plumbing codes are written to prevent the conditions for back-siphonage. Because back-siphonage can occur even through a closed valve, water-using appliances are required to have air gaps, and faucets must be above the water line in bathtubs and sinks. Lawn sprinklers should have antisiphon devices to prevent the backflow of water, but few do. There are no warning signs that your water has been contaminated, but if you discover evidence of back-siphonage, do not attempt to flush out the lines yourself. Call the local health department to have your water tested.

WINTERIZING HOME PLUMBING

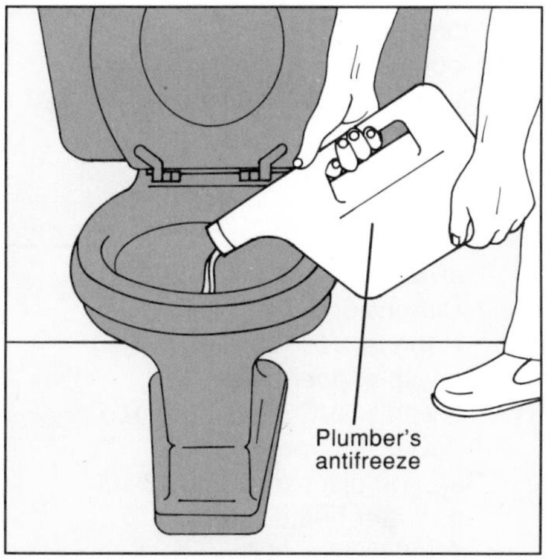

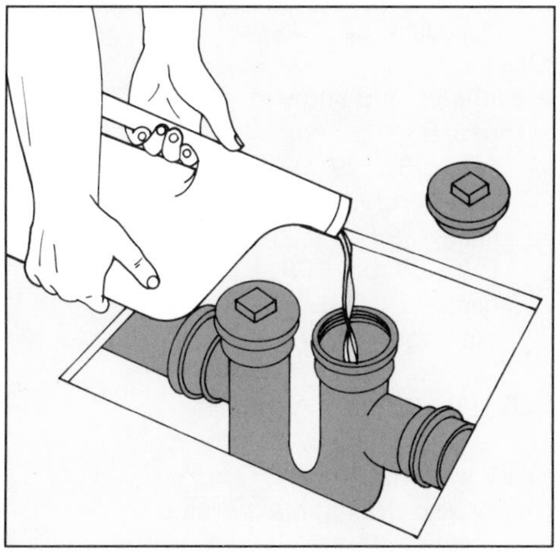

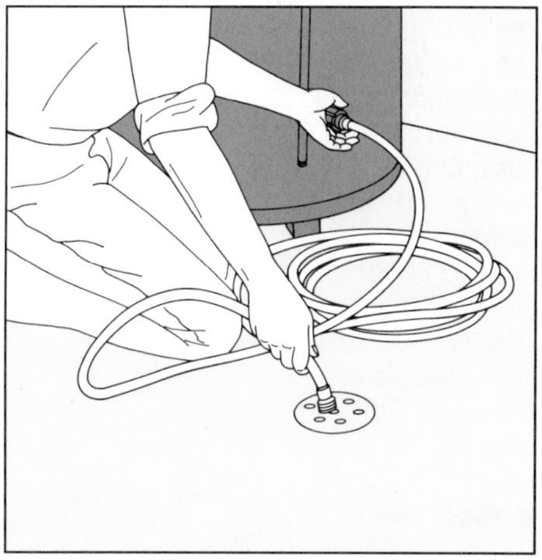

When leaving a house or cottage for a long period during the winter, it may be necessary to take certain precautions to prevent pipes from freezing. First, have the water utility or municipality turn off water at the curb valve near the street. Next, close the main shutoff valve and drain the supply lines by opening all the faucets in the house. Insulation is no help in preventing freeze-ups, but pipes that cannot be drained completely can be protected by a heated cable *(page 13)*. To protect fixture traps, pour plumber's antifreeze in the toilet, kitchen and bathroom sink traps *(above, left)* and in the main house trap, if any *(above, center)*. Turn off the electricity or gas to the water heater and drain its tank *(above, right)*. If you also heat with water, keep the thermostat on a low setting to avoid having to drain the heating system. When restoring the plumbing system to use, be sure to fill the water heater before turning on power or lighting the pilot. You may have to bleed the tank by opening a hot water faucet as it fills. When water flows steadily from the tap, no air remains in the tank.

INDEX

Page references in *italics* indicate an illustration of the subject mentioned. Page references in **bold** indicate a Troubleshooting Guide for the subject mentioned.

A

B

C

D

E

F

G

ACKNOWLEDGMENTS

The editors wish to thank the following:
The American Waterworks Association, Denver, Colo.; Peter J. Banks, Therm-O-Disc Inc., Mansfield, Ohio.; Jacques Bélanger, Darling Duro Ltd., Montreal, Que.; Gilles Bertrand, Gaz Métropolitain, Montreal, Que.; Roy Bradley, Canadian Plumbing and Heating Supplies Ltd., Montreal, Que.; Klaus Bremer, Bremer Electric Inc., Montreal, Que.; Peter Bunce, Crane Canada Inc., Montreal, Que.; Bill Collier, Montebello Plumbing & Hardware, Baltimore, Md.; Robert Déziel and Guy Lapointe, Roger Déziel Inc., Montreal, Que.; Kevin Dryden, Valley Eastman, Mississauga, Ont.; Environmental Protection Agency, Washington, D.C.; Thomas J. Erdman, Dealer Sales Agency, Victor, N.Y.; R.L. Fick and Sons Plumbing Inc., La Mesa, Calif.; Arthur J. Goodhue, Genova Products Inc., Davison, Mich.; Hans Gruenwald, Hudson, Que.; Claude Lesage, Giant Factories Inc., Montreal, Que.; Osborne Plumbing and Heating, San Diego, Calif.; John Sanford and André Sweeney, American-Standard, Toronto, Ont.; Charles Schulman, Canadian Institute of Plumbing and Heating, Montreal, Que., Stephen J. Shafer, D.E./Domestic Engineering Magazine, Chicago, Ill.; Paul Stanfield, Bow Plastics Ltd., Montreal, Que.; Frank S. Stanonik, Gas Appliance Manufacturers Association, Arlington, Va.; WaterTest Corporation, Manchester, N.H.; Ernest Wenczel, State Industries Inc., Ashland City, Tenn.

The following persons also assisted in the preparation of this book: Philippe Arnoldi, Diane Denoncourt, David Donaldson, Claire Dutin, Richard Fournier, Kathleen M. Kiely, Julie Léger, Francine Lemieux, Elizabeth W. Lewis, Michael Mouland, Barbara Peck, Solange Pelland, Natalie Watanabe and John Witczak.

Typeset on Texet Live Image Publishing System.